Society, Environment and Statistics

This interdisciplinary book series demonstrates the significance of statistical science in quantifying and understanding societal and environmental changes while effectively managing resources. The series invites manuscripts that cover a wide range of themes, including original research in traditional fields, emerging topics, practical applications, and connections between novel findings and established works. Particularly encouraged are submissions that showcase methodologies, case studies, good practices, and innovative solutions addressing the complex challenges of the United Nations' Sustainable Development Goals (SDGs). Rigorously peer-reviewed, all contributions adhere to the highest standards of scientific literature.

Florian Dumpert
Editor

Foundations and Advances of Machine Learning in Official Statistics

Springer

Editor
Florian Dumpert
Wiesbaden, Germany

ISSN 2948-2763 ISSN 2948-2771 (electronic)
Society, Environment and Statistics
ISBN 978-3-032-10003-0 ISBN 978-3-032-10004-7 (eBook)
https://doi.org/10.1007/978-3-032-10004-7

This work was supported by Statistisches Bundesamt (3910024289).

This Springer imprint is published by the registered company Springer Nature Switzerland AG
The registered company address is: Gewerbestrasse 11, 6330 Cham, Switzerland

Foreword by Walter J. Radermacher

A book on machine learning in official statistics takes on an ambitious goal, as it attempts to combine two subject areas that are complex in themselves. It could therefore be tempting to confine the discussion to the methodological and technical aspects. This is not the option taken in this publication. Rather, it seeks to take a broad approach, addressing the foundations and the advances, the origins and the future. This is convenient as it allows the authors to understand the integration of machine learning as an evolutionary progress in official statistics, which in turn makes it possible to build on the structures that have emerged from earlier stages and on the experience gained from them. Over the past two centuries, official statistics have contributed to the construction of states and have gone through various ups and downs, good and difficult times, in the tides of politics. Developments in statistics usually progress continuously and slowly. However, there are also times when abrupt changes occur, namely, when several of the driving forces are acting simultaneously, through new data, methods, technologies; or through new societal information needs, crises, framework conditions; or through an interaction between such driving forces. The last time statistics industry experienced a major development spurt of this kind was a quarter of a century ago, after the fall of the Berlin Wall and the end of the Cold War, when in the same period personal computers and the internet began to revolutionise the work in statistics factories and in communication with social users and groups. The term 'statistics industry' is used here because it allows us to refer to similarities in developments such as the switch to a more decentralised organisation of data processing, which on the bright side led to improved efficiency and acceleration, but on the dark side posed risks to the quality of statistics and challenging adjustments to the structure of personnel in the offices. The origins of a modern understanding of official statistics, namely as an infrastructure for public discourse and the common good based on trustworthy information, go back to this time of upheaval, which was both technical and social. Since then, we have become accustomed to defining the quality of statistics from the end, as 'fitness for purpose'. Modern management of quality in a comprehensive 'total' sense, as well as ethical guidelines, has found their way into international standards and codes, such as the UN Fundamental Principles of Official Statistics, the Declaration of Professional Ethics of the International Statistical Institute and the European Statistics Code of Practice. Organisational structures and processes, the business model and governance for the statistics industry have changed fundamentally, and international cooperation has become even more important. In the flow of developments, we are once again passing through a rapid section with significant currents and shallows, which require fast and attentive manoeuvring. The focus is on the interrelated elements in the technology cluster, which includes machine learning as well as cybersecurity or social media, for example. The socio-political context is similarly important in the form of an erosion of social cohesion, changes in lifestyle and, in general, changes in social institutions caused by information flows, new technologies, echoed in turn by political governance through information. This volume addresses a wide range of issues related to the digital transformation, with a view to the possible improvement of efficiency, timeliness and other aspects of quality while maintaining and preserving the core brand of official statistics. All stages in the value chain of statistics are involved, from design and production to communication and application by users. Developing statistical methodology is translated and deployed in practice, where experts in various fields need to be familiarised with it. Both in terms of data collection and communication, new technologies are opening up possibilities that are being explored. Finally, governance and ethics issues are contained in this book, which provide a framework for responsible, secure and fair handling of the new technologies and data. The statistical community as a whole can best address these issues by working together as quickly and on as international a scale as possible. Technology should be seen as a driver, but not the only one. Equally important are the users and their information needs, as well as the experts who, as producers of statistics, are responsible for ensuring high quality. 'Zukunft braucht Herkunft'[1] ('Future needs origin') is taken from an essay by Odo Marquard, a German philosopher. Technologies and crises present themselves in a new configuration, with new challenges, and for this, old recipes are not enough. Without knowledge of the past, without lessons learned from experience, however, one loses one's compass in the present. Striking a balance between openness to the future and orientation to the existing, driving rapid evolutionary development without shaking the supporting pillars that need to be preserved—that is the challenge and art in this epoch of transformation.

Ludwig-Maximilians-University Munich,
Department of Statistics, Germany
March 2025

Walter J. Radermacher

[1] O. Marquard. Zukunft braucht Herkunft: Philosophische Essays. Reclam, 2003.

Foreword by Wesley Yung

I had the pleasure of meeting Florian Dumpert in 2019 when we worked on a project on machine learning in official statistics sponsored by the High Level Group for the Modernization of Official Statistics. When we met, Florian was completing his doctorate, and I thought that given his knowledge of machine learning and his strong theoretical foundations in statistics, he was someone who could move the use of machine learning in official statistics forward. He has proven me right with the workshops that he has organised and this book where he has played the role of editor. This book is an excellent addition to the literature as we navigate the integration of machine learning into the production of official statistics.

National Statistical Offices (NSOs) are at a crossroads with increasing demands for more data in a timelier fashion, at more detailed levels and for more complex concepts. At the same time, NSOs are under external pressures such as competition from data holding organisations producing statistical information, declining response rates to surveys, decreasing trust in governments and budget reductions. In response to these challenges, many NSOs are looking to leverage alternative data such as images, textual information and other unstructured data. The use of these data sources requires data science techniques such as machine learning and has the potential to increase timeliness, reduce or eliminate respondent burden, and increase data quality. As well, NSOs are looking at machine learning to help optimise operations as a way to combat budget pressures. As one can see, the publication of Foundations and Advances of Machine Learning in Official Statistics is very timely.

The use of machine learning in official statistics has to be done carefully and be well thought out. Official statistics are for the public good and have the potential of impacting the lives of millions of citizens. Because of this, official statisticians have developed many frameworks to ensure that any inferences made to the populations of interest are based on scientific methods. The use of scientific methods is included in the United Nations' Fundamental Principles of Official Statistics (Principle #2). To remain true to these principles, machine learning needs to be considered under the existing frameworks or the existing frameworks need to be extended to cover machine learning. Foundational work on this needs to be done as official statisticians cannot allow themselves to be swept up by the hype of machine learning or artificial intelligence. The statistics they produce are too far reaching to implement methods without solid statistical foundations.

The core business of an NSO is to produce estimates of parameters of the populations of interest. These estimates allow data users to make inference to those populations. A core component of these inferential estimates are measures of uncertainty. How to calculate these measures of uncertainty for many machine learning methods is an open question which needs to be resolved if NSOs are to realise the full potential of machine learning models and to produce estimates directly from them. This open challenge is covered in chapters related to resampling and the Total Machine Learning Error framework. Other important aspects related to the adoption of machine learning in the production of official statistics are nicely laid out in the chapters on legal implications and fairness in machine learning. While NSOs may not be able to leverage machine learning models to directly produce inferential estimates at this time, there are other areas which will benefit from their use. See, for example, the chapters touching on streamlining business experiences, building platforms and pipelines, and the chapters related to use cases in Germany and Italy.

Machine learning and data science in general have the potential to be very powerful tools for official statisticians. The current uses are more in what I would call an NSO's ancillary business as opposed to its core business, but they can certainly help relieve some budget pressures. The extension to an NSO's core business requires some research, but it will be well worth it when achieved. As this is a challenge for all NSOs, I would encourage an approach which favours collaborations across NSOs and with the academic community. My hope is that this book will start the ball rolling on this very important research topic.

Statistics Canada, Ottawa, Canada
March 2025

Wesley Yung

Contents

Contributors

Romain Avouac INSEE (National Institute of Statistics and Economic Studies), Montrouge, France

Sandra Barragán Statistics Spain (INE), Madrid, Spain

Felix Beuter Federal Statistical Office of Germany, Wiesbaden, Germany

Bernd Bischl Department of Statistics, LMU Munich, Munich, Germany

Munich Center for Machine Learning (MCML), Munich, Germany

Ludwig Bothmann Department of Statistics, LMU Munich, Munich, Germany

Munich Center for Machine Learning (MCML), Munich, Germany

Anne-Laure Boulesteix Institute for Medical Information Processing, Biometry and Epidemiology, Faculty of Medicine, LMU Munich, Munich, Germany

Munich Center for Machine Learning (MCML), Munich, Germany

Mauro Bruno Istat, Rome, Italy

Elena Catanese Istat, Rome, Italy

Erika Cerasti Istat, Rome, Italy

Frédéric Comte INSEE (National Institute of Statistics and Economic Studies), Montrouge, France

Piet J. H. Daas Department of Methodology, Statistics Netherlands, Heerlen, The Netherlands

Susanne Dandl Department of Statistics, LMU Munich, Munich, Germany

Munich Center for Machine Learning (MCML), Munich, Germany

Massimo De Cubellis Istat, Rome, Italy

Fabrizio De Fausti Istat, Rome, Italy

Marco Di Zio Istat, Rome, Italy

Hendrik Christian Doll Deutsche Bundesbank, Frankfurt am Main, Germany

Florian Dumpert Federal Statistical Office of Germany, Wiesbaden, Germany

Thomas Faria INSEE (National Institute of Statistics and Economic Studies), Montrouge, France

Lars Fiedler IT-Systemhaus der Bundesagentur für Arbeit, Nuremberg, Germany

Sebastian Fischer Department of Statistics, LMU Munich, Munich, Germany

Gerarda Grippo Istat, Rome, Italy

Johannes Gussenbauer Statistics Austria, Wien, Austria

Barbara Hofmann IT-Systemhaus der Bundesagentur für Arbeit, Montrouge, France

Roman Hornung Institute for Medical Information Processing, Biometry and Epidemiology, Faculty of Medicine, LMU Munich, Munich, Germany

Munich Center for Machine Learning (MCML), Munich, Germany

Christoph Kern Department of Statistics, LMU Munich, Munich, Germany

Munich Center for Machine Learning (MCML), Munich, Germany

Emily Kormanyos Deutsche Bundesbank, Frankfurt am Main, Germany

Frauke Kreuter Department of Statistics, LMU Munich, Munich, Germany

Munich Center for Machine Learning (MCML), Munich, Germany

University of Maryland, College Park, MD, USA

Leon Krög Universität Mannheim, Schloss Westflügel, Mannheim, Germany

Theresa Küntzler KfW, Frankfurt am Main, Germany

Giuseppe Lancioni Istat, Rome, Italy

Koen Loogman Master student of FOM Hochschule für Oekonomie & Management, Nuremberg, Germany

Giulio Massacci Istat, Rome, Italy

Elias Minther Federal Statistical Office of Germany, Wiesbaden, Germany

Stefano Mugnoli Istat, Rome, Italy

Francesco Ortame Istat, Rome, Italy

Angela Pappagallo Istat, Rome, Italy

Adrián Pérez-Bote Statistics Spain (INE), Madrid, Spain

Francesco Pugliese Istat, Rome, Italy

Marco J. H. Puts Department of Methodology, Statistics Netherlands, Heerlen, The Netherlands

Alessandra Righi Istat, Rome, Italy

Alberto Sabbi Istat, Rome, Italy

Younes Saidani Federal Statistical Office of Germany, Wiesbaden, Germany

David Salgado Statistics Spain (INE), Madrid, Spain

Luis Sanguiao-Sande Statistics Spain (INE), Madrid, Spain

Patrick Oliver Schenk Department of Statistics, LMU Munich, Munich, Germany

Munich Center for Machine Learning (MCML), Munich, Germany

Tobias Scherl IT-Systemhaus der Bundesagentur für Arbeit Nuremberg, Germany

Hannah Schulz-Kümpel Department of Statistics, LMU Munich, Munich, Germany

Sebastian Seltmann Deutsche Bundesbank, Frankfurt am Main, Germany

Carlos Sáez Statistics Spain (INE), Madrid, Spain

Francesco Sisti Istat, Rome, Italy

Donato Summa Istat, Rome, Italy

Viktoria Szabo Reply, München, Germany

Luca Valentino Istat, Rome, Italy

Susanne Wegner Federal Statistical Office of Germany, Wiesbaden, Germany

Chapter 1
Machine Learning in Official Statistics: A Preface-Like Introduction

Florian Dumpert

Machine learning. What is that supposed to be? Do we need it? Isn't this just something that 'they up there' heard about and now want to have too? Or is it perhaps something that computer nerds claim to need, even though it's just a gimmick? These or something similar were probably the questions in many statistical offices when the first attempts were made to take on machine learning 10–15 years ago. The early adopters certainly did not do this by first designing a strategy and then setting up a project with a steering committee and controlling formal matrix structure. Instead, at least in Germany, there were colleagues at the Federal Statistical Office who, due to their personal interest, were dealing with 'the new stuff', thinking about application examples from their own area of responsibility and just trying it out. Of course, without the IT equipment that we know in many places today. Of course, without thinking about methodological details. Of course, without extensive preprocessing or efficient tuning. None of that was important. What was important was to identify the potential of machine learning for the work of a statistical office. It should be noted, by the way, that early adopters do not necessarily have to be colleagues from the methodology or IT departments. Nor do they necessarily have to be the only ones dealing with such 'new stuff'. On the contrary, if the first attempts and investigations are also supported by managers (and this is done in the full knowledge that it can also fail and produce sunk costs), it can only be beneficial. These and further developments, right up to the point where machine learning is used as a matter of course, are sure to proceed somewhat differently in the various statistical offices and institutes. Dumpert (2024) retraces the journey for Germany, Bruno et al. (2025) a slightly different one for Italy. Nowadays, there is agreement that machine learning (ML) will ultimately be

F. Dumpert (✉)
Federal Statistical Office of Germany, Wiesbaden, Germany
e-mail: florian.dumpert@destatis.de

F. Dumpert (ed.), *Foundations and Advances of Machine Learning in Official Statistics*, Society, Environment and Statistics,
https://doi.org/10.1007/978-3-032-10004-7_1

'completely normal' and no longer anything special. So, what exactly is machine learning? Perhaps this is a futile question. How should one answer it? With a definition from the beginning (Samuel 1959, p. 211)?

> The studies reported here have been concerned with the programming of a digital computer to behave in a way which, if done by human beings or animals, would be described as involving the process of learning. [...] Programming computers to learn from experience should eventually eliminate the need for much of this detailed programming effort.

Or Simon's answer to the question of what this learning is that the machine is supposed to do (Simon 1984, p. 28)?

> Learning denotes changes in the system that are adaptive in the sense that they enable the system to do the same task or tasks drawn from the same population more efficiently and more effectively the next time.

One doesn't really know. Or do you list methods and say that these are machine learning? This in turn leads to the fact that, for example, logistic regression (regularised or not) is sometimes machine learning and sometimes not, depending on the textbook. So that's not a solution either, and everyone probably sees it a little differently. I don't have a solution, and maybe a definition isn't necessary at all. Perhaps because no distinction from a conceptual counterpart, possibly from a supposedly 'classical statistic', is required. If you look at the various sources (and the author hereby indicates that he comes from the field of statistics), a few recurring features can be identified that keep coming up when it comes to machine learning:

- Often (though not always), machine learning is about non-parametric statistical methods with the aim of recognising patterns and relationships.
- Historically, but changing at the moment,[1] the focus—at least in so-called supervised learning—was more on prediction (and less on explanation).
- The methods are used to process certain tasks without being given the solution explicitly.
- They are generally data driven rather than model driven.
- The methods are characterised by the fact that the solution space (meaning the hypothesis space) is often so large that it contains (approximately) all patterns and relationships.

Deep learning is of course also machine learning, and the enigmatic term 'artificial intelligence' (AI) is also often used synonymously with machine learning. In short, this book is about machine learning in the above sense, also AI/ML, and also deep learning.

Machine Learning in Official Statistics

Let's take another look at the (admittedly still quite young) history of machine learning in official statistics: while the early years were characterised by trial and error and getting to know the subject, and also brought about a kind of cultural change, today other topics and questions are coming to the fore that, while all

[1] See e.g., Dandl et al. (2025) in this book.

of them are highly relevant to machine learning, deal with aspects at a different level, a meta-level. This is not to say that there are no more use cases; the opposite is true. But over time, questions have also arisen in the statistical offices that could not be answered at the time. This is not a bad thing either, but rather a testament to the maturing process of machine learning in official statistics. This book helps to capture an interim result of this maturing process, but also to show where action is needed. As is quite natural at meta-levels, this need for action is often not immediately apparent, but only emerges over time and from question to meta-question, etc. These questions are of an incredibly diverse nature and in some cases also typical for official statistics (although not exclusive to it). In the overview of the book below, these questions are also commented on again. By 2018 at the latest, the international community began to address the question of how machine learning and official statistics can be combined. This led, for example, to the UNECE HLG-MOS Machine Learning Project, which was carried out in 2019 and 2020. Under the leadership of Claude Julien, machine learning was addressed with its use cases of coding and classification, editing and imputation, and imagery analysis, along with the topics of quality and integration.[2] In the following years, there were further projects under the leadership of the UNECE and the UK Office for National Statistics (ONS).[3] Topics such as 'From Idea to Valid Solution', 'From Valid Solution to Production' and 'Model Retraining', 'Infrastructure', 'Quality of Training Data', and many use cases were discussed by colleagues from many participating countries and international organisations. In 2023, a Machine Learning for Official Statistics Workshop took place in Geneva,[4] which in turn discussed use cases and provided an opportunity for exchange and discussion on topics such as 'Quality Aspects of Machine Learning in Official Statistics' and 'System-wide Transformation of Statistical Production'. In addition, the Applying Data Science and Modern Methods Group is working on topics such as 'ML based data editing in statistical production'.[5]

The European Commission and Eurostat also recognise that machine learning offers an important contribution to the production of official statistics and are promoting cooperation in the European Statistical System through the 4-year AIML4OS (AI/ML for Official Statistics) project.[6] In addition to questions arising from and about the application domains (earth observation and satellite imagery, editing, imputation, classification and coding, firm-level supply chain networks, large language models, synthetic data), there are overarching work packages on infrastructure, standards, methodology, knowledge transfer, and training. A total of 16 countries are involved in this project. The kick-off meeting of this project

---

[2] https://unece.org/sites/default/files/2022-09/ECECESSTAT20216.pdf

[3] https://statswiki.unece.org/spaces/ML/pages/266142512/Machine+Learning+for+Official+Statistics+Home

[4] https://unece.org/statistics/events/ML2023

[5] https://w3.unece.org/stories/2024/09/mlops_wp/introduction.html

[6] https://cros.ec.europa.eu/dashboard/aiml4os

took place in Wiesbaden in April 2024, right before the International Conference on Foundations and Advances of Machine Learning in Official Statistics,[7] organised by the Federal Statistical Office of Germany. This conference, with its 150 participants from 19 countries and a total of over 40 talks, once again provided an opportunity for intensive exchange and community building. The slides presented at the conference talks are publicly available.[8] At the same time, this conference provided the perfect opportunity for this book. The aim was to give those involved in the conference, but also other dedicated colleagues, the opportunity to summarise their knowledge and experiences, and to make them available to the public in an edited form. In this way, a diverse range of topics in official statistics related to machine learning can now be covered in this book. Some of the topics are very specific (e.g. limited to a specific application), while others provide valuable insights into the activities of an office. Like the conference, the book is truly comprehensive: many areas are discussed, and many countries are represented among the participants. This demonstrates the strong common interest in advancing machine learning in official statistics without forgetting the basics or ignoring essential foundational (statistical) questions—always with the aim of developing official statistics in line with the demands of the times. Of course, machine learning is not an end in itself, but serves to improve official statistics, whether by enabling the use of further data sources, by partially automating processes, or by accelerating tasks. Improving the quality of processes and products is the purpose of using machine learning in official statistics, and some examples of this, along with the questions to be answered for them, can also be found in this book.

Part I: Methodological Aspects

The first part of this book is about methodological aspects. This is an important field and an extremely important one for official statistics: decisions are made on the basis of official statistics. At the same time, the data situation in official statistics is quite different from that in many parts of industry. Here, there are many surveys based on samples, administrative data, and data such as credit card data or remote sensing data. The collection, aggregation, and provision of these data provide a basis for decision-making for the legislative, executive, and judicial powers, as well as for companies and individuals. While sampling and statistical confidentiality have long played a major role in official statistics from a methodological point of view, machine learning, with all its issues, has only come about relatively recently. It therefore cannot look back on a long history of method development. It is possible that a conceptual framework for the methodological issues still needs to be developed before they can be categorised and the right questions identified and formulated (or expressed in formulas). Of course, there are links to sampling theory and statistical confidentiality, as well as to imputation, for example. And yet there are questions that have only found their way into official statistics with the use of

[7] https://www.destatis.de/EN/About-Us/Events/Machine-Learning/_node.html

[8] https://www.destatis.de/EN/About-Us/Events/Machine-Learning/program.html

machine learning. Following Friedrich et al. (2022), you could also say: it's crazy what you can do wrong if you're not clear about these questions (and the associated answers).

The chapter by Puts et al. (2025), i.e., Chap. 2, is groundbreaking for how we think about our work with machine learning in official statistics in the future, how we structure it, and how we approach questions. Like the GSBPM,[9] for example, their Total Machine Learning Error model could develop into a lingua franca. Puts et al. take their inspiration from the Total Survey Error Model and break down machine learning into the individual steps that occur when creating and applying machine learning procedures in official statistics. In doing so, they also address finite and infinite populations, both of which play a role here. Of course, they cannot yet provide answers to all questions. However, this is not the purpose of this chapter. Rather, it is intended to organise our thoughts and show where further research is needed. Unfortunately, one of the authors of this chapter did not live long enough to see the fruits of his labour. Many chapter authors of this book and many more colleagues from official statistics and academia have come to know Piet Daas as an expert in his field and a highly esteemed colleague. It is a high honour for the editor to publish a contribution by Piet Daas in this book.

As already mentioned above, official statistics also differs from many other areas, such as industry, in the fact that the national statistical institutes (NSIs) mainly work with sample data (as well as administrative data). However, data from samples are often not independently and identically distributed, which may be a requirement for many machine learning methods. At least, this applies to many theoretical results and existing implementations. These implementations are therefore not readily applicable to sample data and—possibly much worse—the output quality measures for a procedure, such as generalisation error, are not correctly estimated. In the worst case, they are too optimistic and lead users to believe erroneously that the machine learning procedure is performing acceptably well. In Chap. 3, Schulz-Kümpel et al. describe some challenges in resampling-based performance estimation. They begin by discussing various concepts that fall under the term 'generalisation error' and describe different approaches to estimating these generalisation errors. In addition to good point estimates, confidence intervals for generalisation errors are also important—which should be a matter of course for every statistician. The authors provide recommendations regarding the latter. Another section of Schulz-Kümpel et al. (2025) provides recommendations for generalisation error estimation in nonstandard, especially non-i.i.d., situations that are relevant in official statistics: clustered data, spatial data, presence of sampling probabilities, concept drift, and hierarchical classification.

[9] Generic Statistical Business Process Model, see https://unece.org/statistics/documents/2019/01/standards/gsbpm-v51

Part II: Legal, Ethical, and Quality Aspects

Part II of this book deals with legal, ethical, and quality aspects, which are no less important meta-topics of the use of ML in official statistics. Quality is the core competence and at the same time the unique selling point of official statistics. And without quality, there is no trust in official statistics. In the end, it is quality that counts. This also and possibly especially applies to the use of machine learning as a new technology in official statistics. (At least the possibility of using ML in official statistics on a large scale from a technological point of view is new.) This topic has already been discussed in depth by many colleagues in official statistics, for example, in Yung et al. (2022) and in two special issues of *AStA*, a journal of the German Statistical Society (Dumpert et al. 2023; Burgard et al. 2024). This book continues this discussion.

The contribution by Saidani and Dumpert, Chap. 4, takes up quality dimensions that should be considered when using machine learning (or other statistical algorithmic methods) to further develop production processes and products and substantiates these in the form of concrete quality guidelines (with requirements to be evaluated as fulfilled/not fulfilled) for each quality dimension. However, the quality concept for the use of machine learning does not come out of the blue, but is derived from existing quality concepts and quality frameworks. The quality dimensions derived are accuracy, robustness, explainability, reproducibility, timeliness and punctuality, and cost-effectiveness. MLOps is mentioned as an additional overarching aspect, as also advocated by Barragán et al. (2025) and taken up by Avouac et al. (2025) in this book. At the same time, Saidani and Dumpert (2025) already build a bridge to the concept of fairness as an overarching aspect.

But first, in Chap. 5, Dandl et al. (2025) address the quality dimension of explainability, which is of course not completely separate from the concept of fairness. While Saidani and Dumpert (2025)—where necessary—require the use of explainable methods as a quality guideline, Dandl et al. show possible solutions to fulfil this requirement. They focus on post hoc interpretation methods and particularly highlight the concept of loss-based feature importance, as well as counterfactual and semi-factual explanations. In addition, they introduce `mlr3summary`, an easy-to-use R package that generalises the well-known summary function R and thus provides a practical tool for the interpretability of trained models. Colleagues from the field of official statistics can use this package to address the quality aspect of explainability.

The meaning of the concept of fairness (or of different concepts of fairness, as the field is quite multifaceted) in official statistics is examined by Schenk et al. in Chap. 6. Official statistics is a discipline that often does not deal with automated decision-making with individual effects, but with the use of machine learning in data processing. However, the authors emphasise that official statistics products that may have been created using machine learning can later be used to train systems for automated decision-making, so NSIs play a crucial upstream role for achieving fairness. While so-called protected attributes often play an important role in the fairness consideration in the literature, fairness in official statistics also refers to other subpopulations of interest (e.g. spatial or temporal ones). This highlights the relationship to quality, and the overarching quality aspect of fairness is elaborated in Schenk et al. (2025) through references to other quality dimensions from Saidani and Dumpert (2025).

While fairness already touches on the field of ethics, it should be noted that this book does not contain any further consideration of ethical issues. On the one hand, this is regrettable if one strives for a certain comprehensiveness. Official statistics have indeed dealt with the topic of 'ethics and ML', as shown by Statistics Canada's *Framework for Responsible Machine Learning Processes*[10] and UK Statistics Authority's *Ethical considerations in the use of Machine Learning for research and statistics*.[11] On the other hand, it can be argued that compliance with legal and quality requirements already covers many of the ethical issues; see the two papers by Dumpert et al. (2025a) and Dumpert et al. (2025b).[12]

Even stronger than quality considerations, of course, are the legal requirements already mentioned. These differ from country to country,[13] although, for example, Europe has at least provided a uniform legal framework for the use of machine learning in general with the AI Act. A special consideration of machine learning in the light of legal requirements, in particular data protection and the European AI Act, is provided by Krög's Chap. 7—concluding Part II of this book on legal, ethical, and quality aspects. The chapter focuses on the legal situation in Germany, but in many places, it is at least transferable to other European countries. For example, the author deduces (Krög 2025, Section 7.3.2)

> that ML in official statistics is subject to the AI Act. However, the already existing and planned applications are low-risk AI methods. Consequently, they are not prohibited by the regulation, and the obligations that the regulation places on high-risk AI do not apply to them either. Consequently, the requirements for the Federal Statistical Office as a provider of an AI system are primarily training of the responsible employees in the area of AI literacy.

Part III: Technological Aspects

Part III is dedicated to the technological aspects of machine learning. It is obvious that machine learning methods also require appropriate hardware and software. However, it is not just a question of quantity. It is equally important to manage resources efficiently and effectively, including the administration of data and models. For example (Bruno et al. 2025, Sect. 12.1.2),

> Istat has established an integrated production system with its conventional data acquisition and production infrastructure to facilitate smart statistics production. This system incorporates the Integrated System of Registers, current surveys, and externally manageable processing procedures using agreed methodologies, algorithms, and reliable software. Implementing such a system involves a multi-tiered data processing workflow organization

---

[10] https://www150.statcan.gc.ca/n1/en/pub/89-20-0006/892000062021001-eng.pdf

[11] https://uksa.statisticsauthority.gov.uk/publication/ethical-considerations-in-the-use-of-machine-learning-for-research-and-statistics/pages/1/

[12] Their translations into English can be found here: https://www.destatis.de/EN/Methods/WISTAScientificJournal/_publikationen-articles-en.html

[13] Admittedly, the general quality frameworks also differ, although there is already a strong harmonisation, at least among the member states of the UNECE.

> and adopting standard manufacturing process management models, such as the Generic Statistical Business Process Model (GSBPM).

And Barragán et al. (2025, Sect. 13.4) are absolutely right when they write in their chapter in Part IV:

> Machine learning is a highly data-intensive activity. Therefore, data governance, data management, and data architectures are crucial to implement these methods at scale. The amount of pre-processing tasks in our proofs of concept to prepare data and compute regressors (feature engineering) would clearly benefit if a data architecture with fully fledged metadata is put in place and shared among all surveys. This would reduce the cost for the discovery and development of new ML-based applications.

Chapter 8, the contribution by Avouac et al. from INSEE, the French national statistical institute, takes a very detailed look at this, presents a possible technological solution, and substantiates its applicability with practical examples. The MLOps topic is also extensively examined in this chapter. With Onyxia, Avouac et al. (2025) are introducing an open-source data science platform that enables modern and flexible work with data and (machine learning) models in official statistics. It is becoming apparent that this, in turn, also promotes the consideration of quality aspects and compliance with quality guidelines as described in Part II.

Part IV: Use Cases and Insights

Part IV of this book finally traces developments in statistical offices and deals in detail with typical use cases of official statistics. The contribution of such chapters to the further development of (machine learning in) official statistics should not be underestimated. Of course, these are initially 'only' examples. However, this view falls short. Even if the conditions in the offices, data sets, and further use of results often (and sometimes enormously) differ between the use cases, the presentation of these applications still fulfils important functions:

(i) Demonstrating potential: The NSIs are under pressure to develop further. This is often accompanied by the need to support process steps that were previously carried out purely manually or by means of hard-coded programs with statistical (machine learning) algorithms. Exchanging ideas and possibilities as well as best practices in the form of papers about one's own (successful and unsuccessful) attempts is important for inspiring colleagues in other statistical offices.

(ii) Discovering similarities: It is not always necessary to point out potential. Instead, several NSIs came up with the same ideas and developed solutions for the same or very similar challenges. All of them had their experiences and observations, perhaps wondered about the inadequate performance of their models, and identified difficulties in the data material. An exchange about use cases makes it possible to discover and evaluate any similarities that may exist. Ideally, joint reasons can even be determined on this basis.

(iii) Finally, writing things down often helps to structure one's own thoughts, to question approaches or procedures that may only appear clear, and to evaluate one's own work.

Part IV not only offers a description of use cases but often also reveals the underlying methodology, quality aspects, or technological questions. Many chapters could therefore also have been assigned to Parts I, II, or III.

The automated processing of information transmitted in text form or found on the Internet represents an enormous potential for increasing efficiency in official statistics. It is therefore not surprising that some of the chapters already mentioned and many that follow contain an example of an application in this area. Chapter 9 by Fiedler et al. describes studies by the German Federal Employment Agency to analyse job advertisements. Adaptations of the BERT model are used here. Fiedler et al. (2025) show that domain adaptation is not necessary if a sufficient amount of high-quality data is already available.

In Chap. 10, Statistics Austria and the Federal Statistical Office of Germany, two NSIs in the German-speaking region, explore the possibilities of automated classification of economic activities (NACE) based on German-language textual descriptions. In their contribution, Beuter et al. discuss this NLP problem, including data preprocessing, data augmentation, feature selection, and various classifiers. They also introduce hierarchical performance measures and discuss the classification with respect to these. Finally, Beuter et al. (2025) also examine the use of LLMs to improve the classification. The authors conclude their chapter by emphasising that it seems to be inherent to the NACE classification that at the deeper levels accurate classification becomes difficult, which is, however, also true for human classification because businesses can be complex and not always easy to categorise.

A completely different application is dealt with by Küntzler in Chap. 11. The integration of data plays an important role in the further development of official statistics with regard to analyses and evaluations, but also with regard to reducing the burden on respondents. If data sets are to be merged, i.e., information from two different data sets shall be brought together, and if at the same time there are no common identifiers (because the statistical units observed are at least partly different or because identifiers are practically non-existent or prohibited by law), statistical matching is often the only option. This involves comparing common information (e.g. given by a common set of attributes) of the statistical units. If there is sufficient similarity, the missing attributes in one data set are transferred from the other. The relationships between the variables in the data sets must be modelled appropriately, which is a time-consuming process. Küntzler (2025) offers a machine learning-based alternative and demonstrates its usefulness with two examples.

Chapter 12 gives us a valuable insight into the development of the topics of big data and machine learning at the Italian National Institute of Statistics (Istat). Bruno et al. (the large number of authors alone shows how important the topic is) describe the background and Istat's approach in great detail. They also address challenges that have not yet been mentioned in this book in such explicit terms, namely, the changed skills that colleagues in statistical offices must now possess. Domain knowledge, statistics, and computer science are necessary in order to be able to work appropriately with big data and machine learning. The statistical offices also need close links with academia in order to be able to solve open questions, which often require basic research to answer. Bruno et al. (2025) demonstrate the relevance of big data and machine learning in official statistics by means of some detailed use cases.

Several times already quoted was the contribution of Barragán et al. due to its clear insights. The authors describe their approaches and experiences in improving the processes at the Spanish National Statistical Office (INE). The aim is to improve accuracy, cost-efficiency, timeliness, granularity, response burden reduction, and frequency. In Chap. 13, Barragán et al. (2025) discuss design-based predictive inference, selective and macro editing, statistical classification coding, imputation, integration of administrative data, and time disaggregation of sampling designs. However, they do not simply describe the application, but rather support their use cases with advanced statistical methodology.

The book ends with another example of how LLMs can be used. This example is not directly concerned with the production of official statistics, but rather with understanding how they are used. More specifically, Seltmann et al. in Chap. 14 refer to data sets such as those provided by the research data centres of statistical offices, central banks, other national authorities (ONAs), etc. In order to optimise the data provided, it is first necessary to determine which data set is used for research, how often and in conjunction with which other data sets. Seltmann et al. (2025) use a retrieval-augmented generation pipeline using GPT-3.5 to avoid the problem that citation of data sets is currently hardly standardised and standardised approaches are often not known or are not used for other reasons. This allows the authors to develop a cost-effective and at the same time-effective way to track the use of data sets based on publications.

Some Final Remarks

In addition, a book like this is always a great source of literature and a thorough literature overview, here from different nations as well as fields and varieties of official statistics and science. The authors refer in their contributions to the diverse topics as described above to books, journal articles, official documents, and important and interesting websites. In addition to a treatment of the respective (special) topic, readers of the relevant chapters are thus offered a wide range of additional information and starting points. This is very useful if you want to work on a topic in your own NSI, central bank, ONA, or a comparable institution. To the best of our knowledge, this book offers this overview of primary and secondary sources on the topic of machine learning in official statistics for the first time in this form. For this, I would like to thank the authors of the chapters, who, through their diligent work, laid the foundations for their contributions, who then cast their insights in chapter form and made them available for this book and thus for the public, and who, last but not least, entrusted the editor with their work for compilation. All of this is not self-evident.

References

R. Avouac, T. Faria, F. Comte, A cloud-native data science platform for official statistics, in *Foundations and Advances of Machine Learning in Official Statistics*, ed. by F. Dumpert, Chap. 8 (Springer, Berlin, 2025)

S. Barragán, A. Pérez-Bote, C. Sáez, D. Salgado, L. Sanguiao-Sande, Streamlining business functions in official statistical production with machine learning, in *Foundations and Advances of Machine Learning in Official Statistics*, ed.by F. Dumpert, Chap. 13 (Springer, Berlin, 2025)

F. Beuter, J. Gussenbauer, E. Minther, V. Szabo, S. Wegner, Approaches to automated NACE coding of German business activity descriptions, in *Foundations and Advances of Machine Learning in Official Statistics*, ed. by F. Dumpert, Chap. 10 (Springer, Berlin, 2025)

M. Bruno, E. Catanese, E. Cerasti, M. De Cubellis, F. De Fausti, M. Di Zio, G. Grippo, G. Lancioni, G. Massacci, S. Mugnoli, F. Ortame, A. Pappagallo, F. Pugliese, A. Righi, A. Sabbi, F. Sisti, D. Summa, L. Valentino, Big data and machine learning at Istat, in *Foundations and Advances of Machine Learning in Official Statistics*, ed. by F. Dumpert, Chap. 12 (Springer, Berlin, 2025)

J.P. Burgard, M. Zwick, F. Dumpert, S. Wichert, T. Augustin, N. Storfinger, Vorwort der Herausgeber. AStA Wirtschafts- und Sozialstatistisches Archiv **18**(2), 127–130 (2024)

S. Dandl, B. Bischl, L. Bothmann, Interpretable machine learning for official statistics, in *Foundations and Advances of Machine Learning in Official Statistics*, ed. by F. Dumpert, Chap. 5 (Springer, Berlin, 2025)

F. Dumpert, Maschinelles Lernen im Statistischen Bundesamt – Ein Überblick über die Historie seit 2015 und aktuelle Entwicklungen. WISTA Wirtschaft und Statistik **76**(4), 17–28 (2024)

F. Dumpert, S. Wichert, T. Augustin, N. Storfinger, Editorial issue 3+4, 2023. AStA Wirtschafts- und Sozialstatistisches Archiv **17**(3), 191–194 (2023)

F. Dumpert, J. Reichel, E. Oertel, H. Leerhoff, C. Salwiczek, Ethische Fragen beim Einsatz von KI/ML in der Produktion amtlicher Statistiken – Teil 1: Identifikation. WISTA Wirtschaft und Statistik **77**(1), 15–24 (2025a)

F. Dumpert, J. Reichel, E. Oertel, H. Leerhoff, C. Salwiczek, Ethische Fragen beim Einsatz von KI/ML in der Produktion amtlicher Statistiken – Teil 2: Auseinandersetzung. WISTA Wirtschaft und Statistik **77**(1), 25–36 (2025b)

L. Fiedler, B. Hofmann, K. Loogman, T. Scherl, Domain adaptation of a BERT model for analyzing job advertisements at the German Federal Employment Agency, in *Foundations and Advances of Machine Learning in Official Statistics*, ed. by F. Dumpert, Chap. 9 (Springer, Berlin, 2025)

S. Friedrich, G. Antes, S. Behr, H. Binder, W. Brannath, F. Dumpert, K. Ickstadt, H. A. Kestler, J. Lederer, H. Leitgöb, M. Pauly, A. Steland, A. Wilhelm, T. Friede, Is there a role for statistics in artificial intelligence? Adv. Data Anal. Classif. **16**(4), 823–846 (2022)

L. Krög, Legal implications for the use of machine learning in official statistics, in *Foundations and Advances of Machine Learning in Official Statistics*, ed. by F. Dumpert, Chap. 7 (Springer, Berlin, 2025)

T. Küntzler, An automated machine learning pipeline for statistical matching, in *Foundations and Advances of Machine Learning in Official Statistics*, ed. by F. Dumpert, Chap. 11 (Springer, Berlin, 2025)

M. Puts, D. Salgado, P. Daas, Leveraging machine learning for official statistics, in *Foundations and Advances of Machine Learning in Official Statistics*, ed. by F. Dumpert, Chap. 2 (Springer, Berlin, 2025)

Y. Saidani, F. Dumpert, Quality dimensions and quality guidelines for machine learning in official statistics, in *Foundations and Advances of Machine Learning in Official Statistics*, F. Dumpert, Chap. 4 (Springer, Berlin, 2025)

A.L. Samuel, Some studies in machine learning using the game of checkers. IBM J. Res. Develop. **3**(3), 211–229 (1959)

P.O. Schenk, C. Kern, F. Kreuter, Fairness in machine learning for national statistical organizations, in *Foundations and Advances of Machine Learning in Official Statistics*, ed. by F. Dumpert, Chap. 6 (Springer, Berlin, 2025)

H. Schulz-Kümpel, A.-L. Boulesteix, S. Fischer, R. Hornung, Challenges in resampling based performance estimation, in *Foundations and Advances of Machine Learning in Official Statistics*, ed. by F. Dumpert, Chap. 3 (Springer, Berlin, 2025)

S. Seltmann, E. Kormanyos, H.C. Doll, Building a retrieval-augmented generation pipeline to trace administrative data use in academic papers, in *Foundations and Advances of Machine Learning in Official Statistics*, ed. by F. Dumpert, Chap. 14 (Springer, Berlin, 2025)

H.A. Simon, Why should machines learn? in *Machine Learning – An Artificial Intelligence Approach*, ed. by R.S. Michalski, J.G. Carbonell, T.M. Mitchell (Springer, Berlin, 1984), pp. 25–37

W. Yung, S.-M. Tam, B. Buelens, H. Chipman, F. Dumpert, G. Ascari, F. Rocci, J. Burger, I. Choi, A quality framework for statistical algorithms. Stat. J. IAOS **38**(1), 291–308 (2022)

Part I
Methodological Aspects

Chapter 2
Leveraging Machine Learning for Official Statistics

Marco J. H. Puts, David Salgado, and Piet J. H. Daas

2.1 Introduction

2.1.1 Production of Official Statistics and Machine Learning

Evaluating the impact and utilization of machine learning (ML) in the production of official statistics presents an ongoing challenge. ML is a subfield of artificial intelligence, which aims at "not just to understand but also to *build* intelligent entities" (Russell and Norvig 2010, p. 19). It is thus similar to assessing the impact of intelligent human activity on a production system, which is limitless. ML itself is composed of different subfields about how the process of learning is carried out: supervised learning, unsupervised learning, reinforcement learning, etc. (Alpaydin 2020; Murphy 2013).

Despite the widespread adoption of ML, implementation still has many challenges (O'Neil 2016, discusses this subject). Some of these challenges were already reported by van Delden et al. (2023). Despite the danger and complexity of ML, the compelling "datafication" of our society forces us to look at ML as an addition to our (official) statistical toolbox. Datasets get larger, are more detailed, and become more and more complex in terms of interdependencies between variables and datasets. Because of this, it becomes increasingly difficult to perform (classical) official statistical analysis (i.e., statistical analysis for monitoring the effect of policy change or monitoring the different (socio)economic indicators) on these kinds of data. The

M. J. H. Puts (✉) · P. J. H. Daas
Department of Methodology, Statistics Netherlands, Heerlen, The Netherlands
e-mail: m.puts@cbs.nl

D. Salgado
S.G. for Methodology and Sampling Designs, Statistics Spain, Madrid, Spain
e-mail: david.salgado.fernandez@ine.es

F. Dumpert (ed.), *Foundations and Advances of Machine Learning in Official Statistics*, Society, Environment and Statistics,
https://doi.org/10.1007/978-3-032-10004-7_2

need for algorithmic-based approaches, which can handle larger, more complex, and unstructured datasets, is necessary to be able to perform successful analysis; see Breiman (2001). Without it, we would not be able to extract statistical information from many big data sources, like web scraped data (Daas and van der Doef 2020) and aerial images (Jong et al. 2020).

ML encompasses various techniques, often referred to as "a methodology" within the realm of data science. However, most statisticians, as well as most scientists, would disagree with the usage of this term. In social sciences and econometrics, "methodology," "methods," and "techniques" carry specific and distinct meanings, which we advocate for adhering to, at least when talking about ML from an (official) statistical point of view.

When we look at the term "methods," this becomes most apparent. We have observed that the term "methods" is incorrectly used in many fields that apply ML. It is often used to merely describe the chosen environment in which a study is performed. So, when summarizing the ML algorithms and hyperparameters used, maybe with some kind of rationale, many data scientists assume this describes the "method" used. Such a description, however, falls short when viewed from the statistical standpoint of a methodologist.

So let's start with the basis. Techniques, and how we combine them, are primarily determined by the "why" and "what" questions of the application, such as:

- Why are we doing it?
- Why do we choose certain techniques?
- What are we going to do?
- What is our ground material?
- What is the context?

It is the "how" question that is at the core of the techniques themselves: how do we go about performing certain steps? In addition to the algorithmic description, it describes the (pre- and post-)conditions for applying the technique. From this, we can define a method as

> A method is a systematic procedure of techniques for accomplishing a certain goal. Most of the time these methods are established.

and a technique as

> A technique is a way of carrying out a task. Most of the time these techniques are described as algorithms.

For example, preparing a meal involves cutting vegetables, boiling eggs, and grilling steaks. Recipes can be considered methods. It's also possible to consider a method that takes into account the context in which one prepares a meal, as well as the circumstances that may arise (e.g., a guest may be vegan or allergic to certain ingredients) when preparing a meal. What is the best way to use a specific technique with a specific set of ingredients under what circumstances? Methodology is the subject of this area. When it comes to ML, this implies that we must pay greater attention to the context of methodology when discussing it in relation to its application.

Admittedly, sometimes it is hard to determine if something is a method or a technique. Since methods are procedural, they might be interpreted as an algorithm and the other way around. It's therefore easy to understand why there is confusion about methods and techniques. By definition, they are so closely related that it is easy to confuse them, but for reasons that will become clearer through the rest of this chapter, it is crucial to keep them separate.

A priori every production task is susceptible to being impinged by the growing success of ML and deep learning techniques. At least we distinguish two broad groups of production activities potentially affected by these techniques. On the one hand, we have the inference problem providing us with a set of statistics and indicators describing some fragment of social or economic reality. This is, in our view, the core of the business of official statistical production. On the other hand, complementarily important, we have numerous additional tasks improving the quality of the first goal such as data collection, coding, and statistical dissemination. In both cases, by and large, the use of ML boils down to building a predictive model to be applied to new data, thus constituting a process step in the whole production cycle (see Chapters 23–35 in Snijkers et al. (2023); Dumpert 2023; Measure 2023; Moscardi and Schultz 2023). For example, predicting values of a continuous target variable will be useful in building model-assisted estimators (as a concrete illustrative example of a regression task in the first group). Also, an automatic coding machine will provide a predicted category for a given statistical unit (as a concrete illustrative example of a classification task in the second group).

In this chapter, we shall focus on supervised learning so that different algorithms will be trained, validated, and tested on a given dataset to be then applied to new data (see, e.g., Hastie et al. 2009).

2.1.2 *Production of Official Statistics and Quality*

Quality has been the spinal cord of the production of official statistics making it possible to be used for highly relevant policymaking actions in all countries and the international community (see, e.g., United Nations 2019, and the references therein). Nowadays, quality is a multidimensional concept (Karr et al. 2006) and strongly oriented toward users' needs and purpose (European Statistical System Committee 2022). Measuring and determining quality is thus very important. As a consequence, there exist complementary frameworks developed to assess different aspects of the quality of the production of official statistics (see, e.g., Gootzen et al. 2023, and the references therein).

We may cite the output quality approach of many statistical systems (see, e.g., European Statistical System Committee 2022, in the context of the European Statistical System), which focuses on the assurance of multiple quality dimensions for the statistical outputs. In the light of the use of ML techniques, Puts and Daas (2021) describe how their usage can influence the quality of official statistics in several quality dimensions, namely relevance, accessibility and clarity, coherence and comparability, and accuracy and reliability (see European Statistical System Committee 2022, for definitions, specifically principles 11, 12, 14, and 15). Puts and Daas (2021) argue that the following topics must be worked on regarding the methodology of ML in official statistics:

- Accuracy and reliability
  - Methodology concerning the human annotation of data
  - Sampling the population to obtain representative training sets
  - Using stratification in the context of ML
  - Correcting the bias caused by the ML model
- Accessibility and clarity
  - Data structure engineering and selection to increase the transparency of models
  - Explainability of ML models (explainable AI)
- Coherence and comparability
  - Reducing spurious correlations
  - Methodology for studying causation
  - Dealing with concept drift (representativity over time)

Taking the European Statistics Code of Practice (ESCoP; European Statistical System Committee 2022) as a starting point, Saidani et al. (2023) provide an overview of all the attempts made to create an ML quality framework for official statistics.[1] Similar to Puts and Daas (2021), they also conclude that the ESCoP principles 11–15 (statistical output) are crucial in defining the quality of ML algorithms, with the additional consideration of principles 7–10 (statistical processes). They add to this an extra quality dimension, called robustness. As producers of official statistics, we need to make sure that the models are robust for changes in the population and trends; we should ensure that our estimates stay unbiased and accurate, even if the outside world changes rapidly.

However, in the following, we shall focus on the Total Survey Error Model (TSEM henceforth) (Groves and Lyberg 2010) as the starting point for our proposal to adapt this framework to the use of ML in the production of official statistics. As stated above, we shall simplify the use of these techniques to the construction of a predictive model either of continuous, semicontinuous, or categorical variables trained, validated, and tested on a given dataset to be applied to new data for any purpose (thus covering both the core and additional groups of tasks).

The TSEM is a comprehensive framework used in survey research and practice to understand and quantify the various sources of error that can affect the accuracy of survey estimates (thus we are focusing on the reliability quality dimension). The approach acknowledges that no survey is perfect, and errors can arise at different

[1] See also Saidani and Dumpert (2025) in this book.

stages of the survey process. The goal of the model is to identify, measure, and minimize these errors to improve the overall quality of survey data. This framework implicitly assumes that design-based inference is used to construct estimators and their accuracy assessment (confidence intervals, variance estimation, etc.).

In the TSEM, the concepts of population unit and target variable are central, depicted by the so-called representation line and measurement line (see Fig. 2.1). As such, the model identifies measurement-related errors and representation-related errors. The measurement-related errors are associated with what is being measured, whereas the representation-related errors are about the population and its units and how they are included in the sample.

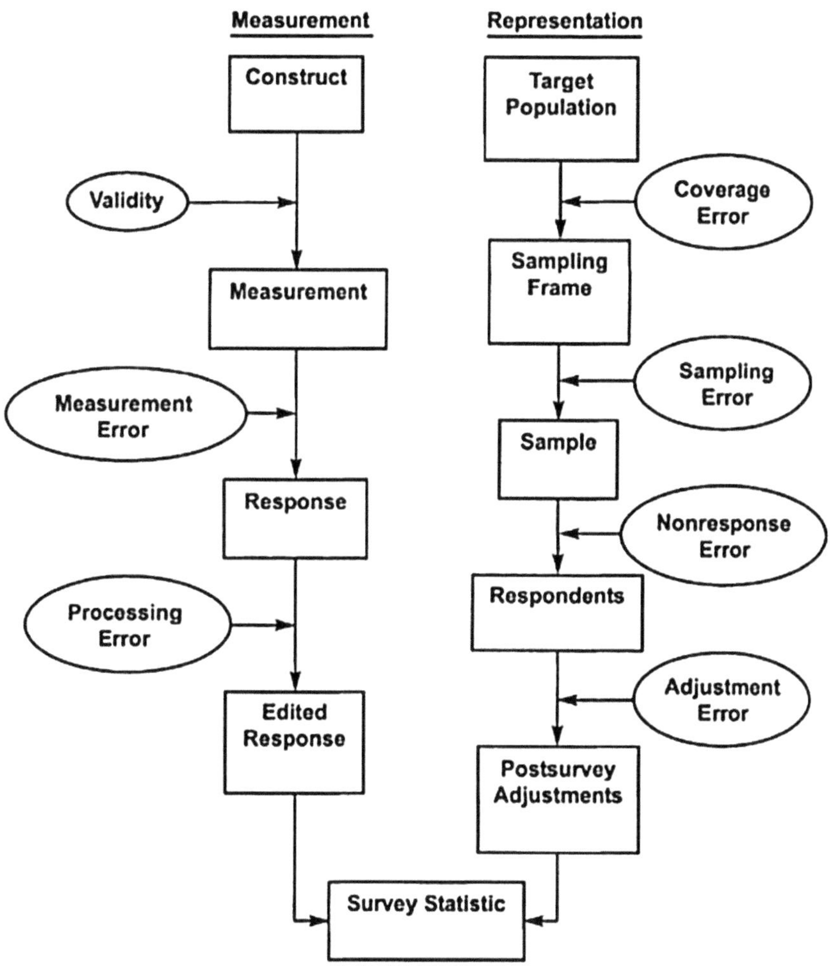

Fig. 2.1 The total survey error model (source: Moya 2020, slightly modified)

We give a brief overview of the definitions of the different sources of error in the TSEM: The measurement-related errors are:

- **(Construct) validity error:** Construct validity errors are related to whether a survey instrument is effectively measuring the theoretical construct or concept it is intended to measure. It assesses the degree to which the survey questions align with the underlying construct of interest.
- **Measurement error:** Measurement errors occur when there is a discrepancy between the true value of a variable and the value obtained through measurement. This can result from respondent misunderstanding, misreporting, or errors in the survey instrument.

  It is caused by respondent biases, unclear survey questions, or issues with data collection instruments.
- **Processing error:** Processing errors involve mistakes made during the data collection and data processing stages, including errors in data entry, coding, and editing.

  It is caused by human error in the handling and processing of survey data.

The representation-related errors are:

- **Coverage error:** Coverage errors arise when the sampling frame does not accurately represent the target population. It includes individuals who should be in the population but are not in the frame (undercoverage) or individuals in the frame who are not in the population (overcoverage). It is caused by an incomplete or inaccurate sampling frame.
- **Sampling error:** Sampling errors occur due to the inherent variability that arises when a sample is used to estimate the characteristics of a population. It is the difference between a sample statistic and the true population parameter. It is caused by a random chance in the selection of the sample.
- **Nonresponse error:** Nonresponse errors result from differences between respondents and nonrespondents. It occurs when individuals chosen for the survey do not participate, and their characteristics differ from those who do participate. It is caused by refusals, inability to reach respondents, or other factors leading to nonparticipation.

Puts et al. (2022) argued that such a model can also be valid for ML. First of all, the training set is a sample of the population, the measurements are imperfect (e.g., the result of annotations) and can contain errors, and the parameters of a trained model can be seen as estimates. They argued that, due to the presence of both measurement errors and representation errors in the training dataset with respect to the population, the model will generate errors, and most of these errors will result in biases when applied to new data.

In addition, Saidani et al. (2023) claim translation of TSEM errors into ML presents a challenge. This observation is fundamentally accurate. TSEM was originally designed for evaluating survey methodology errors, but adapting it to ML presents a number of challenges. Moreover, quantifying the various errors outlined in the TSEM is difficult, and, of course, achieving this task would exceed the scope of this chapter.

In this document, we propose including a TSEM-like view on the construction of a supervised learning model as a means to assess the quality of the model. It will be referred to as the Total Machine Learning Error (TMLE) model. Although the framework will be completed to the greatest extent possible for the time being, there might be some gaps that will need to be filled, thus providing further opportunities for research and refinement in the future.

2.1.3 Internal and External Validity

Validation is an important step in assessing the quality of a model. This is commonly done by splitting the dataset into a training set and a test set.

We would like to argue that splitting the data into a training set and a test set is not enough to prove the validity of the model. Validity can be described as internal or external (Onwuegbuzie and McLean 2003). The internal validity refers to the fact that the model learned the relationships that were present in the training set, whereas the external validity refers to validity beyond that point: The model is able to predict outside the context of the given population.

Since the predictive model has to predict future data, which comes from a future population, the external validity of a model is much more important than the internal validity. This external validity can be checked by (1) using a dataset from a different year and (2) using a dataset from a different location than the training set.

Throughout the chapter, it should be kept in mind that our goal is to reach an external valid model.

2.1.4 Outline of the Chapter

The remainder of this chapter is structured as follows. After the introduction, the next section focuses on populations and samples. Here, the viewpoint of Deming on this topic provides valuable insights. This is followed by a description of the TMLE model and the various phases discerned. In Sect. 2.4 an overview is given of the consequences of the TMLE-model which provides valuable insights. This section is followed by an overview of some ML applications that both inspired and benefited from the development of the TMLE-model. The chapter ends with a discussion in which the most important topics for future research are listed.

2.2 Deming's Machine: Populations and Samples

For our purposes, let's start with an example used by Deming (1942) (later on suggested as an exercise in his well-known textbook about survey sampling (Deming 1950)). An industrial business person owns an industrial machine producing bolts with a set of technical specifications (weight, size, lengths, resistance to temperature, etc.). Every, say, $N = 100$ units are packed up in a box to be sold to retailers. For evident reasons, this person can pose two complementary and different questions in this situation: (a) how many defective bolts (failing technical specifications) exist in each box of $N = 100$ bolts, and (b) how many defective bolts are produced on average by the machine? Notice that these are a priori independent concerns rightfully relevant in this situation. This simple example will allow us to introduce relevant statistical and mathematical concepts which will be the basis for our proposed TMLE-model for the production of official statistics.

Firstly, notice that in question (a) no random element exists in the formulation of the concern: For each box, say, U_i, with a specific known number of bolts, say, N_{U_i}, there exists a fixed but unknown number of defective bolts, say, $N_{U_i^{(D)}} \leq N_{U_i}$. All we want is to know $N_{U_i^{(D)}}$ to monitor the output quality of our production system. There is no reference to any probability distribution underlying the box, and quantities such as total, mean, and variance must be understood as numerical aggregation figures much in the line of exploratory data analysis. The characteristics of each bolt, i.e., the target variables,[2] are fixed but unknown numbers.

However, question (b) contains an implicit reference to an underlying random process or random experiment every time a bolt is produced by the machine. The question is meaningful only when randomness is recognized to be present in the generation process of each bolt so that the result may be different, i.e., sometimes defective and sometimes working. The interest is focused not only on the generation mechanism in the past but also genuinely on (immediately) future instances of the generation process.

This distinction allows us to formalize and motivate the following definitions, which are implicitly used at all times in similar related statistical analyses. In the context of question (a) a finite population U is a set of identifiable units u_k which we usually denote by their labels k so that $U = \{1, \ldots, N\}$. In this line, a sample s is just a subset of units selected from U, i.e., $s \subset U$ (see, e.g., Cassel et al. 1977). Ordered and/or with-replacement samples (Koop 1974) can be further expressed in rigorous mathematical language, but the key underlying concept is the same: A finite population is a collection of population units; no probability measure is involved in the definition; and it is a set-theoretic concept. The characteristics of interest $\mathbf{y}_k$ of these population units are just expressed by numbers, i.e., $y_{qk} \in \mathcal{D}_q$ for each variable $q = 1, \ldots, Q$, where $\mathcal{D}_q$ stands for the numerical set (range) of possible

[2] In our simplified case, the defectiveness binary indicator $\delta_k(U_i^{(D)}) \in \{0, 1\}$.

values of variable y_q. For simplicity's sake, in the following we shall concentrate on a unidimensional variable y so that we can drop the subscript q.

In the context of question (b), more subtleties are needed. The underlying randomness (and probability space) enters into play by defining the population as the probability distribution function $F_{\boldsymbol{\theta}}$ of variable y, thus now turned into a random variable Y. This is indeed the concept of infinite population introduced by Fisher (1922). This distribution function $F_{\boldsymbol{\theta}}$ basically concentrates on the random generation mechanism of the value y_k for each bolt k, in particular, for the defectiveness indicator variable $\delta(U^{(D)})$. The concept of population unit in this context is much subtler since the mathematical definition of a random variable does not make any reference to such a concept. It is a modeling assumption. In our simplified modeling scenario for the machine, we may conceive of each variable $\delta(U^{(D)})$ as a realization of the same binary random variable $\delta(U^{(D)}) \simeq F_{\boldsymbol{\theta}}$. In this way, every time the random experiment is conducted (generation of a bolt), the random variable is realized (as when we toss a coin), and a new value $\delta_k(U^{(D)})$ is generated. This motivates the following definition of a sample (Casella and Berger 2002): "the random variables $Y_1, \ldots, Y_n$ are called a *random sample of size n* from the population $F_{\boldsymbol{\theta}}$ if $Y_1, \ldots, Y_n$ are mutually independent random variables and the marginal distribution function of each variable Y_k is the same function $F_{\boldsymbol{\theta}}$." Alternatively, they are called independent and identically distributed random variables. Following these definitions, a population unit in the context of question (b) amounts to each instance the random experiment is conducted, giving rise to a new bolt, thus identified with this bolt. Notice how the concept of infinite population naturally fits in this description: We have, on the one hand, already generated bolts, but, on the other hand, we may also generate as many as we want since they come from a random experiment.

In sum, the difference between questions (a) and (b) stems from the modeling assumption of conceiving target variable values y_k as realizations of random variables or not, i.e., whether there exists an underlying random generation mechanism or not. In this line of reasoning, we can observe the two fundamental concepts to assess quality in the production of statistics using the TSEM, namely population and variable. Assessing the quality of both concepts amounts to assessing the quality of the final statistics. This is what the TSEM does for the production of official statistics in the context of question (a), thus motivating the design-based inference paradigm (see, e.g., Tillé 2020, and multiple references therein), where no assumption for an underlying random generation mechanism is made for the target variables.

This analysis does not mean that more complex modeling assumptions cannot be made or alternative inferential paradigms cannot be followed (Chambers and Clark 2012; Little 2012). Our proposal concentrates on how to fit the increasing use of ML models into the classical quality assessment framework provided by the TSEM.

2.3 The Total Machine Learning Error Model

To propose a TMLE-model, we shall assume that we are still providing answers to question (a) in Deming's machine scenario, which we understand as the traditional realm of the production of official statistics (typically by statistical offices). This is in contrast to question (b), which we understand as the natural realm of the analysis of official statistics (typically by policymakers, analysts, researchers, and stakeholders in general). This is our reading of the difference between enumerative and analytic surveys by Deming and Stephan (1941) and Deming (1942, 1953).

The challenge we face is twofold. Firstly, we intend to provide a proposed TMLE-model for supervising statistical learning models independently of their specific use in a business function in the production process (see Barragán et al. 2025, in this book for some examples). Secondly, we propose a combination with the TSEM to produce statistics in a general fashion.

Like the TSEM, the TMLE-model deals with the total error on an ML model as a result of different errors introduced during the process of creating the training set and assessing the model with the test set to be subsequently applied to our target population to produce the statistics of interest. In the TSEM the basic assumption is the existence of true values for the target variables (Groves and Lyberg 2010). Likewise, for statistical learning models we shall assume the existence of a true statistical model expressing the random generation mechanism of a target variable Y from auxiliary variables $\mathbf{X}$, so that with little loss of generality, we may write $Y = f(\mathbf{X}; \mathbf{\Theta}) + \epsilon$, where $\mathbf{\Theta}$ represents any parametrization of the functional dependence f. The model output will be basically an estimated function $\hat{f}$ with an estimated set of parameters $\widehat{\mathbf{\Theta}}$ so that $\widehat{Y} = \hat{f}\left(\mathbf{X}; \widehat{\mathbf{\Theta}}\right)$. The choice of $\hat{f}$, and the resulting parameters $\widehat{\mathbf{\Theta}} = \mathbf{\Theta} + \epsilon_{\Theta}$, will lead to errors in the final predictions of the model. Given the model $\hat{f}$, it is in our opinion essential to identify the error $\epsilon_{\mathbf{\Theta}}$, since it will deliver a substantive contribution to the total error of the estimation of Y.

In the TMLE-model, we try to identify all the sources of errors that will occur during this process impinging on the quality of all predicted values $\widehat{y}_k$ for the units $k \in U$ in our target population U and their inclusion in the final statistics for the target population of interest. To accomplish this, the model is defined in two phases. The first phase, the training phase, tries to estimate the optimal set of parameters, $\widehat{\mathbf{\Theta}}$, based on the selected training set (see Sect. 2.3.1), whereas, during the application phase, the model is used to find an estimation of the target variable $\widehat{Y}$ (see Fig. 2.4 for an overview of the total model; Sect. 2.3.3). But first, we will focus on the estimation of the parameters $\widehat{\mathbf{\Theta}}$ with its error term $\epsilon_{\widehat{\Theta}}$.

2.3.1 *The Training Phase*

The model also makes use of a measurement line and a representation line, as in the TSEM. The basic scheme for the training phase is depicted in Fig. 2.2.

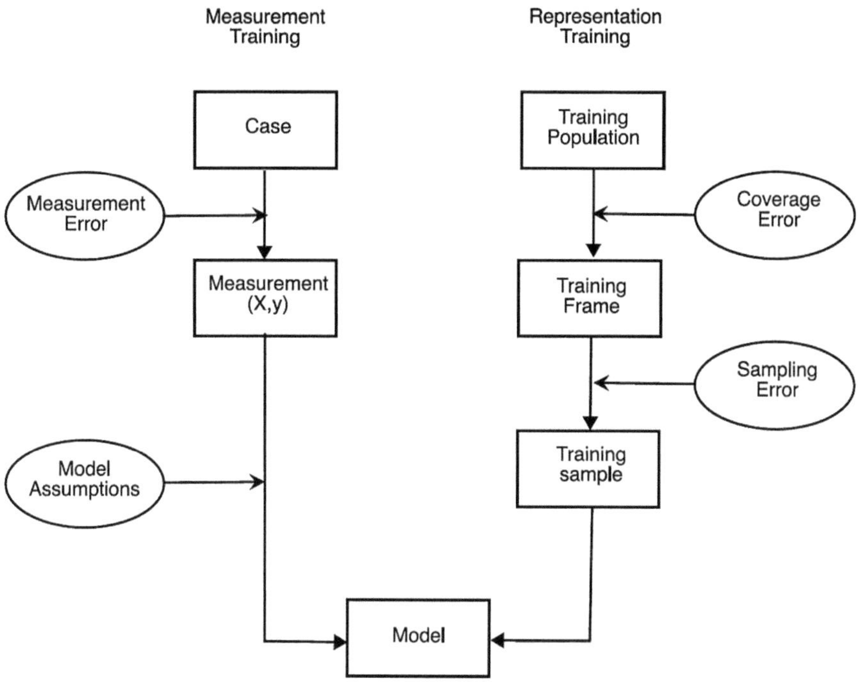

Fig. 2.2 The total machine learning error model: training phase

2.3.1.1 The Measurement Line

The measurement line represents the evolution of variable values from its conception to its influence in the final model. It consists of two important steps within the training of an ML algorithm.

Case

The first block at the upper left corner references each instance in the training set. The case still does not have any features, since these will be described in the next block (measurement). A case is best described as a pointer to an event or an object in the real world that we have to make a prediction about. In the next step, we will observe and measure this object or event.

Measurement

In the process of measuring, the first errors will occur. We state here that these measurement errors are often neglected in data science and can therefore be an important source of errors. We often hear many data scientists refer to the training set as the ground truth (true values in the original TSEM's terminology), and this is—by definition—wrong. In the case of supervised learning, for example, a training set has features and a target variable. Firstly, since the features (**X**) can have measurement errors, they cannot be equal to their true values. We can write with little loss of generality $\mathbf{x} = \mathbf{x}^{(0)} + \boldsymbol{\epsilon}_{\mathbf{x}}$, where $\mathbf{x}^{(0)}$ denotes the true values (ground truth), $\mathbf{x}$ denotes the actually measured, distorted, values, and $\boldsymbol{\epsilon}_{\mathbf{x}}$ denotes the measurement error of variables $\mathbf{x}$.

Secondly, the model target variable Y can also have errors. In general, we can write $y = y^{(0)} + \epsilon_y$, where $y^{(0)}$ stands for the true value (ground truth), y denotes the actually measured, distorted, values, and ϵ_y stands for the measurement error of variable y. In the special case of a classification task for, say, I categories, this leads to misclassifications (e.g., by a human annotator), which can be expressed by $\mathbf{y} \neq \mathbf{y}^{(0)}$, where $y_i, y_i^{(0)} \in \{0, 1\}^I$:

$$\mathbf{y}^{(0)} = \begin{bmatrix} y_1^{(0)} \\ y_2^{(0)} \\ \vdots \\ y_I^{(0)} \end{bmatrix}, \tag{2.1}$$

where

$$\begin{cases} y_i^{(0)} = 1 & \text{if the case belongs to class } i, \\ y_i^{(0)} = 0 & \text{otherwise.} \end{cases}$$

Let T be the transition matrix, where T_{ij} describes the transition probabilities from the ground truth class i to the annotator's class j. The error in classifications can now be expressed in terms of the expected values of y and this transition matrix:

$$\mathbb{E}\left[\mathbf{y}\right] = \mathbb{E}\left[\mathbf{y}^{(0)} + \boldsymbol{\epsilon}_y\right] = T\mathbf{y}^{(0)}. \tag{2.2}$$

It is actually easy to show that this transition matrix T is, in fact, equal to a normalized confusion matrix: a matrix of the True Positive rate (TPR), True Negative rate (TNR), False Positive rate (FPR), and False Negative rate (FNR).

This matrix gives, per actual condition, the probability that the annotator decides that a certain case ends up in a certain class. The normalized confusion matrix is an empirical result of this stochastic matrix (see also Feng and Tang 2024). As such, a normalized confusion matrix can be used to identify the measurement errors during the annotation process.

Model (Measurement Perspective)

Based on the preceding distorted dataset ($\mathbf{x}$, y), we have to form the model. At the moment that we have measurements (even with its measurement errors), we can start to train the model. Another source of errors we can identify is best described as the model assumptions. As is usual for models, the ML model is an abstraction of reality. It will describe how certain variables are related to each other. This is based on the way the ML model assumes how these variables ($\mathbf{X}$, Y) are related and how the features $\mathbf{X}$ are located in the feature space. Also, the choices made by the data scientist about which features he/she selects and how they are coded or transformed belong to the realm of model assumptions.

Hence, errors due to model assumptions are:

- Errors due to the assumptions on how the features $\mathbf{X}$ are represented in the feature space (e.g., encodings)
- Errors due to the selection or actual availability of features $\mathbf{X} \neq \mathbf{X}^{(0)}$, where $\mathbf{X}^{(0)}$ stands for the true features generating the target variable Y and $\mathbf{X}$ denotes the actual set of features used in the modeling exercise
- Errors due to the assumptions in the functional dependence f, i.e., $f \neq \tilde{f}$, where f stands for the true functional dependence and $\tilde{f}$ denotes the actually assumed dependence (to be estimated)

All these assumptions degrade the model, but one should not forget that the idea of modeling is to generalize and the only way we can generalize is by making assumptions. The famous quote from G.E.P. Box is highly relevant in this context. Box stated, "All models are wrong, but some are useful. The question you need to ask is not 'Is the model true?' (it never is) but 'Is the model good enough for this particular application?" (Box et al. 2009, p. 61). In the end, it is all about the usefulness of the model, and a model with fewer erroneous assumptions is expected to be better.

After the model is defined and a functional dependence $\hat{f}$ is chosen, a parameter space is determined, and errors arise due to the optimization of model parameters $\mathbf{\Theta}$, i.e., $\widehat{\mathbf{\Theta}} \neq \mathbf{\Theta}^{(0)}$, where $\mathbf{\Theta}^{(0)}$ stands for the values for the parameters generating the smallest errors in target variable Y from the features $\mathbf{X}$ and $\widehat{\mathbf{\Theta}}$ denotes the estimated (by optimization search, usually) values of these parameters. In many cases, a model is under-determined, leading to many equal optimal solutions ($\widehat{\mathbf{\Theta}}$ is in fact one instance from a set of solutions). This makes the solution not unique and could mean that, although the error is equal for different $\widehat{\mathbf{\Theta}}$, there are more stable points in the parameter space.

2.3.1.2 The Representation Line

The representation line deals with errors regarding the concepts of population, frame, and sample (specific dataset for training, testing, and application). For ML models, these concepts are much subtler than in the TSEM, where the conceptions of set-theoretic population, frame, and sample are used (see Sect. 2.2). ML models are statistical models, and the focus is on the joint distribution $F(Y, \mathbf{X})$ of the target variable and the features, which are usually decomposed as $F(Y, \mathbf{X}) = F_{\Theta}(Y|\mathbf{X})\, F(\mathbf{X})$ so that the focus is actually placed on the conditional dependence $F_{\Theta}(Y|\mathbf{X})$ (see, e.g., Hastie et al. 2009). In consequence, the conception of population, frame, and sample must be stated in terms of the generating distribution function, i.e., as an infinite population in the traditional jargon.

The dichotomy in these concepts lies much in line with the dichotomic distinction between enumerative and analytic surveys by Deming (1942). Indeed, in this work he introduced the concept of question A and question B types, noticing that each type has distinct implications for inference and decision-making. The production of official statistics engages mainly in type A questions, which involve actions based on existing data, intended to characterize and make decisions about known populations (set-theoretic conception). Conversely, question B types go beyond the limitations of current data, focusing on future measurements of unknown entities (statistical conception). To understand the inherent challenges of making inferences about unseen data, which is also an essential aspect of ML, we must first understand this dichotomy.

This dichotomy is not even novel in the current practice. Model-assisted estimation (see, e.g., Särndal et al. 1992) makes an intelligent and profuse use of the dichotomy to introduce linear regression models (question B type) in the design-based inference paradigm (question A type).

Understanding the dynamics of populations is essential to understanding the difference between these two types of questions. Deming's 1941 paper, "On the Interpretation of Censuses as Samples" (Deming and Stephan 1941), challenged the conventional view of a census as a complete enumeration of a population (set-theoretic definition). According to him, even a 100 % census represents a sample from a broader, infinite population. He argued that if one needs to make an inference on a very small cohort of the complete population, even a 100 % sample could be too small to make that inference. Even though the sampling error will be zero, the more general type B questions cannot be answered with absolute precision based on this data (hence the need for statistical inference, the identification of errors, and uncertainty quantification). We need to consider the broader context when answering type B questions, i.e., we need to focus on the generating distribution function underlying the random phenomena behind the data generation.

In this line of thought, ML models and algorithms are generally designed to assist in solving problems similar to Deming's type B questions, aiming to predict (often future) instances of unobserved units based on observed data. As opposed to type A questions, which focus on actions based on existing information, type B questions involve anticipation and inference. Having understood this distinction, one can better understand the essence of ML, where predictive models are trained not simply to describe past events (although they can be used in this sense (see, e.g., Barragán et al. 2025, in the same book)) but also to extrapolate patterns and relationships for informed decisions about data that has not yet been seen. As we shall argue, the problem of concept drift, data drift, or model drift originates here.

To illustrate our proposal let us focus on a concrete example where the dichotomy between set-theoretic and statistical concepts can be clearly observed. Let us try to construct a model-assisted estimator like the GREG estimator (see, e.g., Särndal et al. 1992) but using a CART-type regression tree (see, e.g., Murphy 2013) instead of a linear regression model.[3] Basically, we need to build a regression tree model (which involves both training and testing) to be applied to a concrete dataset to provide estimates for population totals of a continuous target variable Y. To be more specific, we may think of Y as the turnover target variable in a periodic business statistics and the features $\mathbf{X}$ as the set of auxiliary variables available for the whole target population of interest (set-theoretic), as in the GREG scenario.

In terms of Deming's machine analogy, each box corresponds to actual data from a given reference time period. For concreteness' sake, we may think of the machine as corresponding to the country's social and economic conditions producing these data for all periods. To build such a CART-assisted estimator amounts to investigating the statistical functional dependence $Y = f(\mathbf{X}; \boldsymbol{\Theta}) + \epsilon$ behind the data generation mechanism $F_{\boldsymbol{\Theta}}(Y|\mathbf{X})$ to provide estimates for, say, the population totals of the target population U at a given reference time period t. As usual in the production of official statistics, we have a (set-theoretic) probabilistic sample s_t selected according to a sampling design $p(\cdot)$ from the (set-theoretic) target frame U_{Ft}. For simplicity, we shall assume $U_{Ft} = U_t$, so there are no coverage errors affecting the target sample. We also consider there is no nonresponse. Errors from the measurement line are also considered non-present so that we can focus on the representation errors in the ML model. The CART-assisted estimator of Y_t will be

$$\widehat{Y}_t^{CART} = \sum_{k\in U_t} \widehat{\widehat{y}}_k + \sum_{k\in s_t} \frac{y_k - \widehat{\widehat{y}}_k}{\pi_k}, \tag{2.3}$$

where π_k stands for the first-order inclusion probability of unit k and $\widehat{\widehat{y}}_k$ is the design-based estimate[4] of the CART-predicted value of variable Y for unit k. The CART-predicted value $\widehat{y}_k$ can be written as (see, e.g., Murphy 2013)

$$\widehat{y}_k = \sum_{m=1}^{M} w_m I_{\widehat{R}_m}(\mathbf{x}_k),$$

where $\{\widehat{R}_m\}_{m=1}^M$ denotes the binary disjointly split regions of the feature space according to the estimated model and $w_m = \frac{1}{n_m}\sum_{k\in\widehat{R}_m} y_k$. Notice that for concreteness' sake, we are focusing on the continuous variable case (turnover).

---

[3] The argument remains valid for a linear regression model, but by using CARTs we hope to underline the different concepts involved, usually away from usual practice.

[4] The double-hat notation is becoming clear below to distinguish between model-based and design-based predicted values.

Now since we are taking a (set-theoretic) probabilistic sample s_t, as in the GREG estimation, we need to provide the design-based estimator so that

$$\widehat{\widehat{y}}_k = \sum_{m=1}^{M} \left(\frac{\sum_{k \in \widehat{R}_m} y_k / \pi_k}{\sum_{k \in \widehat{R}_m} 1/\pi_k} \right) I_{\widehat{R}_m}(\mathbf{x}_k).$$

We shall use this example to introduce the concepts of training and target populations, training and target frames, and training, test, and target samples. See, e.g., Nalenz et al. (2024) for a deeper insight in this direction.

Training Population

Let us denote by $F_{\Theta}^{tr}(Y|\mathbf{X})$ the data generation distribution function for those statistical units U^{tr} used for training the model. This infinite population $F_{\Theta}^{tr}(Y|\mathbf{X})$ will be, in our TMLE-model, the training population. In our analogy using Deming's machine, the training population is a machine generating data later to be used just for training.

To assess its effect on the final statistics to be produced, we need to investigate its relationship with the final target population U_t of interest. The difficulty arises because we need to compare this (set-theoretic) target population U_t of interest with the (statistical) concept of training population $F_{\Theta}^{tr}(Y|\mathbf{X})$. We can consider U_t as the realization of the underlying data generation distribution function $F_{\Theta}(Y|\mathbf{X})$ for the target population so that the finite population U_t can indeed be conceived as a sample of this infinite target population (i.e., the so-called superpopulation approach (see, e.g., Cassel et al. 1977)).

In this line, the more different $F_{\Theta}^{tr}(Y|\mathbf{X})$ is from $F_{\Theta}(Y|\mathbf{X})$, the less accurate the final statistics will be. Although representativity as a comparison between two sets for estimating purposes constitutes a slippery concept (Kruskal and Mosteller 1979a,b,c, 1980), a mathematical definition in the set-theoretic realm (Bethlehem 2009) can be provided in terms of the difference between (empirical) distribution functions for a given variable. In this same (to be made precise) we can talk of the representativity of the training population with respect to the target population in terms of the distance between $F_{\Theta}^{tr}(Y|\mathbf{X})$ and $F_{\Theta}(Y|\mathbf{X})$. The differences arise because of the changing dynamics of populations. For example, if data from the past are used, a judgment is implicitly made about the stability in time in the relationship between Y and $\mathbf{X}$ for the target population at stake.

Training Frame

Once the data generation distribution function $F_{\Theta}^{tr}(Y|\mathbf{X})$ is in place, data must be actually generated (Deming's machine must produce the bolts) so that we have a (set-theoretic) frame population U_F^{tr} from which statistical units (instances) will be taken to train the model. This data-generating mechanism may be affected by different factors producing overcoverage, undercoverage, imbalance, etc. Errors can arise when the relationship between Y and $\mathbf{X}$ is not extensively and properly covered throughout all generated instances. In mathematical terms, this means that it is impossible to obtain an accurate estimation of the training data generation distribution function $F_{\Theta}^{tr}(Y|\mathbf{X})$ from the frame population U_F^{tr}.

Target Sample

The actual dataset used for training is composed of a selection of instances from U_F^{tr}. This selection may have been executed in many different ways, producing either a non-probability or (ideally) a probability sample s_{tr}.

Notice that even selecting a representative sample s_{tr} with respect to the target frame U_F^{tr}, i.e., even having a small distance between the (empirical) distribution functions of the target variable in U_F^{tr} and s_{tr}, does not guarantee the final quality of the model if yet $F_{\Theta}^{tr}(Y|\mathbf{X})$ is very different to $F_{\Theta}(Y|\mathbf{X})$. However, to meet this final requirement, this notion of (set-theoretic) representativity of s_{tr} with respect to U_F^{tr} is necessary.

Model (Representation Perspective)

In our previous sections, we discussed both the coverage error and the sampling error for the training phase. To create the optimal training set from the representation perspective to build the model, we need to ensure that $F_{\Theta}^{tr}(Y|\mathbf{X})$ is very close to $F_{\Theta}(Y|\mathbf{X})$. The concept of representativity, as we have seen, is indeed slippery and involves diverse subtleties.

New methodology needs to be developed to solve this representativity problem. As a result of this new methodology, we believe that problems like concept drift will be minimized to the best of our ability, as the final model will be robust to the dynamics occurring in the finite population under consideration.

2.3.2 *The Testing Phase*

For the testing phase, the model also makes use of a measurement line and a representation line, as in the training phase. The basic scheme for the combination of the training and testing phases is depicted in Fig. 2.3.

A cautious reader may think that, in order to account for over- and underfitting, an infinite population $F_{\Theta}^{test}(Y|\mathbf{X})$ may be needed for the testing phase so that both the measurement and representation lines can be defined in a similar fashion. However, this is in contrast to usual (good) practice. The available data are divided for training and testing so that indeed we are ensuring $F_{\Theta}^{test}(Y|\mathbf{X}) = F_{\Theta}^{tr}(Y|\mathbf{X})$ (unless this

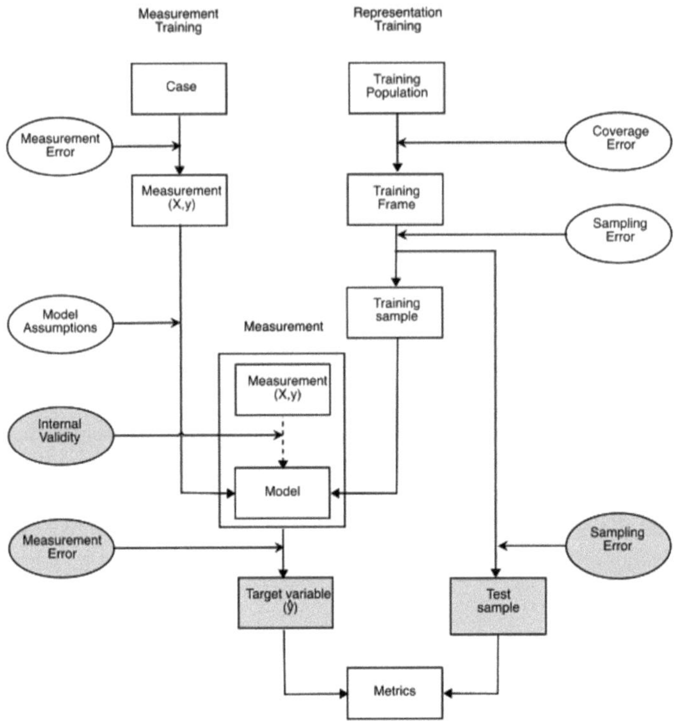

Fig. 2.3 The Total Machine Learning Error model: training and testing phases. The training phase is highlighted in white, and the testing phase is highlighted in gray

division is executed in a highly non-ignorable way). This has a direct consequence for the scope of the validity of the tested models. This kind of validity is not reachable at the moment that $F^{test}_{\boldsymbol{\Theta}}(Y|\mathbf{X}) \neq F^{tr}_{\boldsymbol{\Theta}}(Y|\mathbf{X})$.

It also has immediate consequences for both the measurement and the representation lines in the testing phase, as well as consequences for the scope of the validity of the models tested in this fashion.

2.3.2.1 The Measurement Line

The measurement line in the testing phase still represents the evolution of variable values since its generation to its influence in the final model, but now focused on those units used in the testing phase.

Measurement

When measuring variables for the testing phase, the same situation occurs as in the training phase: Measurement of variables takes place with errors. Measured values are not in general equal to the corresponding true values.

Within set bounds for statistical uncertainty, internal validity is indeed ensured when both training and testing data come from the same training target frame, as stated above. This may not be the case, for example, when training data are taken from preceding time periods and testing data are used for the next period in the time series.

Target Variable

Once the model has been trained and tested, even with measurement errors, we can produce the predicted values $\widehat{y}$ since the functional dependence $\widehat{f}$ has been estimated, as well as the parameters $\widehat{\boldsymbol{\Theta}}$.

Metrics (Measurement Perspective)

Once predicted values can be produced, the model performance evaluation can be undertaken with the corresponding metrics. The basis for these kinds of validation measures often lies in the well-known confusion matrix. It is important to underline that the model quality in its final application to the target population of interest will depend on the choice of metrics. The best reference to an overview of the confusion matrix and all its derived metrics can be found on Wikipedia (Wikipedia contributors 2024).

2.3.2.2 The Representation Line

Once the training/testing splitting is executed ensuring that both underlying infinite populations are the same, we can assume that the testing population and testing frame are the same as in the training phase so that only the samples will indeed be different: This is the unique novel element.

Test Sample

The actual dataset used for testing is composed of a selection of instances from U_F^{tr} according to the mentioned training/testing splitting strategy. This selection may have also been executed in many different ways, producing either a non-probability or a (ideally) probability sample s_{test}.

Under these assumptions, representativity properties of s_{tr} and s_{test} are shared, and thus sampling errors are equally present in the testing phase, although it is common that the test set is much smaller than the training set. Particular training/testing splitting strategies may introduce differences between s_{tr} and s_{test}. Cross-validation, out-of-bag procedures, and similar techniques are highly convenient and adequate in this sense.

Metrics (Representation Perspective)

Model performance indicators and metrics are then computed for s_{test}. Notice that when performance indicators and metrics on the test set are considered adequate, the whole model is retrained in the whole training/testing dataset, thus for the same underlying infinite population.

2.3.3 *The Application Phase*

Ultimately, one of the goals of creating an ML model is to use it on a set of new data. In this chapter, we will focus on the use of the predictor to count how many units in the populations have a certain treat. Of course, this is not the only application for the use of machine learning. One can imagine that we will use machine learning for, for instance, imputation, editing, or determining weights. However, for the sake of the argument, we will now look at using the predictor directly for counting.

The application phase is shown in the TMLE-model in Fig. 2.4. The training phase (implicitly including the testing phase) is highlighted in white, whereas the prediction/application phase is highlighted in gray.

The central aspect of error assessment is that the model now includes two infinite populations, namely the training population $F_{\Theta}^{tr}(Y|\mathbf{X})$, and the infinite target population $F_{\Theta}(Y|\mathbf{X})$ and the realized finite target population U, which is the population for statistical analyses.

At this point, the original TSEM (Fig. 2.1) should be taken into account both for the variables y (measurement line) and for the units $k \in U$ (representation line). As a final step, following Eq. (2.3), we will make an estimate based on collected target variable values y and the predictions $\widehat{y}$ that have been made by the model. From a general point of view, notice that this final step represents indeed the inclusion of the predicted values in our working dataset so that we can further either produce estimates (e.g., following Eq. (2.3) or similar) or use in an intermediate process step (e.g., for coding, error detection, etc.).

It is important to note that we are now discussing the external validity of the model on the measurement side of the model. This is because we are applying the model to a population that is completely different from the one on which we trained and tested it. If the model does not perform, as well as it did on the test set, it is not externally valid. As a form of validity, external validity is stronger than internal

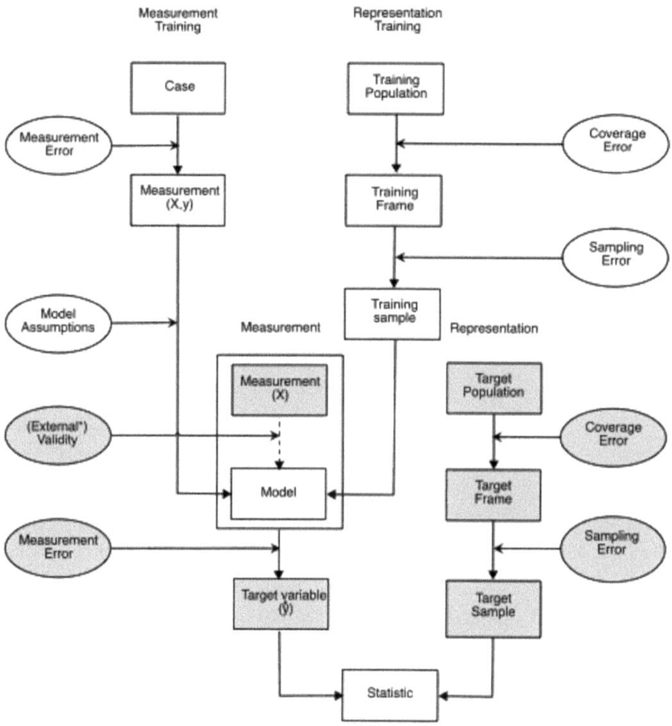

Fig. 2.4 The Total Machine Learning Error model: training/testing and application phases. The training/testing phases are highlighted in white, and the application/prediction phase is highlighted in gray

validity, and when applying ML algorithms, we should strive for an as good as possible external validity. In our opinion, this is also a better way to test the model. Instead of taking a sample of the same population as the training set, we should take a sample from a different (part of the) population. This would provide us with a much better assessment of the model.

2.4 Summary of the Model

Having a comprehensive understanding and meticulous rectification of the inherent measurement errors prevalent in a variety of tasks is a critical aspect of ML endeavors. In addition to inaccuracies in feature measurement, these errors can also be attributed to the complexities associated with misclassifications of target variables, which can significantly impair ML accuracy and reliability. By meticulously acknowledging and adeptly mitigating these multifaceted errors, practitioners can substantially enhance the robustness and effectiveness of their models, resulting in greater trust and utility in their applications across various domains and scenarios.

A pivotal juncture occurs in the course of the representation dimension, as the focus shifts toward the intricacies of creating training populations, frames, and samples. Deming's seminal contributions to population analysis are underpinning this transition, whose distinction between type A and type B questions provides invaluable insight into population analysis' nuance. It is evident from Deming's distinction that it is paramount that training data accurately capture the subtle nuances and complexities inherent in the infinite population being studied. In addition, a growing emphasis is being placed on cultivating representative training frames that reflect the true essence of the broader population landscape as a resounding call to action among discussions regarding finite populations and their broad-ranging implications for model generalization.

Our focus is increasingly on rigorously evaluating model performance by applying a dedicated test dataset as we proceed through the testing phase. As practitioners navigate a labyrinth of measurement errors, they face a multitude of challenges, among others, the complexity of internal validity concerns. In light of these challenges, robust evaluation methodologies that go beyond mere performance metrics, examining the intricate nuances of model behavior and efficiency across a variety of contexts and scenarios, are imperative. Practitioners can increase their confidence and reliability in their ML endeavors by embracing the imperative of assessing external validity and ensuring that models can be applied to entirely new datasets.

In applying a model to previously unobserved data, we are presented with the ultimate frontier of ML deployment, wherein real-world applications and insights are derived from unobserved data, resulting in the fruits of labor. Iterative learning culminates in this phase, when models move from theory to practical application, enabling them to impact diverse domains and industries tangibly. There is a strong emphasis on external validity in this context, underlining the importance of models demonstrating robustness and generalizability across a variety of scenarios and contexts. In contrast to the confines of finite populations, Deming's principles promote models that transcend the confines of finite populations and embrace the dynamic complexity of the broader landscape by drawing parallels with the TSEM. For ML to reach greater heights of innovation and impact, practitioners must maintain a commitment to excellence and a steadfast commitment to advancing the frontiers of knowledge to navigate the rapidly evolving landscape with confidence and efficacy.

Based on these considerations, we would like to propose the following best practices:

- **Understand measurement errors:** Thoroughly investigate and address measurement errors in feature variables and target variables during model training.
- **Construct representative training frames:** Strive to create training frames that accurately represent the characteristics and complexities of the infinite population, considering diversity and coverage issues.

- **Evaluate model performance:** Use test sets to evaluate model performance, focusing on the external validity. Having a test set that is not part of the training population helps in determining the external validity.
- **Continuously monitor and refine models:** Keep checking the validity of the training population that was used to create the model. Make sure that the target population is still a sample of the assumed infinite population. Iterate, when necessary on model development, incorporating feedback from evaluations on test sets and real-world applications to improve performance and robustness.

In the context of ML pipelines, we would like to emphasize the importance of integrating these insights into every stage of the development life cycle. From data collection and preprocessing to model training and evaluation, incorporating robust mechanisms for addressing measurement errors and ensuring representativeness in training data is paramount. This entails implementing rigorous validation protocols, leveraging diverse datasets to capture the full spectrum of population characteristics, and continually refining models to enhance their generalizability and reliability (e.g., using audit sampling—see Zhang 2023).

Moreover, fostering a culture of transparency and accountability is essential, wherein practitioners actively document and communicate the limitations and assumptions underlying their models. By promoting open dialogue and collaboration, we can collectively identify and address potential biases and distortions, thereby fostering greater trust and confidence in ML applications. In our opinion, a concrete course of action in this direction would be to document and understand the data generation mechanisms in all data sources used for the production of official statistics. This clearly demands a close dialogue between data holders and the statistical offices.

Furthermore, investing in ongoing research and development efforts aimed at elucidating the intricate dynamics of population analysis and model evaluation is crucial. This entails exploring novel methodologies for assessing external validity, refining sampling strategies to minimize bias and variance to optimize the two fold connection between sampling designs and ML model construction, and advancing techniques for quantifying and mitigating measurement errors.

Ultimately, by integrating these best practices into ML pipelines, we can try to create a way to develop more robust and reliable models that better fit the high standards we try to uphold within the field of official statistics.

2.5 Applying Machine Learning Models: Some Classification Examples

As described above, creating a good externally valid ML model is essential when applying this kind of algorithm in the context of official statistics. Here, we will first describe the approach followed in the study performed to detect innovative companies, followed by the detection of online platforms. In the end, some recent insights gained during a detailed study of the creative industry are discussed. All studies use website text to identify different types of companies in the Netherlands. These studies were performed by some of the coauthors and nicely illustrate the insights gained during our study of ML methodology.

2.5.1 Detecting Innovative Companies

Producing an overview of innovative companies in a country is a challenging task. Traditionally, this is done by sending a questionnaire to a sample of companies. This approach, however, focuses on large companies and completely misses small companies, such as start-ups. Therefore, an alternative approach was investigated by determining if a company is innovative by studying the text on its website. An ML model was developed based on the texts of the websites of companies included in the Community Innovation Survey of the Netherlands. The latter is a survey carried out every 2 years that focuses on the detection of innovative companies with ten or more working people. All websites of the innovative companies were included (a total of 3,340) in addition to a similar sized random sample of the non-innovative companies (3,302) (Daas and van der Doef 2020). This provided the training frame which, according to the population topics discussed above, is very likely representative. It was found that the ML model developed was able to reproduce the results from the Community Innovation Survey, with an accuracy of 93 %, a precision of 99 %, and a recall of 87 %. It was also able to detect innovative companies with less than ten employees, such as start-ups (Daas and van der Doef 2020). Manual checking was performed to determine the accuracy of the model on the classification of small companies and confirmed its external validity (regarding this subpopulation).

As a next step, the model was applied to a very large, more representative, dataset of websites, of around 400,000. When a new model was trained on a random sample of 20,000 websites, from the dataset of around 400,000 classified websites, it became clear that even though the new model was trained on a more representative part of the (finite) population, the accuracy of the model had decreased to 88 %. This revealed that misclassification errors can build up during model development.

The downside of the original model was its instability over time. Here, a fairly rapid decline was observed on the same set of websites scraped at various points in time (Daas and Jansen 2020). This is referred to as concept drift. Studies to reduce this issue were performed and found to be a challenging topic. Application of models based on large language models provided the most successful "solution" thus far with an accuracy of 86–87 % for the various datasets over time. However, the external validity of that model, on new unseen data, was not that high. It had an accuracy of 72 %. This example shows that a high internal validity does not have to result in an equally high external validity. Hence, the need for robust evaluation methodologies that go beyond mere performance metrics (see Sect. 2.4) is obvious.

2.5.2 Detecting Online Platforms

Obtaining reliable information from a small or rare subpopulation is a challenging topic. Approaches commonly used to find rare or so-called hard-to-identify groups are a screening survey, network sampling, area sampling, or a combination. An example of a rare subpopulation is online platform businesses. To get a complete overview of this subpopulation, an ML model was developed to identify these kinds of platforms. This required a training frame. Hence, business statistics experts of Statistics Netherlands were asked to manually provide examples of such websites. This resulted in a set of 569 online platforms and 303 non-platform organizations, with very similar characteristics, which were additionally identified during this process. To the latter, a random sample of 266 non-platform organizations, from the websites linked to the Business Register, was additionally added. This is not a lot of data for ML, but it was found to be sufficient for the task at hand.

The combined frame contained 50 % platform and 50 % non-platform websites used for model development. This resulted in an ML model that was able to identify online platforms with an accuracy of 82 % on the test set, a precision of 84 %, and a recall of 79 % (Daas et al. 2024). After applying the model to the entire population of websites linked to the business register and manual inspection of random samples, it became clear that the model seriously overestimated the number of online platforms. In other words, the external validity of the model was low, and its findings were biased. By sending questionnaires to a large sample of businesses, this problem was initially solved, and the answers obtained were used to validate the ML model-based findings (Daas et al. 2024).

The initial study revealed that online platforms merely compose 0.22 % of the total population of businesses with a website in the Netherlands. This made clear why false positives were such a huge problem. Subsequently, various steps were studied with the aim to considerably reduce the number of false positives detected by the ML model (Gubbels et al. 2024). The combination of steps that worked well is listed in Table 2.1.

Apart from producing binary labels, the model used could also produce the probability of a website being an online platform. These probabilities have a value between 0 and 1. Using the "probabilities" of the model was found to increase the online platform estimates, hence increasing the bias (Table 2.1, Row 2). This makes clear that, under these circumstances, the model did not produce actual probabilities. This did not affect the accuracy.

Table 2.1 Effect of various model-based approaches on online platform detection

	Type of model	True Pos.	Est. Pos.	Bias	Accuracy
1	Log. reg.	69	2,991	0.098	0.901
2	Log. reg. prob.	69	7,657	0.255	0.901
3	Log. reg. cal. prob.	69	637	0.019	0.985
4	Ensemble cal. prob.	69	306	0.007	0.993

Subsequently, a method was applied to correctly calibrate the probabilities of the model used. This method corrects for the intrinsic prevalence of the model, i.e., the prevalence caused by the ratio of positives and negatives on which the ML model was originally trained (Puts and Daas, 2021). Since the number of online platforms is very low in the population and the model was trained on a much higher ratio of positive and negative cases (either 50–50% or 30–70%), it can be expected that correcting for this prevalence may seriously reduce the number of positive cases estimated. Applying the calibration method revealed that this was indeed the case; see Table 2.1, Row 3. As a consequence, the accuracy considerably increased.

When the results of multiple trained and calibrated models, up to 10, were combined, the bias was reduced even further; see Table 2.1, Row 4. The accuracy also increased somewhat. The bias is, however, not completely removed by the combination of correction methods applied. This is not unexpected for a model detecting rare events. We think there are two ways to even further improve this approach. The first is by increasing the number of models included in the ensemble. The second is by improving the representativeness of the websites included in the training frame used and those in the dataset used to produce the findings shown in Table 2.1. This obviously relates to the discussion on infinite and finite populations in Sect. 2.2.

2.5.3 *Detecting the Creative Industry*

Next, we discuss an ML study on the detection of businesses belonging to the creative industry in the Dutch municipality of Eindhoven. Since the creative industry is very difficult to define, it was interesting to study the topic with a data-driven approach. The big question in this study was—initially—if such businesses could be identified with an ML model trained in the texts on the websites of positive and negative cases. This required examples and, hence, local experts were asked to provide a list of websites of businesses belonging to the creative industry.

At the start of the study, a list of 110 positive websites was provided. Considering the number of positive cases included in the training frame in the cases discussed above, this is a very low amount. However, assuming that "a website of a business belonging to the creative industry" is a rare event, the idea emerged that a (small) random sample of the websites linked to the business register, excluding the websites already in the positive set, could provide a nearly perfect list of noncreative industry examples. Such an approach is referred to as Positive and Unknown (PU) learning (Bekker and Davis 2020) and might provide a solution. Subsequently, random samples were drawn from the websites of the business register and combined with the 110 positive cases, and various models were trained. Be aware that the selection procedure used assured that no websites already included in the positive set were selected from the business register linked list of websites. After some trial and error, a multilayer perceptron classifier was trained that seemed to produce fairly accurate results, on the test set. It had an accuracy of 86 %, a precision of 79 %, and a recall of 91 %.

Applying the model to the population resulted in a probability distribution that was composed essentially of two clearly distinct peaks: a large one of noncreative websites (with an average of around 0.05) and a smaller group of potential creative websites (with an average probability of 0.99). However, a manual inspection of a sample of 370 relatively high-scoring websites revealed that only 52 % of those websites actually belonged to the creative industry. The external validity of the model was obviously poor.

Since random samples were drawn for manual inspections, the idea emerged to add the findings for the websites to the training frame. Subsequently, in the next iteration, the combination of 110 positive and the 370 manually classified cases (including both negative and positive cases) was combined with various amounts of randomly sampled websites from the websites linked to the business register. Here again, the selection procedure used ensured that no websites already included in the positive and manually classified set were selected from the business register list of websites. This procedure resulted in a second multilayer perceptron classification model trained on much more data. It was found to an accuracy of 85 % on the test set, with a precision of 88 % and a recall of 80 %.

Applying this second model to the population revealed a probability distribution composed of three distinct groups: (i) a very large group with an average probability of 0, (ii) a small group with an average probability of 0.1, and (iii) a small group with an average probability of 1. Manually inspecting random samples from each group revealed that each group contained, respectively, 2 %, 40 %, and 87 % websites belonging to the creative industry. Hence, the second model was much better able to discern creative industry websites compared to the first model. This leads to the conclusion that including the 370 manually inspected websites in the training frame, the 100 positive cases, and a random sample of unknown cases made the resulting frame much more representative of the population studied. We think that following such an iterative approach is a very interesting way to create high-quality (and more representative) training frames with a fairly low manual effort. It reminds us of the Deming cycle of continuous improvement. Because this study was performed during the development of the TMLE-model described above, this also indicated that having such a frame in mind while performing an ML-based study also helps to improve the quality of its findings.

2.6 Discussion

From the above, it is clear that the methodology of applying ML in official statistics is just in its infancy. Compared to the questions posed in Puts and Daas (2021) this document already sheds some light on the methodology concerning the human annotation of data, sampling the population to obtain representative training sets, dealing with concept drift, and correcting the bias caused by the ML model. However, there are several additional and important considerations—from a methodological perspective—that became apparent while writing this document. These are at various stages of development and are described in the paragraphs below. Some of them are quite fundamental, and all should be thoroughly investigated.

(1) The terminology used by statisticians and data scientists differs. Historically, emerging fields use their own terminology. This is usually not problematic, since this terminology is typically only used within the newly defined paradigm. However, if the emerging field is adopted in another field, with its own paradigm, it will result in a loss of common ground. This does not necessarily have to result in a total misunderstanding between the fields. The diverging terminology within different fields has a striking similarity to the term "false friends," which are words written exactly the same in two different languages but with a divergent meaning. Even though multilingual families have to deal with "false friends" on a daily basis, they can function in peace and harmony without many misunderstandings. An outstanding example is provided by the term "bias," more often than not applied in expressions such as *biased sample* (samples are not biased; only estimators can be biased), *bias in an instance value* (confused with a measurement error for a given statistical unit), and *biased estimate* (confused with an estimate error for a given sample), etc. Why would this be a problem in science? The reason is evident: The older field (in this case Statistics) will consider the terms used in the younger field (in this case ML) as incorrect. To deal with "false friends," however, we should acknowledge their different meanings and keep them under consideration when communicating (like in bilingual families). Part of the purpose of this chapter was to acknowledge that methodology in statistics means something different compared to ML and subsequently describe the field of ML in the terminology used within the field of official statistics.

(2) Ensure homogeneity in the construct measured. Developing an ML model starts with a training frame including cases of the best possible quality. To enable this, there is usually a "human in the loop," for instance, to ensure that the cases included are correctly classified or to check the findings of the model on new data. Including human checking is challenging because the findings of multiple humans need to be consistent, the so-called inter-annotator agreement, which requires considerable effort. Efficient ways to verify the construct measured by ML models need to be developed.

(3) Representativity in the context of ML. This fundamental and challenging question touches the heart of statistical inference. There is a definitive need to develop more theory in this area to ensure that ML models, as accurately as possible, measure the concept of interest on new cases. Which steps should be taken, in which order, to enable this? For example, should stratification be approached in a different way? Should sampling algorithms be updated or used in a particular way? Should sampling designs be reconsidered to optimize the use of the sample in ML models instead of to minimize design-based mean square errors? In many of the questions included in this list, the notion of representativity in the context of ML is paramount and should therefore be resolved.

(4) How to deal with ML and type A and type B questions? Deming distinguishes these types of questions. Type A questions are action-based on existing information and more in line with the production of official statistics, while type B questions aim to predict instances of unobserved units and go beyond the limitations of current data. Here, one could simply decide to just focus on ML and type A questions, but one needs to be aware that ML models developed for answering type B questions are expected (when properly developed) to better deal with new data and changing conditions. Both types are highly relevant, and ML models and their appropriate usage dealing with each or both types of question should be studied.

(5) Bias(es) resulting from errors made during training of the model (especially for type B questions). Many of the errors occurring during model development will result in biased estimates. Reducing the errors as much as possible during model development is a way to decrease their disruptive effect. The TMLE-model is a good starting point here, and the focus should now shift to identify, measure, and correct each of the errors or any combination of them. Of course, one should bear in mind that this is not as evident as it seems. For instance, errors introduced by model assumptions are almost impossible to quantify.

(6) Representativity problems resulting from differences in the population composition of the training frame and the infinite population. We saw that machine learning involves answering a type B question (see the previous point). As stated, this has its consequences regarding the representativity of a training set. The main challenge is to abstract from the finite population to generate cases that do not occur (yet) in the finite population. From a technical point of view, data augmentation could deliver a solution from this. Nevertheless, procedures are needed to ensure that the cases in the training frame include the relevant, as well as possible, features occurring in the infinite population.

(7) How to approximate the infinite population? A possible approximation of the infinite population can be done by taking one simple random sample: the finite population. We will, however, introduce parts in the feature space that are not well represented in this final population. The ill-covering finite population (with respect to the infinite population) can be seen not only as a coverage error but also as a sampling error. Acknowledging this error is already an important step. However, the question remains how we could approximate the infinite population? To answer this question, we need to have a better understanding of representativity, both in general and in the context of machine learning. Advances in understanding the data generation mechanisms in real situations, quantifying uncertainty (both bias and variance), and gaining insight into the underlying (many times unknown) selection probabilities for units in datasets are needed. Deming suggested approximating the infinite population by taking several finite populations, separated in time to ensure independence between the "samples," but this approach is not always feasible. Consequently, the question remains and needs to be answered.

(8) Sampling with replacement and bootstrap methods in machine learning. An important insight that emerged is the need to keep in mind the infinite population when developing ML models. This perspective influences how samples are used during model training. An approach to achieve a better approximation of the underlying distribution is sampling with replacement. However, this raises questions about the role of bootstrap methods in ML. In particular, it is worth investigating whether commonly used techniques such as bagging or boosting are suitable for achieving a robust understanding of the infinite population. Furthermore, should adaptations or entirely new methods be developed to ensure bootstrapping techniques contribute effectively to improving model quality?

(9) Develop a procedure that ensures that only the most important features (variables) are included in the model. When larger training frames are being used in ML model development, we have observed that increasing numbers of features become included in models. There is a definite need for an approach that reduces this effect and ensures that only the "best" features are selected. Such a procedure could also improve the accuracy and stability of the model over time.

(10) Develop a procedure that ensures that the ML model is both internally and externally valid and as stable as possible over time. From the above, it has become obvious that the ultimate goal when developing an ML model is creating a model with a high external validity. This is linked to topic (6). There is a definitive need for a procedure that ensures a model is obtained with both a high internal and an external validity. The findings of topic (6) will certainly help here.

These topics go beyond the focus of many traditional ML practitioners and highlight the importance of a statistical view on ML and the need for ML methodology. We are convinced that the TMLE-model, in particular, the sources or errors identified, will help users to get a better grip on the challenging application of ML in a statistical context and also when applying ML in general. In addition, the document gives an overview of the topics that need to be studied in more detail and, as such, sets the stage for future research in this interesting and challenging area.

Acknowledgments First of all, we would like to extend our heartfelt thanks to Yvonne Gootzen for her invaluable contributions to this work. We are deeply grateful for her involvement.

We also wish to acknowledge Luuk Gubbels and Sanne Peereboom, whose internship work provided valuable insights that contributed to Sect. 2.5. Delorian Canlon and Sourav Bhattacharjee are gratefully acknowledged for stimulating discussions that considerably improved the document. We also acknowledge the work by S. Barragán, L. Sanguiao, and C. Sáez, underlying many of the theoretical concepts presented here.

References

E. Alpaydin, *Introduction to Machine Learning* (MIT Press, Cambridge, 2020)

S. Barragán, A. Pérez-Bote, C. Sáez, D. Salgado, L. Sanguiao-Sande, Streamlining business functions in official statistical production with machine learning, in *Foundations and Advances of Machine Learning in Official Statistics*, ed. by F. Dumpert, Chap. 13 (Springer, Berlin, 2025)

J. Bekker, J. Davis, Learning from positive and unlabeled data: a survey. Mach. Learn. **109**, 719–760 (2020)

J. Bethlehem, *Applied Survey Methods: A Statistical Perspective* (Wiley, Hoboken, 2009)

G.E. Box, A. Luceño, M.D.C. Paniagua-Quiñones, *Statistical Control By Monitoring and Adjustment* (John Wiley & Sons, Hoboken, 2009)

L. Breiman, Statistical modeling: the two cultures. Stat. Sci. **16**(3), 199–215 (2001)

G. Casella, R.L. Berger, *Statistical Inference* (Duxbury Press, 2002)

C.-M. Cassel, C.-E. Särndal, J.H. Wretman, *Foundations of Inference in Survey Sampling* (Wiley, Hoboken, 1977)

R.L. Chambers, R.G. Clark, *An Introduction to Model-Based Survey Sampling with Applications* (Oxford University Press, Oxford, 2012)

P. Daas, J. Jansen, Model degradation in web derived text-based models, in *Paper for the 3rd International Conference on Advanced Research Methods and Analytics (CARMA)* (2020), pp. 77–84

P. Daas, S. van der Doef, Detecting innovative companies via their website. Stat. J. IAOS **36**(4), 1239–1251 (2020)

P. Daas, W. Hassink, B. Klijs, On the validity of using webpage texts to identify the target population of a survey: an application to detect online platforms. J. Off. Stat. **40**(1), 190–211 (2024)

W.E. Deming, On a classification of the problems of statistical inference. J. Am. Stat. Assoc. **37**(218), 173–185 (1942)

W.E. Deming, *Some Theory of Sampling* (Wiley, Hoboken, 1950)

W.E. Deming, On the distinction between enumerative and analytic surveys. J. Am. Stat. Assoc. **48**(262), 244–255 (1953)

W.E. Deming, F.F. Stephan On the interpretation of censuses as samples. J. Am. Stat. Assoc. **36**(213), 45–49 (1941)

F. Dumpert, Machine learning in German official statistics, in *Advances in Business Statistics, Methods and Data Collection*, ed. by G. Snijkers, M. Bavdaž, S. Bender, J. Jones, S. MacFeely, J.W. Sakshaug, K.J. Thompson, A. van Delden, Chapter 23 (Wiley, Hoboken, 2023), pp. 537–560

European Statistical System Committee, Quality Assurance Framework of the European Statistical System, version 2.0. https://ec.europa.eu/eurostat/documents/64157/4392716/ESS-QAF-V2.0-final.pdf (2022)

Y. Feng, W. Tang, Confusion matrix design for decision making with prediction models. SSRN, 4950758 (2024)

R.A. Fisher, On the mathematical foundations of theoretical statistics. Philos. Trans. R. Soc. London. Seri. A, Contain. Pap. Math. Phys. Char. **222**, 309–368 (1922)

Y.A. Gootzen, P. Daas, A. van Delden, Quality framework for combining survey, administrative and big data for official statistics. Stat. J. IAOS **392**, 439–446 (2023)

R.M. Groves, L. Lyberg, Total survey error: past, present, and future. Public Opin. Q. **74**(5), 849–879 (2010)

L. Gubbels, M. Puts, P. Daas, Bias correction in machine learning-based classification of rare events, in *Symposium on Data Science and Statistics* (2024)

T. Hastie, R. Tibshirani, J. Friedman, *The Elements of Statistical Learning*, 2nd edn. (Springer, Berlin, 2009)

T.D. Jong, S. Bromuri, X. Chang, M. Debusschere, N. Rosenski, C. Schartner, K. Strauch, M. Boehmer, L. Curier, Monitoring spatial sustainable development: semi-automated analysis of satellite and aerial images for energy transition and sustainability indicators (2020). https://arxiv.org/abs/2009.05738

A.F. Karr, A.P. Sanil, D.L. Banks, Data quality: a statistical perspective. Stat. Methodol. **3**(2), 137–173 (2006)

J. Koop, Notes for a unified theory of estimation for sample surveys taking into account response errors. Metrika **21**, 19–39 (1974)

W. Kruskal, F. Mosteller, Representative sampling, I: non-scientific literature. Int. Stat. Rev. **47**, 13–24 (1979a)

W. Kruskal, F. Mosteller, Representative sampling, II: scientific literature, excluding statistics. Int. Stat. Rev. **47**, 111–127 (1979b)

W. Kruskal, F. Mosteller, Representative sampling, III: the current statistical literature. Int. Stat. Rev. **47**, 245–265 (1979c)

W. Kruskal, F. Mosteller, Representative sampling, IV: the history of the concept in statistics, 1895–1939. Int. Stat. Rev. **48**, 169–195 (1980)

R.J. Little, Calibrated Bayes, an alternative inferential paradigm for official statistics. J. Off. Stat. **28**, 309–334 (2012)

A. Measure, Six years of machine learning in the bureau of labor statistics, in *Advances in Business Statistics, Methods and Data Collection*, ed. by G. Snijkers, M. Bavdaž, S. Bender, J. Jones, S. MacFeely, J.W. Sakshaug, K.J. Thompson, A. van Delden, Chapter 24 (Wiley, Hoboken, 2023), pp. 561–572

C. Moscardi, B. Schultz, Using machine learning to classify products for the commodity flow survey. in *Advances in Business Statistics, Methods and Data Collection*, ed. by G. Snijkers, M. Bavdaž, S. Bender, J. Jones, S. MacFeely, J.W. Sakshaug, K.J. Thompson, A. van Delden, Chapter 25 (Wiley, Hoboken, 2023), pp. 573–591

C. Moya, Introduction to survey and causal inference methods for social sciences. https://bookdown.org/cristobalmoya/iscs_materials/ (2020). Accessed 18 Jun 2025. Licensed under CC BY-SA 4.0

K.P. Murphy, *Machine Learning: A Probabilistic Perspective* (MIT Press, Cambridge, 2013)

M. Nalenz, J. Rodemann, T. Augustin, Learning de-biased regression trees and forests from complex samples. Mach. Learn. **113**, 3379–3398 (2024)

C. O'Neil, *Weapons of Math Destruction: How Big Data Increases Inequality and Threatens Democracy* (Crown Publishing Group, New York, 2016)

A.J. Onwuegbuzie, J.E. McLean, Expanding the framework of internal and external validity in quantitative research. Res. Sch. **10**(1), 71–89 (2003)

M. Puts, P. Daas, Machine learning from the perspective of official statistics. Surv. Stat. **84**, 12–17 (2021)

M. Puts, A. da Silva, L.D. Consiglio, I. Choi, D. Salgado, C. Clarke, S. Jones, A. Baily, ONS-UNECE Machine Learning Group 2022. Quality of Training Data. Theme Group Report. Technical report, UNECE (2022). https://statswiki.unece.org/download/attachments/338329823/The%20quality%20of%20Training%20data_final.pdf?version=1&modificationDate=1671449307809&api=v2

S. Russell, P. Norvig, *Artificial Intelligence: A Modern Approach*, 3rd edn. (Prentice Hall, Hoboken, 2010)

Y. Saidani, F. Dumpert, Quality dimensions and quality guidelines for machine learning in official statistics, in *Foundations and Advances of Machine Learning in Official Statistics*, ed. by F. Dumpert, Chap. 4 (Springer, Berlin, 2025)

Y. Saidani, F. Dumpert, C. Borgs, A. Brand, A. Nickl, A. Rittmann, J. Rohde, C. Salwiczek, N. Storfinger, S. Straub, Qualitätsdimensionen maschinellen Lernens in der amtlichen Statistik. AStA Wirtschafts- und Sozialstatistisches Archiv **17**(3), 253–303 (2023)

C.-E. Särndal, B. Swensson, J. Wretman, *Model Assisted Survey Sampling* (Springer, Berlin, 1992)

G. Snijkers, M. Bavdaž, S. Bender, J. Jones, S. MacFeely, J.W. Sakshaug, K.J. Thompson, A. van Delden, *Advances in Business Statistics, Methods and Data Collection* (Wiley, Hoboken, 2023)

Y. Tillé, *Sampling and Estimation from Finite Populations* (Wiley, Hoboken, 2020)

United Nations, *UN National Quality Assurance Frameworks Manual for Official Statistics* (United Nations, 2019)

A. van Delden, J. Burger, M. Puts, Ten propositions on machine learning in official statistics. AStA Wirtschafts- und Sozialstatistisches Archiv **17**(3), 195–221 (2023)

Wikipedia contributors, Confusion matrix – Wikipedia, the free encyclopedia (2024). https://en.wikipedia.org/wiki/Confusion_matrix. Online; Accessed 21 Feb 2024

L.-C. Zhang, Audit sampling as a quality standard for multisource official statistics. Span. J. Stat. **5**(1), 67–83 (2023)

Chapter 3
Challenges in Resampling-Based Performance Estimation

Hannah Schulz-Kümpel, Anne-Laure Boulesteix, Sebastian Fischer, and Roman Hornung

3.1 The Generalization Error

3.1.1 Two Variants of the Generalization Error: Definitions and Interpretation

Ensuring transparency, documenting assumptions, and communicating uncertainty of statistical findings are crucial in official statistics. These practices are not only necessary for the accurate interpretation of results but also foster trust in the data and the institutions responsible for producing them. Especially given the increasing complexity of machine learning models, this transparency is necessary for policymakers, researchers, and the public to make informed decisions. In this context, quantities that evaluate model performance, particularly on new, unseen data, are an important tool. The state-of-the-art approach to make inferences about

H. Schulz-Kümpel · S. Fischer
Department of Statistics, LMU Munich, Munich, Germany
e-mail: hannah.kuempel@stat.uni-muenchen.de; sebastian.fischer@stat.uni-muenchen.de

A.-L. Boulesteix
Institute for Medical Information Processing, Biometry and Epidemiology, Faculty of Medicine, LMU Munich, Munich, Germany

Munich Center for Machine Learning (MCML), Munich, Germany
e-mail: boulesteix@ibe.med.uni-muenchen.de

R. Hornung (✉)
Institute for Medical Information Processing, Biometry and Epidemiology, Faculty of Medicine, LMU Munich, Munich, Germany

Munich Center for Machine Learning (MCML), Munich, Germany
e-mail: hornung@ibe.med.uni-muenchen.de

F. Dumpert (ed.), *Foundations and Advances of Machine Learning in Official Statistics*, Society, Environment and Statistics,
https://doi.org/10.1007/978-3-032-10004-7_3

model performance on unobserved data is to utilize resampling techniques, the particulars of which will be described in Sect. 3.1.2.

In this chapter, we explore the concept of generalization error, which serves as a pivotal metric for evaluating the performance of machine learning models on unseen data based on resampling the available data. To be more precise, the term "generalization error" (GE) is used as an umbrella term for two very similar quantities that provide insight into how well a model is expected to perform on new, unseen instances, thereby offering a crucial measure of its reliability and predictive power in real-world scenarios.

Specifically, for a prediction function $\hat{f}_{\mathcal{D}}$ that results from fitting a model on data $\mathcal{D}$ with n observations, the GE is often (see Bates et al. 2024 and Schulz-Kümpel et al. 2025) used to refer to one of the following two quantities:

$$\mathcal{R}_P(\hat{f}_{\mathcal{D}}) = \mathbb{E}[L(\boldsymbol{y}^*, \hat{f}_{\mathcal{D}}(\boldsymbol{x}^*))|\mathcal{D}] \tag{3.1}$$

$$\mathbb{E}\big[\mathcal{R}_P(\hat{f}_{\mathcal{D}})\big] = \mathbb{E}_{\mathcal{D}}[\mathbb{E}[L(\boldsymbol{y}^*, \hat{f}_{\mathcal{D}}(\boldsymbol{x}^*))|\mathcal{D}]], \tag{3.2}$$

where L denotes some loss function and $(\boldsymbol{y}^*, \boldsymbol{x}^*)$ a new, unseen data point drawn from the same distribution as the observations in $\mathcal{D}$.

We refer to Eqs. (3.1) and (3.2) as *risk* ($\mathcal{R}_P(\hat{f}_{\mathcal{D}})$) and *expected risk* ($\mathbb{E}\big[\mathcal{R}_P(\hat{f}_{\mathcal{D}})\big]$), respectively.

Interpreting GE Quantities

Generally, one may say that the GE quantifies *the expected loss, i.e., some quantification of the distance between observation and model prediction for a new, unseen data point.* More accurate, however, would be the following distinct interpretations of the (expected) risk; see also Remark 1 of Schulz-Kümpel et al. (2025).

- The *risk* $\mathcal{R}_P(\hat{f}_{\mathcal{D}})$ informs on how well suited a specific model that has been fit on specific data $\mathcal{D}$ will be on average for predictions on data stemming from the same data generating process as $\mathcal{D}$.
- Meanwhile, the *expected risk* $\mathbb{E}\big[\mathcal{R}_P(\hat{f}_{\mathcal{D}})\big]$ informs, more generally, on how well-suited models that have been fit using a certain algorithm on data of size n will be on average for predictions on data stemming from the same data-generating process. As such, it may be interpreted as informing on the suitability of a certain algorithm for a certain kind of data.

Of course, to make inferences about quantities whose definitions contain unobserved data points, one must in practice generate observations of loss between predictions from a fitted model and corresponding observations that were not necessarily used to fit the said model, given that one has only one data set. As previously mentioned, this is achieved through resampling the available data set and repeatedly refitting the model. Section 3.1.2 explains this procedure and the resulting complexities in detail.

3.1.2 Inference on the GE Through Resampling

The state-of-the-art approach to perform inferences on the GE is to utilize resampling, which refers to a variety of techniques to repeatedly produce pairs of train and test sets. This allows for the repeated application of a model-fitting algorithm to each training set and the evaluation of the resulting model on the corresponding test set. The following table provides an overview of the most common resampling techniques.

Due to the inherent challenges associated with performing inference on the generalization error—particularly when constructing confidence intervals—many other resampling methods have been proposed in the literature (Jiang et al. 2008; Nadeau and Bengio 2003; Bates et al. 2024; Noma et al. 2021; Dietterich 1998). These methods are largely combinations and modifications of the approaches outlined in Table 3.1. For a more comprehensive overview of these methods, the reader is referred to Schulz-Kümpel et al. (2025).

Table 3.1 Common resampling techniques

Resampling procedure	No. of refits	Explanation	Reference
Holdout	1	The data $\mathcal{D}$ is split only once into $\mathcal{D}_{\text{train}}$ and $\mathcal{D}_{\text{test}}$	James et al. (2021), Chap. 5.1.1.
Subsampling	K	The holdout procedure is repeated K times	Shao and Wu (1989), as *delete-d jackknife*
Cross-validation (CV)	K	The data $\mathcal{D}$ is split into K folds of size $\lfloor n/K \rfloor$. Each fold is used as a test set $\mathcal{D}_{\text{test}}^{(i)}$ once, where all the remaining folds are used as the training data $\mathcal{D}_{\text{train}}^{(i)}$, $i = 1, \ldots, K$	James et al. (2021), Chap. 5.1.3.
Leave-one-out cross-validation	n	Apply cross-validation with $K = n$	James et al. (2021), Chap. 5.1.2.
Bootstrap	R	Sample R data sets of size m (often $m = n$) with replacement and use each sample as the training data and the corresponding out-of-bag data as the test set	James et al. (2021), Chap. 5.2.
Insample	1	Use the whole data $\mathcal{D}$ as train and test data	

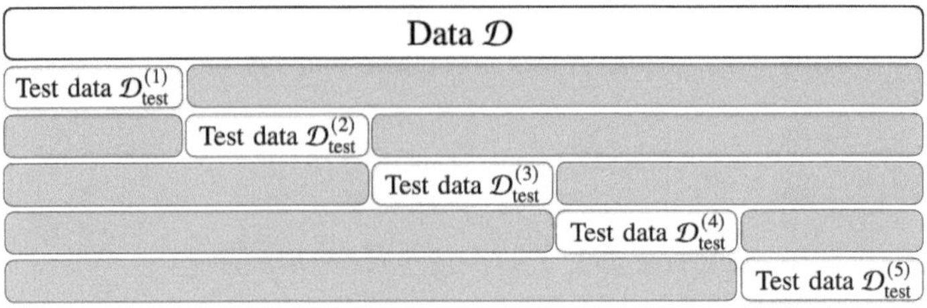

Fig. 3.1 Visualization of fivefold cross-validation (CV). Gray areas depict the training data $\mathcal{D}_{\text{train}}^{(i)}$, $i = 1, 2, 3, 4, 5$, corresponding to the test data in each fold (visualized as a row)

Let us consider the common K-fold cross-validation (CV) point estimate for the GE to illustrate resampling-based inference in this context. To aid readability, Fig. 3.1 provides a visualization of fivefold CV.

In this example, given original data $\mathcal{D}$ that is independent and identically distributed (i.i.d.), pointwise losses between each observation in every test set and the corresponding prediction based on a model fit on the corresponding training set are computed, yielding n overall observations of pointwise losses. The arithmetic mean over those loss observations then gives the point estimate:

$$\widehat{\text{GE}}^{(\text{CV})} := \frac{1}{n} \sum_{i=1}^{K} \sum_{(x,y)\in\mathcal{D}_{\text{test}}^{(i)}} L\big(y, \hat{f}_{\mathcal{D}_{\text{train}}^{(i)}}(x)\big) . \tag{3.3}$$

Note that while Eq. (3.3) is commonly used as a point estimate for any GE quantity, it might technically be more suitable as a point estimate for the average risk over the training sets of size $n(1 - \frac{1}{K})$, because the average of pointwise losses is taken over K models fit on data of that size.

Indeed, all resampling-based inference, and thereby all inference on the generalization error, is in some way affected by either a lower number of observations available for training, repeated fitting of an algorithm, or both. Additionally, utilizing the same observations for training and testing within a resampling procedure, such as K-fold CV, results in dependencies between the pointwise losses. The latter point especially affects the construction of confidence intervals (CIs) for the GE, even when the original observations arise from an i.i.d. setting; see Sect. 3.2.2.

A different challenge arises in nonstandard data settings, especially those where the data are not i.i.d. Here, standard resampling techniques often produce suboptimal and, in particular, biased GE estimates. In such cases, specialized resampling techniques are required that explicitly account for the relationship between the new data, to which the prediction model will be applied, and the given training data.

Next, Sect. 3.2 will provide more detail on the complicated, but important, inference problem of deriving CIs for the GE and present two well-performing methods from the current literature. Subsequently, Sect. 3.3 will discuss specialized resampling techniques for point estimation of the GE in different nonstandard settings, most of which deviate from the i.i.d. assumption.

Important Takeaways for the Rest of the Chapter

1. The generalization error (GE) is an umbrella term for the risk and expected risk, important performance metrics for ML models that quantify the expected loss between observation and model prediction for a new, unseen data point.
2. Inference about the GE is based on resampling methods.
3. Due to the dependence structure that results from resampling the available data, deriving confidence intervals for the (expected) risk is complex even in i.i.d. data settings.
4. While point estimation for GE is relatively straightforward in i.i.d. data settings, the same does not hold when the data is not i.i.d.

3.2 Constructing Confidence Intervals for the GE

This section explores methods for constructing confidence intervals for the generalization error of models fit on independent and identically distributed (i.i.d.) data. After discussing the importance of confidence intervals in this context, we discuss simple techniques like Holdout and K-fold CV, which provide basic but useful estimates. Finally, we introduce more advanced methods—Corrected Resampled-T and Conservative-Z—that offer improvements over these simpler techniques and provide specific recommendations for practitioners on when to employ each method.

3.2.1 Why and How Confidence Intervals for the GE Matter

Point estimates for the GE play a central role in assessing the performance of machine learning models as they inform on how accurately a model will predict on new data observed in the future. However, reporting a single point estimate of expected pointwise loss—say an error rate of 5% for a classification model—conceals the fact that this number is itself uncertain. Having observed slightly different data, one may have arrived at a different estimate. Additionally, when applying a stochastic model such as XGBoost (see Chen et al. 2023), where randomness arises from sources like hyperparameter tuning, the exact point estimate will likely be replicable only if a fixed seed is set, even on the same data. Without a measure of this variability, one risks overconfidence in the results—which should especially be avoided in official statistics, where model-based decisions must be transparent, reproducible, and statistically sound; see Sect. 3.2.1.2.

Confidence intervals offer a solution by quantifying the uncertainty around point estimates for the GE. Instead of stating that a model's (expected) risk is 5%, we might report that it lies between 3.8% and 6.2% with 95% confidence. More precisely, a reported CI may be interpreted as follows: *If we could have applied the same procedure for computing a GE point estimate and CI to 100 data sets of the same size and drawn from the same distribution as the data we used to compute the CI, (on average) 95 of those CIs would have covered the true GE.* Such an interval communicates the inherent uncertainty that stems from only "running an experiment once," in our case applying the estimation procedure once to the single data set available. Thereby, it allows for more cautious and informed decision-making.

3.2.1.1 What Makes a Good Confidence Interval

As detailed in Sect. 3.1, computing CIs for the generalization error is a complex problem, with many different procedures having been proposed in the literature. When making recommendations on which methods may be generally considered reliable in Sect. 3.2.3, this assessment is based on the following three properties.

High Coverage Probability

Obviously, a reliable CI should contain the true (expected) risk at the rate of the stated confidence level (e.g., 95%). Even within the classic setting of i.i.d. data to which common model-agnostic methods are currently limited, a crucial point is whether a given method maintains accurate coverage across a broad range of model classes. While reliable CI should cover the GE for different model classes—ranging from simple linear regressions to deep neural networks—only few methods manage to provide stable coverage for ML models with even moderate stochasticity, e.g., stemming from hyperparameter tuning; see Schulz-Kümpel et al. (2025).

Narrow Width

Generally, an extremely wide CI may indicate correspondingly high uncertainty about the estimate. However, some methods for computing CIs for the generalization error achieve high coverage probability via overly wide CIs. Of course, accepting a higher width for the benefit of reliable coverage is generally legitimate—such CIs are referred to as conservative—but achieving a balance between coverage and precision is paramount. Consequently, producing generally narrow CIs at reliable coverage rates is a desirable property for any considered method.

Computational Feasibility

Excellent performance of CI methods in terms of coverage and width is insufficient when the underlying procedure is not computationally efficient enough to be applied in practice. This is especially true in official statistics where complex models are regularly applied to large data sets. Due to the necessity of resampling for computing CIs for the generalization error, the driving force behind computational efficiency is how many training-test splits—and, thereby, refits of the given model—are required by a method.

Section 3.2.3 will detail two subsampling-based CI methods originally proposed by Nadeau and Bengio (2003) that perform very well in terms of all three previous metrics. Meanwhile, the nested-cross-validation procedure suggested more recently by Bates et al. (2024) is an example of a method that performs very well only for less stochastic model classes, at least at resample specifications that lead to practical computational efficiency.

3.2.1.2 Confidence Intervals in Quality Reporting

Just as uncertainty quantification for population estimates (such as childlessness rates and life satisfaction) is already common in quality reporting in official statistics, regularly including confidence intervals for the GE of any applied model in quality reports is an excellent tool for further increasing transparency and conveying reliability of models trained on i.i.d. data. Beyond communicating uncertainty about the performance of a single model, reporting such CIs allows for both assessing the suitability of a modeling approach for routinely collected data over time and comparing the suitability of different models for the same application.

In fact, while this chapter is focused on the construction of CIs for the expected loss on new data for a single model, all discussed methods may also be used to directly construct a CI for the expected difference in loss between two models. Specifically, for a reference model $\hat{f}_{\mathcal{D}}^{\text{ref}}$ and alternative model $\hat{f}_{\mathcal{D}}^{\text{new}}$, this is achieved by replacing $L(\boldsymbol{y}^*, \hat{f}_{\mathcal{D}}(\boldsymbol{x}^*))$ in Eqs. (3.1) and (3.2) with $L(\boldsymbol{y}^*, \hat{f}_{\mathcal{D}}^{\text{new}}(\boldsymbol{x}^*)) - L(\boldsymbol{y}^*, \hat{f}_{\mathcal{D}}^{\text{ref}}(\boldsymbol{x}^*))$. Here, a CI containing 0 would indicate little evidence for a significant difference in performance between these two models.

Lastly, the wide variety of methods available for computing CIs for the generalization error warrants the documentation of which specific method was used, which loss function was chosen, the number of resamples employed, and any other relevant parameter settings. Recommendations for these specifications are given in Sect. 3.2.3 and have recently been implemented in the R-package `mlr3inferr`; see Fischer and Schulz-Kümpel (2025).

3.2.2 Basic Approaches: Holdout and K-Fold Cross-Validation

Generally, any resampling scheme naturally gives rise to a point estimate for the GE that is simply the arithmetic mean of all pointwise losses computed on a test set using the predictions based on the corresponding test set, as explained in Sect. 3.1.2. The following defines such point estimates for Holdout resampling, where the data is split only once, and K-fold CV, where the data is split into K folds (see Table 3.1 for a detailed description of common resampling methods):

$$\widehat{\mathrm{GE}} := \begin{cases} \frac{1}{n_{\mathrm{test}}} \sum_{(x,y)\in\mathcal{D}_{\mathrm{test}}} L\big(y, \hat{f}_{\mathcal{D}_{\mathrm{train}}}(x)\big) & \text{for Holdout resampling} \\ \frac{1}{n} \sum_{i=1}^{K} \sum_{(x,y)\in\mathcal{D}_{\mathrm{test}}^{(i)}} L\big(y, \hat{f}_{\mathcal{D}_{\mathrm{train}}^{(i)}}(x)\big) & \text{for K-fold CV resampling.} \end{cases}$$

While such point estimates are usually (nearly) unbiased estimators of their targets, obtaining an unbiased estimate of their variance is notoriously difficult, if not theoretically impossible; see Bengio and Grandvalet (2004). This difficulty arises due to dependencies between the pointwise losses computed on different test sets, which do not affect the point estimates due to the properties of the expected value but significantly complicate the variance estimation. Of course, of all common resampling methods, only the Holdout is not affected by this dependence since it only produces one test set; but it comes with its own drawback, which is detailed below.

In any case, there is the option to take the same approach to variance estimation as with the point estimates: simply viewing the pointwise losses computed on all test sets as one sample and computing the sample variance. This results in confidence intervals, but with formal targets that are neither the risk nor the expected risk. For Holdout and K-fold CV, however, these targets are sufficiently close to the formal quantity targets referred to as GE for such intervals to have been suggested in the literature. Specifically, for Holdout and K-fold CV, respectively, this approach results in the following variance estimates:

$$\hat{\sigma}^2 := \begin{cases} \frac{1}{n_{\mathrm{test}}-1} \sum_{(x,y)\in\mathcal{D}_{\mathrm{test}}} \Big(L\big(y, \hat{f}_{\mathcal{D}_{\mathrm{train}}}(x)\big) - \widehat{\mathrm{GE}}\Big)^2 & \text{for Holdout resampling} \\ \frac{1}{n} \sum_{i=1}^{K} \sum_{(x,y)\in\mathcal{D}_{\mathrm{test}}^{(i)}} \Big(L\big(y, \hat{f}_{\mathcal{D}_{\mathrm{train}}^{(i)}}(x)\big) - \widehat{\mathrm{GE}}\Big)^2 & \text{for K-fold CV resampling} \end{cases}$$

and produces CIs of the form

$$\left[\widehat{\mathrm{GE}} \pm z_{1-\frac{\alpha}{2}} \frac{\hat{\sigma}}{\sqrt{N}}\right], \tag{3.4}$$

where N equals n_{test} for Holdout and n for K-fold CV and $z_{1-\frac{\alpha}{2}}$ denotes the $(1-\frac{\alpha}{2})$-quantile of the normal distribution.

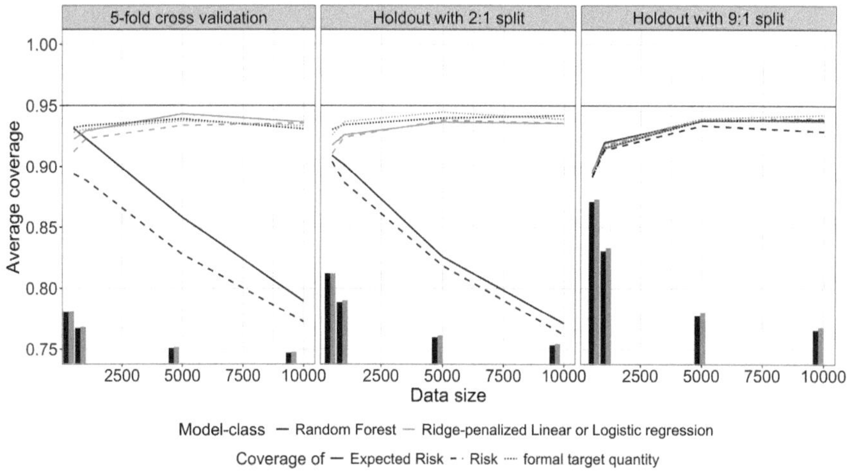

Fig. 3.2 Coverage of advanced, subsampling-based CI methods, illustrated using a subset of the results from the benchmark study of Schulz-Kümpel et al. (2025). See Figure 3 of that work for more results. The squared error was chosen as loss for regression and 0-1 loss for classification. The horizontal black line represents the target confidence level, with the other lines showing the average coverage for three possible targets, colored by model class. The median widths of the corresponding CIs are represented by the bars at the bottom, without absolute values as they serve purely for comparison between the depicted CI methods

For the Holdout, the formal target quantity corresponding to Eq. (3.4) is simply $\mathbb{E}[L(\boldsymbol{y}^*, \hat{f}_{\mathcal{D}_{\text{train}}}(\boldsymbol{x}^*))|\mathcal{D}_{\text{train}}]$: the risk conditional on the training data instead of the entire data set. Meanwhile, for K-fold CV, Eq. (3.4) formally covers the *K-fold test error* $\frac{1}{n}\sum_{i=1}^{K}\sum_{(x,y)\in\mathcal{D}_{\text{test}}^{(i)}}\mathbb{E}\left[L\left(y, \hat{f}_{\mathcal{I},\mathcal{D}_{\text{test}}^{(i)}}(x)\right) \mid \mathcal{D}_{\text{train}^{(i)}}\right]$ as proven by Bayle et al. (2020). In this instance, the issue of dependencies between test sets is bypassed by averaging over the risks conditional on the different training sets.

Figure 3.2 illustrates an important caveat regarding these basic CI methods: while their formal target quantities are mathematically similar to the (expected) risk, their values can differ substantially in practice. Comparing coverage rates for the three possible targets reveals, at least for some resampling specifications, notable discrepancies between them across different model classes.

The first thing that becomes apparent is that the Holdout with a training-test split of 90%:10% (9:1) of the observations provides, for large data, solid coverage of both the formal target quantity and the GE. However, this coverage comes with the drawback of comparatively very wide CIs. Additionally, it has long been established that the Holdout point estimate generally performs worse than that based on K-fold CV and other more sophisticated resampling procedures; see James et al. (2021). Still, when extremely large data ($\gg$100.000) is available, Holdout with a 9:1 split is certainly a valid choice for computing CIs for the generalization error.

Meanwhile, the CIs based on fivefold CV and Holdout with a 2:1 training-test split cover their respective formal target quantities as well as the GE for ridge-penalized linear or logistic regression reliably, while the coverage of the GE for a random forest model actually notably decreases for data of larger size. This is the result of the pessimistic bias between GE and formal target quantities resulting from conditioning on data of smaller size, which is especially pronounced when models still have a steep learning curve for a rising number of observations as the random forest does.

3.2.3 Advanced Subsampling-Based Methods: Corrected Resampled-T and Conservative-Z

The Corrected Resampled-T and Conservative-Z, proposed by Nadeau and Bengio (2003), are subsampling-based methods that offer highly reliable alternatives to the basic approaches discussed in Sect. 3.2.2. While, in contrast to the basic approaches, formal target quantities for these methods have not been theoretically obtained in the literature thus far, Corrected Resampled-T and Conservative-Z have demonstrated excellent empirical coverage across model classes in the setting of i.i.d. data; see Schulz-Kümpel et al. (2025). Additionally, these two methods may be applied to estimate the GE with non-decomposable metrics such as AUC or F1 in addition to the pointwise loss setting of Eqs. (3.1) and (3.2).

For both methods, the confidence intervals are constructed around the following point estimate to which the subsampling procedure, where the Holdout procedure is repeated K times (see Table 3.1), gives rise:

$$\widehat{\mathrm{GE}}^{\mathrm{sub}} := \frac{1}{K}\sum_{i=1}^{K}\frac{1}{n_{\mathrm{test}}}\sum_{(x,y)\in\mathcal{D}^{(i)}_{\mathrm{test}}} L\big(y, \hat{f}_{\mathcal{D}^{(i)}_{\mathrm{train}}}(x)\big),$$

where $\mathcal{D}^{(i)}_{\mathrm{test}}$ and $\mathcal{D}^{(i)}_{\mathrm{train}}$ denote the test and training set corresponding to the ith Holdout conducted during the subsampling procedure.

Then, the Corrected Resampled-T method utilizes the quantile of the t-distribution with $K-1$ degrees of freedom and the variance estimate

$$\hat{\sigma}^2_{cort} := \frac{1}{K}\sum_{i=1}^{K}\left(\left(\frac{1}{n_{\mathrm{test}}}\sum_{(x,y)\in\mathcal{D}^{(i)}_{\mathrm{test}}} L\left(y, \hat{f}_{\mathcal{D}^{(i)}_{\mathrm{train}}}(x)\right)\right) - \widehat{\mathrm{GE}}^{\mathrm{sub}}\right)^2,$$

which is adjusted by multiplying with the term $(1/K + n_{\mathrm{test}}/(n - n_{\mathrm{test}}))$.

Meanwhile, the Conservative-Z method utilizes *paired subsampling* to directly estimate the standard error for a CI. In this resampling procedure, the available data is split into two sets of the same size R times, and regular subsampling is conducted on both of these sets. Then, estimates of the following forms are utilized for the standard error estimate:

$$\widehat{P}^{conz}_{r,t} := \frac{1}{K}\sum_{i=1}^{K}\frac{1}{n_{\text{test}}}\sum_{(x,y)\in\mathcal{D}^{(i)}_{\text{test},r,t}} L\left(y, \hat{f}_{\mathcal{D}^{(i)}_{\text{train},r,t}}(x)\right), \quad r = 1,\ldots,R,\ t \in \{1,2\},$$

where $\mathcal{D}^{(i)}_{\text{test},r,t}$ and $\mathcal{D}^{(i)}_{\text{train},r,t}$ denote the test and training set corresponding to the ith Holdout conducted during the rth subsampling procedure on the tth (of two) original data split.

These preliminaries now result in the following squared standard error estimates:

$$\widehat{\text{SE}}^2 := \begin{cases} \left(\frac{1}{K} + \frac{n_{\text{test}}}{n-n_{\text{test}}}\right)\cdot\hat{\sigma}^2_{cort} & \text{for Corrected Resampled-T} \\ \frac{1}{2R}\sum_{r=1}^{R}\left(\widehat{P}^{conz}_{r,1} - \widehat{P}^{conz}_{r,2}\right)^2 & \text{for Conservative-Z.} \end{cases}$$

and the corresponding CIs are given by

$$\left[\widehat{\text{GE}}^{\text{sub}} \pm q_{1-\frac{\alpha}{2}}\widehat{\text{SE}}\right], \tag{3.5}$$

where $q_{1-\frac{\alpha}{2}}$ equals the $(1-\frac{\alpha}{2})$-quantile of the normal distribution for the Conservative-Z method and that of the t-distribution with $K-1$ degrees of freedom for the Corrected Resampled-T method.

Recommended Specifications of Corrected Resampled-T and Conservative-Z

Corrected Resampled-T and Conservative-Z require more parameter specifications than the basic methods of Sect. 3.2.2. We recommend choosing the following based on the benchmark study of Schulz-Kümpel et al. (2025).

Conservative-Z:

- *For data of moderate size* ($\geq$*500)*: $K = 5$ for both subsampling for the point estimate and in the paired subsampling scheme for standard error estimation. For the latter, repeating paired subsampling $R = 10$ times for variance estimation is recommended, resulting in 105 overall repetitions.[1]
- *For data of small size (*$<$*500)*, the same specifications are valid, although both K and R may be slightly increased here (see Schulz-Kümpel et al. 2025) if computational capacity allows.

Within the (paired) subsampling Holdout with a split of 9:1 is recommended.

(continued)

[1] This procedure requires $(2R + 1)K$ model refits; see Table 2 of Schulz-Kümpel et al. (2025).

Corrected Resampled-T: We recommend the use of this method for data of moderate size (≥500) only. Here, we suggest choosing $K = 25$ for the subsampling with a Holdout split of 9:1, resulting in 25 overall repetitions.

Figure 3.3 illustrates the reliable performance of both Conservative-Z and Corrected Resampled-T, with parameter specifications for data of moderate size as recommended above, when covering the generalization error. As the name suggests, the Conservative-Z method produces wider CIs that do lead to an even more stable coverage over the investigated model classes and data sizes. However, due to the high number of repetitions, i.e., model refits, it should also be expected to be about four times as computationally expensive as the Corrected Resampled-T method. Notably, both methods also provide reliable coverage for more random model classes such as XGBoost (Chen et al. 2023) that other approaches struggle with, with Conservative-Z again proving slightly more stable.

When choosing between Corrected Resampled-T and Conservative-Z, those factors should be weighed against each other.

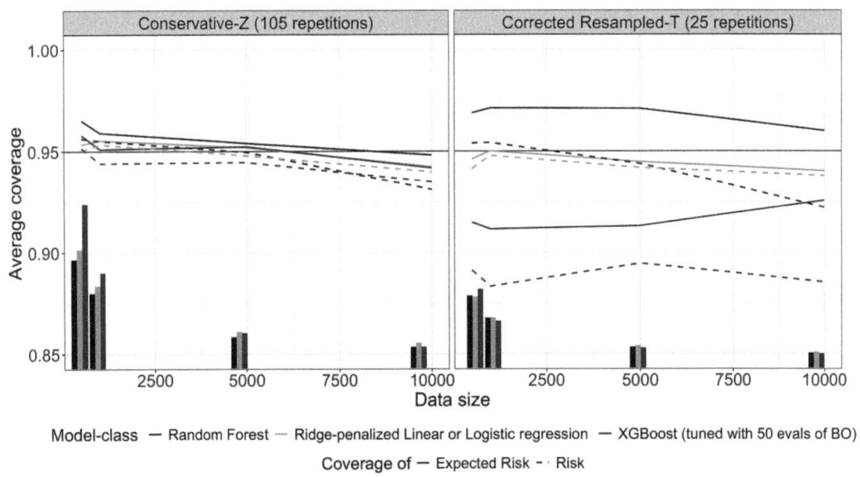

Fig. 3.3 Coverage of advanced, subsampling-based CI methods, illustrated using a subset of the results from the benchmark study of Schulz-Kümpel et al. (2025). See Figure 3 of that work for more results. The squared error was chosen as loss for regression and 0-1 loss for classification. The horizontal black line represents the target confidence level, with the other lines showing the average coverage for three possible targets, colored by model class. Again, the median widths of the corresponding CIs are represented by the bars at the bottom, without absolute values as they serve purely for comparison between the depicted CI methods. Therefore, the height of the bars should not be compared to the height of those in Fig. 3.2, as the basic methods generally result in much wider CIs than those of Corrected Resampled-T and Conservative-Z

Important Takeaways for Computing CIs for the GE

1. When fitting models to extremely large i.i.d. data ($\gg$100,000), the Holdout method with a 9:1 split is a solid and computationally very efficient choice to compute confidence intervals for the generalization error.
2. When fitting models to i.i.d. data of moderate size, we recommend using either Conservative-Z (with $R = 10$ and $K = 5$) or the Corrected Resampled-T (with $K = 25$), both with a Holdout split of 9:1 throughout for computing confidence intervals for the generalization error.
3. When fitting models to i.i.d. data of small size, only the Conservative-Z method is recommended. Here the parameters R and K may be slightly increased if that is computationally feasible.
4. The recommended methods for computing confidence intervals for the generalization error are implemented in the R-package `mlr3inferr`; see Fischer and Schulz-Kümpel (2025).

3.3 GE Estimation in Nonstandard Data Settings

3.3.1 Considered Data Settings and Need for Specialized Resampling

The discussions in the previous section focused exclusively on i.i.d. data, which is the most fundamental and common type of data in machine learning research and practice. Despite its prominence, the assumption of i.i.d. data is not always valid in official statistics, where data often exhibit more complex structures. Applying methodologies designed for i.i.d. data to these nonstandard data settings can lead to suboptimal, in particular biased, results.

In this section, we turn our attention to the resampling-based estimation of the GE in various nonstandard data settings. In particular, we depart from the previous discussion by excluding inference considerations related to the GE. It is important to acknowledge that the inference methods described previously are likely unsuitable for the nonstandard settings addressed here.

We consider the following nonstandard settings: clustered data, spatial data, unequal sampling probabilities, concept drift, and hierarchically structured outcomes. With the exception of the latter, these settings do not conform to the i.i.d. assumption. The content of this section draws heavily upon the work of Hornung et al. (2025), which aimed to offer practical guidance for estimating the GE under these challenging conditions with minimal bias and variance. This publication provided an extensive review of existing literature and introduced original simulation studies aimed at filling knowledge gaps, thereby providing the most comprehensive guidance possible.

A fundamental principle shared across the resampling techniques discussed in this section is that, during each iteration, the test data should reflect the new observations to which the model will be applied. More precisely, the relationship between the test and training data should mirror the relationship between the data the final model will be applied to and the data from which the final model was developed. Additionally, it is crucial that the training data selected in each iteration is of similar size and composition as the data used to develop the final model. This principle, which is both intuitive and practical, offers a valuable framework for devising strategies to estimate the GE in nonstandard settings not covered in this section.

3.3.2 Clustered Data

Clustered data frequently appear in official statistics, particularly in surveys and censuses, where clusters may be, for example, households or apartment buildings. A cluster is generally defined as a collection of observations that are interconnected in some manner. Two different types of clusters can be distinguished: those comprising measurements from different individuals and those consisting of measurements from a single individual. In the latter scenario, it is common for all measurements to have the same value for the response variable.

It has been shown repeatedly in the literature that standard resampling procedures introduce an optimistic bias when applied to clustered data (Brenning and Lausen 2008; Saeb et al. 2017; Gholamiangonabadi et al. 2020; Kunjan et al. 2021; Tougui et al. 2021). This bias arises because measurements within the same cluster tend to be more similar than measurements from different clusters. In standard resampling, the training and test data for each iteration contain observations from the same clusters. Consequently, when training the prediction model, information specific to the clusters present in both the training and test sets is used, making the model more suited to the test data than to new, previously unseen clusters. Therefore, it is recommended to use grouped resampling, which differs from standard resampling by performing the resampling at the cluster level rather than at the individual observation level.

In grouped resampling, each iteration involves training and testing on different clusters. A common variant of this approach is the leave-one-subject-out CV (Gholamiangonabadi et al. 2020; Kunjan et al. 2021), also referred to as leave-one-object-out CV (Bischl et al. 2024). An exception where standard resampling is appropriate occurs in applications aimed at predicting the response variable values for new observations from clusters already represented in the data used to train the final model. Here, an overlap of clusters between training and test data is beneficial as this mirrors the overlap of clusters between the data used to train the final model and the data to which the final model will be applied.

Empirical analyses and simulation studies to date, as discussed earlier, indicate a substantial underestimation of the GE by standard resampling. However, these considered scenarios where the response values within the clusters were equal or where the clusters were large; in both cases, the underestimation is expected to be particularly strong. In contrast, Hornung et al. (2025) considered clusters of varied sizes with different response values per cluster and found only a slight underestimation of the error, which increased with the presence of cluster-constant covariates. As a general guideline, however, grouped resampling is advisable when dealing with clustered data if the prediction goal involves estimating the response values of observations from new clusters.

3.3.3 Spatial Data

Reflecting the nature of official statistics, spatial data are widespread in this field. These data can be divided into continuous and discrete types. For continuous spatial data, the precise location of each observation is known, whereas for discrete spatial data, only the broader spatial region (e.g., county or federal state) is identified.

Spatial data typically exhibit spatial correlations, as observations located near each other tend to be more similar than those farther apart. In supervised machine learning contexts, this implies that both the values of the covariates and that of the response variable show correlations that depend on the distances between observations. Previous studies have highlighted that neglecting the spatial structure and prediction objectives during GE estimation can lead to over-optimistic estimates; see, for instance, Heikkinen et al. (2012), Wenger and Olden (2012), Roberts et al. (2017), Schratz et al. (2019), and Schratz et al. (2022).

Spatial CV methods are recommended for addressing these issues. These methods ensure that the test data are geographically separated from the training data in each iteration. Various spatial CV methods exist, many of which are equipped with user-selectable parameters. The choice and configuration of these methods should consider the spatial structure of the data and the specific prediction goals.

A basic method involves a one-time division of the data into spatially separated training and test sets. This is appropriate when the spatial relationship between new data (for which predictions are needed) and the training data is well known.

For scenarios where new data is geographically distant, incorporating a buffer zone between training and test sets is advisable. This zone, excluded from both training and testing, acts as a spatial buffer. Buffer zones, applicable to various spatial CV methods, should only be used when predictions for new areas are desired, with their width reflecting the anticipated distance between the new data and the training data. The empirical correlogram's range parameter, indicating the minimum distance for uncorrelated observations, often guides the buffer zone's minimum width (Brenning 2005).

Beyond simple data splitting, another CV variant is the rectangular tiles approach, where the observation area is divided into rectangular sections for use as CV folds. The tile width should align with the application of the prediction model, guided by the empirical correlogram for predictions in new areas, similar to buffer zone considerations.

If observations are unevenly distributed, the rectangular tiles approach may lead to tiles with widely varying observation counts. In such cases, the k clustered groups method, which clusters spatial coordinates to define CV folds, is preferred. An alternative variant clusters based on covariate values instead of coordinates, which is suitable when the prediction involves new covariate values not present in the training data. To prevent optimistic bias with this approach, spatial correlation still needs to be taken into account. This can be achieved, for example, by incorporating the spatial coordinates into the clustering process.

In summary, spatial CV is essential for GE estimation with spatial data, and various methods are available. The choice and setup of these methods should be informed by the spatial structure of the data and the prediction objectives.

3.3.4 Unequal Sampling Probabilities

In official statistics surveys, it is common for observations to be selected with unequal probabilities. This can have various reasons. For instance, there might be an interest in oversampling certain subgroups (such as minorities) to ensure their adequate representation in the sample. Sampling strategies that deviate from uniform selection probabilities for all observations are termed "non-simple random sampling" (NSRS) schemes.

The implementation of NSRS schemes results in samples that do not accurately represent the population. This discrepancy must be addressed in parameter estimation to prevent bias and increased variance. For estimating population parameters, such as the mean value or totals, the Horvitz-Thompson theorem is commonly applied. In the case of estimating the mean value, the following formula is used:

$$\hat{Y}_{HT} = \frac{1}{N} \sum_{i=1}^{n} \frac{1}{\pi^{(i)}} y^{(i)}, \tag{3.6}$$

where N is the population size, n is the sample size, $y^{(i)}$ represents the value of the ith observation, and $\pi^{(i)}$ is the probability that observation i gets selected into the sample. We have $\pi^{(i)} = n\pi^{*(i)}$, where $\pi^{*(i)}$ is the probability of selecting observation i in a single draw.

Recently, it has been shown that the Horvitz-Thompson theorem can also be used for unbiased estimation of the GE in the presence of NSRS schemes (Holbrook et al. 2020). Let $\mathbf{y}^{test}$ represent the response value vector in a test data set of size n_{test} and

$\hat{f}_{\mathcal{D}}^{\,test}$ the corresponding predicted values. The GE, in relation to a loss function L, can be estimated without bias as follows:

$$\hat{\rho}_{L,HT}(\mathbf{y}^{test}, \hat{f}_{\mathcal{D}}^{\,test}) = \frac{1}{Nn_{test}} \sum_{i=1}^{n_{test}} \frac{1}{\pi^{*(i)}} L(y^{(i)}, \hat{f}_{\mathcal{D}}(x^{(i)})), \tag{3.7}$$

where $\pi^{*(i)}$ is again the probability of selecting observation i in a single draw, with $y^{(i)}$ and $\hat{f}_{\mathcal{D}}(x^{(i)})$ being elements of $\mathbf{y}^{test}$ and $\hat{f}_{\mathcal{D}}^{\,test}$, respectively. In the context of CV, this estimator is computed for each test fold and then averaged across all folds. It is important to note that the lack of representativeness of the sample also has an impact on the learned model. However, the present discussion focuses on GE estimation, and readers are referred to Breidt and Opsomer (2017) for a comprehensive review of sampling-consistent learning approaches.

The simulation conducted in Hornung et al. (2025) has illustrated that while the bias of the standard GE estimator, not corrected by the Horvitz-Thompson theorem, depends on the model class and specification, the Horvitz-Thompson adjusted estimator consistently yields unbiased estimates. Therefore, the Horvitz-Thompson corrected estimator is recommended when dealing with NSRS.

3.3.5 Concept Drift

The term "concept drift" refers to the situation where the distribution of data changes over time. This change presents a challenge because the predictive performance of prediction models deteriorates if these models are not updated or retrained regularly. Furthermore, it is crucial to account for concept drift when estimating the GE to prevent highly biased estimates. In the field of official statistics, where prediction models are often employed over extended periods, concept drift is a common occurrence. For instance, societal aging over time (European Commission 2023) or shifts in governmental policies or regulations (World Bank Group 2019), as exemplified by the recent COVID-19 pandemic, can cause concept drift.

Concept drift can be categorized into different types. Pure covariate drift occurs when only the distribution of the covariates, $\mathbb{P}_x$, changes, while the conditional distribution of the response given the covariates, $\mathbb{P}_{y|x}$, remains unchanged. This type of drift is considered less problematic since prediction models rely on $\mathbb{P}_{y|x}$. The other types of concept drift include pure response drift, where only $\mathbb{P}_{y|x}$ changes while $\mathbb{P}_x$ stays constant, and full concept drift, where both $\mathbb{P}_{y|x}$ and $\mathbb{P}_x$ change. Additionally, concept drift can be characterized based on the nature of the change: abrupt change, continuous change, and recurring concepts. Abrupt change describes a sudden shift in the data distribution, continuous change refers to a steady evolution or a gradual shift to a new distribution, and recurring concepts describe temporary changes where the distribution eventually reverts to its original state. In practice, concept drifts may not fit neatly into a single category, and mixed forms are possible.

Several statistical tests exist for detecting concept drift, focusing on changes in the prediction model's error rate or in the data distribution. To accommodate concept drift in model training, various approaches are adopted. Incremental learning involves regular retraining of the model using all historical data. For (infinite) data streams, windowing techniques can be applied, using data from a specific time horizon for training, allowing models to adapt more flexibly to distribution changes. Alternatively, models can be adjusted to new data distributions without complete retraining, offering an even more flexible response to distribution changes.

The literature on GE estimation in the context of concept drift is relatively sparse. A commonly employed method is "prequential validation" (Dawid 1984), also known as "time-series CV" (Hyndman and Athanasopoulos 2018), where the data are divided into temporally consecutive folds for training and testing in a sequential manner. Another approach, out-of-sample validation, involves a single division of data into training and test sets, with the test set positioned at the end of the observation period, closely aligning with real-world application scenarios. This method, along with prequential validation, may incorporate a temporal buffer zone between training and test data to estimate the GE to be expected in the slightly more distant future.

In the simulation study by Hornung et al. (2025), the GE was estimated largely unbiasedly using out-of-sample validation. The introduction of a buffer zone allowed the GE to be estimated slightly into the future with little bias. However, attempts to estimate the GE far into the future are discouraged, advocating for frequent model retraining or updating due to potential unanticipated distributional changes. Both prequential validation and ordinary CV resulted in biased GE estimates. In the simulation, an incremental drift was used. However, as argued in Hornung et al. (2025), it is likely that out-of-sample validation is also suitable for other types of drift. A disadvantage of out-of-sample validation is that it is associated with a fairly high variance for small data sets. However, the bias of prequential validation and ordinary CV was much more pronounced. Depending on the model class, also pure covariate drift led to a deterioration in predictive performance in the simulation, which is why the common assumption that prediction models are robust to this type of concept drift does not seem to be generally valid.

3.3.6 Hierarchical Classification

Hierarchical classification problems are characterized by the fact that each observation belongs to a hierarchy of nested classes rather than to a single class. For instance, the profession "Pet Groomer" can be hierarchically classified within the ISCO-08 classification system for occupations developed by the International Labour Organization (International Labour Organization 2012) as follows: (1) Major Group "5 Services and Sales Workers," (2) Sub-major Group "51 Personal Services Workers," (3) Minor Group "516 Other Personal Services Workers," and

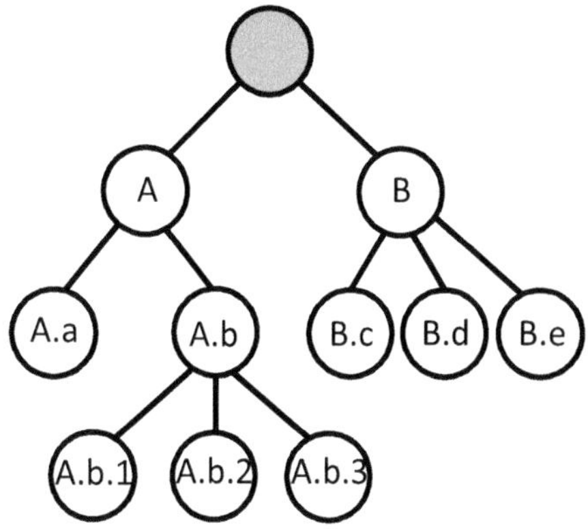

Fig. 3.4 Example of a category tree in hierarchical classification problems. Nodes higher up the tree represent more general classes—a modified version of a figure originally developed for Hornung et al. (2025)

(4) Unit Group "5164 Pet Groomers and Animal Care Workers." Such classification systems are prevalent in official statistics.

Methodologically, there are different types of hierarchical classification structures, but this paper focuses on tree-structured classification problems like ISCO-08, where each observation is assigned to exactly one class at each level. Figure 3.4 illustrates such a category tree.

It is crucial to consider the hierarchical structure when developing prediction models. There are several approaches that can be broadly divided into two groups: local classifier approaches and "big bang" approaches. Local classifier approaches divide the hierarchical classification problem into multiple individual tasks, employing a standard classifier for each task. A subcategory of these approaches uses a multi-class classifier at each node within the tree hierarchy. In the example shown in Fig. 3.4, one classifier would differentiate among classes A and B; another among classes $A.a$ and $A.b$; a third between classes $B.c$, $B.d$, and $B.e$; and a fourth between classes $A.b.1$, $A.b.2$, and $A.b.3$. Conversely, "big bang" approaches tackle the hierarchical classification problem in its entirety, facilitating the modeling of parameter relationships across different tree levels. Typically, hierarchical classification employs a top-down strategy, whereby each observation is navigated from the top downward through the hierarchy, assigning it to one of the subsequent subclasses at each node. A limitation of this method is the propagation of errors; an incorrect classification at an early stage results in subsequent misclassifications.

There are a variety of evaluation metrics designed to assess the predictive performance of hierarchical classification models. These take into account various aspects of the hierarchical structure, acknowledging that a prediction might be wrong at one level yet correct at another (Kiritchenko et al. 2005), that certain classes are more closely related within the hierarchy than others (Wang et al. 1999), and that misclassifications at higher hierarchy levels are more consequential than those at lower levels (Blockeel et al. 2002; Cesa-Bianchi et al. 2004; Wu et al. 2019).

Prior to Hornung et al. (2025), guidelines for selecting appropriate resampling methods for hierarchical classification appeared to be absent. Drawing on findings from Kohavi (1995), Hornung et al. (2025) postulated that stratified CV in hierarchical classification might exhibit lower bias and variance compared to traditional CV. In the simulation study presented in Hornung et al. (2025), ordinary CV slightly underestimated the predictive performance, whereas stratified CV was not associated with bias except for very small data sets. Another exception was macro-averaged performance metrics, where stratified CV also led to biased results, possibly due to the specific design of the simulation. The study did not find differences in variance between stratified and ordinary CV.

Important Takeaways for GE in Non standard Data Settings

1. Clustered data: Grouped cross-validation (CV), with fold allocation based on clusters rather than observations, provides unbiased GE estimates.
2. Spatial data: Spatial CV techniques yield realistic GE estimates, with the choice and configuration of a suitable technique depending on prediction goals and data structure.
3. Unequal sampling probabilities: The Horvitz-Thompson correction allows for unbiased GE estimation.
4. Concept drift: Out-of-sample validation, where the test data is from the end of the observation period, provides realistic GE estimates.
5. Hierarchical classification: In contrast to stratified CV, ordinary CV may slightly underestimate performance.
6. General principle: Resampling techniques should, among other things, take into account the relationship between training and new data.

References

S. Bates, T. Hastie, R. Tibshirani, Cross-validation: what does it estimate and how well does it do it? J. Am. Stat. Assoc. **119**(546), 1434–1445 (2024)

P. Bayle, A. Bayle, L. Janson, L. Mackey, Cross-validation confidence intervals for test error. Adv. Neural Inf. Proces. Syst. **33**, 16339–16350 (2020)

Y. Bengio, Y. Grandvalet, No unbiased estimator of the variance of k-fold cross-validation. J. Mach. Learn. Res. **5**, 1089–1105 (2004)

B. Bischl, R. Sonabend, L. Kotthoff, M. Lang, (eds.) *Applied Machine Learning Using mlr3 in R*, Chapter 3 (CRC Press, Boca Raton, 2024)

H. Blockeel, M. Bruynooghe, S. Džeroski, J. Ramon, J. Struyf, Hierarchical multi-classification, in *Workshop Notes of the KDD'02 Workshop on Multi-Relational Data Mining* (2002), pp. 21–35

F.J. Breidt, J.D. Opsomer, Model-assisted survey estimation with modern prediction techniques. Stat. Sci. **32**(2), 190–205 (2017)

A. Brenning, Spatial prediction models for landslide hazards: review, comparison and evaluation. Nat. Hazards Earth Syst. Sci. **5**(6), 853–862 (2005)

A. Brenning, B. Lausen, Estimating error rates in the classification of paired organs. Stat. Med. **27**(22), 4515–4531 (2008)

N. Cesa-Bianchi, C. Gentile, L. Zaniboni, Incremental algorithms for hierarchical classification, in *Advances in Neural Information Processing Systems*, ed. by L.K. Saul, Y. Weiss, L. Bottou (eds.), vol. 17 (MIT Press, Cambridge, 2004), pp. 233–240

T. Chen, T. He, M. Benesty, V. Khotilovich, Y. Tang, H. Cho, K. Chen, R. Mitchell, I. Cano, T. Zhou, M. Li, J. Xie, M. Lin, Y. Geng, Y. Li, J. Yuan, *xgboost: Extreme Gradient Boosting* (2023). https://CRAN.R-project.org/package=xgboost. R package version 1.7.3.1

A.P. Dawid, Present position and potential developments: some personal views statistical theory the prequential approach. J. R. Stat. Soc. Ser. A **147**(2), 278–290 (1984)

T.G. Dietterich, Approximate statistical tests for comparing supervised classification learning algorithms. Neural Comput. **10**(7), 1895–1923 (1998)

European Commission, The impact of demographic change – in a changing environment. Commission Staff Working Document SWD(2023) 21 final (2023). https://commission.europa.eu/system/files/2023-01/theimpactofdemographicchangeinachangingenvironment2023.pdf

S. Fischer, H. Schulz-Kümpel, *mlr3inferr: Inference on the Generalization Error* (2025). https://mlr3inferr.mlr-org.com. R package version 0.1.0, https://github.com/mlr-org/mlr3inferr

D. Gholamiangonabadi, N. Kiselov, K. Grolinger, Deep neural networks for human activity recognition with wearable sensors: leave-one-subject-out cross-validation for model selection. IEEE Access **8**, 133982–133994 (2020)

R.K. Heikkinen, M. Marmion, M. Luoto, Does the interpolation accuracy of species distribution models come at the expense of transferability? Ecography **35**(3), 276–288 (2012)

A. Holbrook, T. Lumley, D. Gillen, Estimating prediction error for complex samples. Canadian J. Stat. **48**(2), 204–221 (2020)

R. Hornung, M. Nalenz, L. Schneider, A. Bender, L. Bothmann, F. Dumpert, B. Bischl, T. Augustin, A.-L. Boulesteix, Evaluating machine learning models in non-standard settings: an overview and new findings. Stat. Sci. (2025). Accepted

R.J. Hyndman, G. Athanasopoulos, *Forecasting: Principles and Practice* (OTexts, Melbourne, 2018)

International Labour Organization, International standard classification of occupations: ISCO-08. Technical report, 2012. https://webapps.ilo.org/ilostat-files/ISCO/newdocs-08-2021/ISCO-08/ISCO-08%20EN%20Vol%201.pdf

G. James, D. Witten, T. Hastie, R. Tibshirani, *An Introduction to Statistical Learning: with Applications in R* (Springer, Berlin, 2021)

W. Jiang, S. Varma, R. Simon, Calculating confidence intervals for prediction error in microarray classification using resampling. Stat. Appl. Genet. Mol. Biol. **7**(1), Article8 (2008)

S. Kiritchenko, S. Matwin, A.F. Famili, Functional annotation of genes using hierarchical text categorization, in *Proceedings of the BioLINK SIG: Linking Literature, Information and Knowledge for Biology (Held at ISMB-05)* (2005)

R. Kohavi, A study of cross-validation and bootstrap for accuracy estimation and model selection, in *Proceedings of the 14th International Joint Conference on Artificial Intelligence* (1995), pp. 1137–1143

S. Kunjan, T.S. Grummett, K.J. Pope, D.M. Powers, S.P. Fitzgibbon, T. Bastiampillai, M. Battersby, T.W. Lewis, The necessity of leave one subject out (LOSO) cross validation for EEG disease diagnosis, in *Proceedings of the 14th International Conference on Brain Informatics, Virtual Event* (2021), pp. 558–567

C. Nadeau, Y. Bengio, Inference for the Generalization Error. Mach. Learn. **52**(3), 239–281 (2003)

H. Noma, T. Shinozaki, K. Iba, S. Teramukai, T.A. Furukawa, Confidence intervals of prediction accuracy measures for multivariable prediction models based on the bootstrap-based optimism correction methods. Stat. Med. **40**(26), 5691–5701 (2021)

D.R. Roberts, V. Bahn, S. Ciuti, M.S. Boyce, J. Elith, G. Guillera-Arroita, S. Hauenstein, J.J. Lahoz-Monfort, B. Schröder, W. Thuiller, D.I. Warton, B.A. Wintle, F. Hartig, C.F. Dormann, Cross-validation strategies for data with temporal, spatial, hierarchical, or phylogenetic structure. Ecography **40**(8), 913–929 (2017)

S. Saeb, L. Lonini, A. Jayaraman, D.C. Mohr, K.P. Kording, The need to approximate the use-case in clinical machine learning. Gigascience **6**(5), gix019 (2017)

P. Schratz, J. Muenchow, E. Iturritxa, J. Richter, A. Brenning, Hyperparameter tuning and performance assessment of statistical and machine-learning algorithms using spatial data. Ecol. Model. **406**, 109–120 (2019)

P. Schratz, M. Becker, M. Lang, A. Brenning, mlr3spatiotempcv: Spatiotemporal resampling methods for machine learning in R (2022). https://arxiv.org/abs/2110.12674

H. Schulz-Kümpel, S. Fischer, R. Hornung, A.-L. Boulesteix, T. Nagler, B. Bischl, Constructing confidence intervals for 'the' generalization error – a comprehensive benchmark study. J. Data-Centric Mach. Learn. Res. **6**, 1–73 (2025)

J. Shao, C.F.J. Wu, A general theory for jackknife variance estimation. Ann. Stat. **17**(3), 1176–1197 (1989)

I. Tougui, A. Jilbab, J. El Mhamdi, Impact of the choice of cross-validation techniques on the results of machine learning-based diagnostic applications. Healthcare Inf. Res. **27**(3), 189–199 (2021)

K. Wang, S. Zhou, S.C. Liew, Building hierarchical classifiers using class proximity, in *Proceedings of the 25th International Conference on Very Large Data Bases* (1999), pp. 363–374

S.J. Wenger, J.D. Olden, Assessing transferability of ecological models: an underappreciated aspect of statistical validation. Methods Ecol. Evol. **3**(2), 260–267 (2012)

World Bank Group, *The Changing Nature of Work*. World Development Report. World Bank (2019)

C. Wu, M. Tygert, Y. LeCun, A hierarchical loss and its problems when classifying non-hierarchically. PLOS ONE **14**(12), e0226222 (2019)

Part II
Legal, Ethical, and Quality Aspects

Chapter 4
Quality Dimensions and Quality Guidelines for Machine Learning in Official Statistics

Younes Saidani and Florian Dumpert

4.1 Introduction

Official statistics enjoys special privileges over other statistics providers: Its work is not subject to market forces and profitability concerns, and respondents are in many cases legally required to respond to its surveys, thus enabling statistical offices to continuously collect and publish data on topics of public interest. In turn, it is expected that official statistics produces data of high quality in order to fulfil its important role as a reliable data provider for parliament, government, administration, society and industry, among other stakeholders. ‘The high quality of data from official statistics is both an aspiration and a unique selling point’, as emphasised in the recommendations of the Statistical Advisory Committee for the 21st legislative period in Germany.[1] After all, ‘bad quality erodes trust [in official statistics] very, very fast’,[2] as Walter Radermacher, former President of the German Federal Statistical Office and Director-General of Eurostat, once remarked. The quality of statistical data has therefore always been of great importance in official statistics. But what does it mean for data to be ‘of high quality’? In short, it has to be fit for use. In Europe, the Statistics Code of Practice (CoP) (European Statistical System Committee 2017) defines 16 key principles for the institutional environment, statistical processes and statistical outputs, which are used to assess and safeguard

[1] https://www.destatis.de/DE/Ueber-uns/Leitung-Organisation/Statistischer-Beirat/empfehlungen.pdf?__blob=publicationFile, p. 3. Translated into English by the authors.
[2] Translated into English by the authors.

Y. Saidani (✉) · F. Dumpert
Federal Statistical Office of Germany, Wiesbaden, Germany
e-mail: Younes.Saidani@destatis.de; Florian.Dumpert@destatis.de

F. Dumpert (ed.), *Foundations and Advances of Machine Learning in Official Statistics*, Society, Environment and Statistics,
https://doi.org/10.1007/978-3-032-10004-7_4

quality in statistical offices. The Quality Assurance Framework (QAF) (European Statistical System Committee 2022) provides further guidance on how to implement the high-level principles from the CoP by offering more detailed methods, tools and good practices.

4.2 The Need for Tailor-Made Quality Guidance for ML

Among other things, the CoP and the QAF explicitly require official statistics to be 'constantly striving for innovation' (QAF, Indicator 7.1) to improve the quality of its products. Yet both documents are derived from and based on the requirements and challenges of 'traditional' statistics production. This poses a practical challenge, since innovative production methods can differ substantially from traditional ones, potentially reducing the usefulness of existing quality frameworks in one of three ways:

1. Certain quality dimensions may not be applicable to new methods at all.
2. They may be applicable in principle but differ with regard to the methodological details.
3. New methods may present new challenges that are not covered by existing quality dimensions.

Consequently, upon adopting new methods, there is a need to assess their compatibility with existing official statistics quality frameworks and to offer accompanying quality guidance in case it is needed.[3] Machine learning (ML)—an example of such an 'innovation'—has matured from future technology to industry standard. The term refers to a collection of approaches that (according to one definition) differ from traditional statistical methods with regard to their intended use: While 'classical', research-oriented statistics often focuses on hypothesis testing, ML algorithms mostly aim to optimally predict the properties of new observations, often with the aim of process automation. ML methods such as tree-based approaches (including random forests and boosting), support vector machines and neural networks offer great potential for classification and coding tasks, error detection and correction as well as the imputation of missing values. As a result, they have been actively taken up and piloted in official statistics. Yet the adoption of ML is marred with methodological, technological and regulatory challenges—among them the question if and how existing quality guidance from official statistics can be applied to ML methods. The further development and refinement of the concept of quality with regard to machine learning by official statistics, as well as its practical

[3] This is a frequently chosen approach in official statistics; see, for example, Kowarik and Six (2022) for the acquisition and usage of big data; Puts and Daas (2021) for an additional view on machine learning; Gootzen et al. (2023) for combining survey, administrative and big data; and Ascari et al. (2020) for the quality of multisource statistics.

implementation, is also important with regard to ethical questions, as the work of Dumpert et al. (2025a,b) shows.

4.3 Existing Quality Frameworks for Official Statistics Are Useful but Insufficient

The quality principles in the Code of Practice are grouped into three levels: the institutional environment, the statistical processes and the statistical products. Of these, the last two are primarily of interest, as the principles on institutional and organisational factors are too abstract and not directly affected by the use of machine learning procedures. The *principles for statistical processes* define European standards, guidelines and practices for the management, efficiency and innovation of processes. They also aim to maintain and strengthen the credibility of official statistics. The *principles for statistical products* are primarily focussed on user needs. They aim to ensure that statistics meet the needs of European institutions, governments, research organisations, businesses and the public in general. Table 4.1 provides an overview of all nine principles for statistical processes and products.

In order to assess the usefulness of existing official statistics quality frameworks for the application of machine learning methods, we now turn to the indicators from the Quality Assurance Framework supplementing each principle and evaluate them with regard to their relevance for ML, noting along the way which indicators might be positively or adversely affected by widespread ML use.

4.3.1 The QAF Indicators and Their Relevance for Machine Learning

Sound Methodology

All seven indicators of this quality principle are relevant for the use of machine learning methods. The first one states that the methods used must generally follow international standards (7.1). Deviations therefrom must hence be explained in quality and methodological reports for ML methods. Furthermore, the indicators describe that standard concepts, definitions and classifications are to be used uniformly and that these classification systems must be consistent at the national and European level (7.2 and 7.4). When machine learning methods are used to support classification tasks, they should be compatible with existing classification systems, thereby facilitating the exchange of training, validation and test datasets. In addition, ML methods can be used to support regular updates to the statistical registers such as the business register (7.3), for example, by imputing missing values or by automating classification. The remaining indicators state that in order to produce high-quality statistics and ensure sound methodology, graduates from relevant academic disciplines must be recruited (7.5) and that employees must receive

Table 4.1 Principles for statistical processes and products from the European Code of Practice

Principles for statistical processes	
7. Sound methodology	Sound methodology underpins quality statistics. This requires adequate tools, procedures and expertise
8. Appropriate statistical procedures	Appropriate statistical procedures, implemented throughout the statistical processes, underpin quality statistics
9. Non-excessive burden on respondents	The response burden is proportionate to the needs of the users and is not excessive for respondents. The statistical authorities monitor the response burden and set targets for its reduction over time
10. Cost-effectiveness	Resources are used effectively
Principles for statistical products	
11. Relevance	European statistics meet the needs of users
12. Accuracy and reliability	European statistics accurately and reliably portray reality
13. Timeliness and punctuality	European statistics are released in a timely and punctual manner
14. Coherence and comparability	European statistics are consistent internally, over time and comparable between regions and countries; it is possible to combine and make joint use of related data from different data sources
15. Accessibility and clarity	European statistics are presented in a clear and understandable form, released in a suitable and convenient manner, available and accessible on an impartial basis with supporting metadata and guidance

continuous vocational training (7.6). In addition, cooperation with the scientific community should be maintained to continuously improve the methodology (7.7). Knowledge acquisition and knowledge transfer are especially crucial when dealing with the rapidly developing field of machine learning. Although machine learning methods are now finding their way into university curricula, active training is needed for those employees in statistical offices that are not recent graduates. A high level of ML competency can only be created through the targeted recruitment of specialists and large-scale training programmes. In order to promote dialogue with the scientific community, the experience already built up in official statistics must be shared and discussed at scientific events.

Appropriate Statistical Procedures

Of the seven indicators of this principle, one is directly related to machine learning methods: 'Statistical processes are routinely monitored and revised as required' (8.3). This includes the design of survey and sampling plans, on which machine learning methods can have an impact via their use in estimation procedures. It also covers methods and procedures for data collection and processing. It is particularly in these phases of the Generic Statistical Business Process Model (GSBPM) that machine learning offers great potential for increasing data quality and saving staff costs, for example, through automated coding or imputation. In order to conform to quality standards, ML algorithms must thus be adequately deployed, monitored and, if necessary, regularly revised.

Non-excessive Burden on Respondents
Two of the six indicators are relevant for machine learning: The burden on respondents is to be 'limited to what is absolutely necessary' (9.1), including through 'measures that enable the linking of data sources' (9.6). This is an area in which machine learning methods are routinely applied and offer much potential (e.g. as part of a web scraping pipeline in price statistics). However, it should be noted that in some EU member states such as Germany, the merging of data sources is subject to the provisions of the Federal Statistics Act (see § 13a and § 6 (5) BStatG); data from freely accessible sources, for example, may be linked with official data.

Cost-Effectiveness
This principle is subdivided into four indicators, two of which are relevant to the use of machine learning. First, the instruction to optimise the productivity potential of information and communications technology for the statistical processes (10.2) provides a strong mandate for appraising the use of ML algorithms in all areas of official statistics. Second, any such algorithms should be developed as 'standardised solutions that increase effectiveness and efficiency' (10.4), although depending on the particular application in question this may not be straightforward or indeed possible.

Relevance
This quality principle does not contain any direct guidance on the use of ML algorithms. It may, however, be positively affected, since machine learning methods often allow analyses that were previously not possible or increase the speed with which statistics are produced and hence published, thus increasing their relevance to the wider public.

Accuracy and Reliability
Two of the three indicators describe aspects that are absolutely essential when using machine learning methods: the assessment and validation of source data (12.1) and the measurement of sampling and non-sampling errors (12.2). Machine learning models are trained using source data, for which a model with high predictive power is estimated. The better the source data represents the phenomenon of interest, the higher the quality of model predictions. When using machine learning methods, particular attention must therefore be paid to evaluating the source data used in order to ensure accurate results. It is standard practice to describe and document the quality of the model using suitable metrics (e.g. accuracy and precision).

Timeliness and Punctuality
Two of the five indicators for this principle are relevant to machine learning. When using machine learning methods, the timeliness of data publications and deliveries (e.g. to Eurostat) must continue to 'meet European and other international release standards' (13.1). Beyond merely meeting existing deadlines, however, statistics production can also be accelerated, as ML algorithms can replace manual work, automate process steps and thus improve timeliness. They can also often allow 'preliminary results of acceptable aggregate accuracy' to be released at an earlier stage (13.5). For example, Salgado et al. (2023) use a gradient boosting procedure to impute survey data that has not yet been received, thus enabling an increasingly accurate early estimate for the Spanish industrial turnover index.[4]

Coherence and Comparability
This principle with its five indicators relates to machine learning methods in two ways: On the one hand, statistics must be 'internally coherent and consistent' (14.1). On the other hand, they must also remain comparable over time—especially if there are methodological changes (14.2). Deviations must be documented and made accessible to users. If machine learning methods are used as a supplement or substitute for traditional methods, it is essential that cross-statistical relationships are maintained so as not to jeopardise comparability. In addition, such methodological changes should be communicated transparently in articles or method papers.

Accessibility and Clarity
The seven indicators for this principle deal with data access and the description of statistics, which should be designed as user friendly as possible. Two indicators are relevant for the use of machine learning methods: Users must be provided with all the necessary information required for 'proper interpretation and meaningful comparisons' (15.1). In addition, the methods must be explained transparently, if only in brief. The use of ML algorithms must therefore be communicated in quality reports and method papers; the corresponding metadata (15.5) must also contain methodological information.

4.3.2 Summary

A closer look at official statistics quality frameworks reveals that a number of quality principles contained in the CoP and many of their respective quality indicators from the QAF are either potentially affected by machine learning methods or can be used to derive quality requirements for their usage. Most of these quality principles are thus broad enough to cover most methodological characteristics of ML. Yet they are also not specific enough to provide useful guidance for the use of ML methods in official statistics: The principles for statistical processes (7 to

[4] See also Sect. 13.3.1 of Barragán et al. (2025) in this book.

10) are too abstract, and the principles for statistical products (11 to 15)—referring to the quality of statistical publications rather than that of intermediate results, the generation of which is where ML methods are most often employed—are too indirect. Discussing the potential need for new quality criteria when integrating new data and methods in official statistics, de Broe et al. (2021) similarly conclude that 'the methodological quality aspects related to statistical learning and big data clearly contain new elements compared to the well-established ESS [European Statistical System] output quality dimensions. Those new elements can be seen as extensions of the well-established dimensions rather than completely new quality dimensions of their own' (ibid., p. 357). On the one hand, this underscores the need for developing quality guidance specifically tailored to ML in order to enable adequate quality management (see also Julien 2020 and Dumpert 2021). On the other hand, it shows that such new guidance can—and in fact should—build on the existing quality frameworks for official statistics, highlighting relevant links where possible, providing further details where useful and offering extensions where needed. This chapter aims to contribute to the development of such new quality guidance for machine learning methods in official statistics. It is based on previous collaborative work in German by Saidani et al. (2023).

4.4 A Four-Step Approach Towards Comprehensive Quality Guidance for ML

Attempting to establish an extension of existing quality frameworks for machine learning methods is a multistep process. Similar to existing quality guidance—which starts with high-level quality principles (in the CoP) that are then broken down into quality indicators and further supplemented by so-called methods (in the QAF)—we suggest the following four-step structure for ML, starting at the abstract level and then moving into specifics:

1. Quality dimensions that define what 'quality' means for ML, i.e. clarify the concept and its main components
2. Quality guidelines for statistical processes that describe what needs to be considered during ML development in order to ensure quality along the above dimensions
3. Quality indicators and metrics for statistical results generated using ML that permit a quantitative evaluation of quality along the above dimensions during development and production
4. Standards and recommendations for quality documentation that aid in communicating the quality of ML used in statistics productions in an appropriate, standardised manner

All four steps are necessary components of a holistic quality guidance for ML in official statistics. Quality dimensions provide the conceptual background, quality guidelines are like checklists and guide the design of processes, and quality indicators formulate adequate metrics for evaluating quality. Last but not least, given the regulatory background and increasing user requirements, standardised quality documentation ensures transparency about the ML methods used and their effect on the quality of statistical products. Developing such dimensions, guidelines, metrics and documentations requires the collaboration of machine learning practitioners, subject matter statisticians and quality officers. Given the plethora of methods that can be considered 'machine learning', they must strike a balance between being overly general—and thus not useful—and too specific, and thus only applicable to certain methods. They must also consider that ML is a rapidly evolving field and thus allow for changing best practices. Last but not least, theoretical musings are only useful if they are implemented in practice; thus ensuring adoption of new standards in statistical offices—the difficulty of which must not be underestimated due to cultural and behavioural obstacles—is of utmost importance.[5] Besides structuring the task at hand, this chapter describes high-level quality dimensions tailored to the methodological peculiarities of ML, thereby expanding on previous suggestions, and two cross-cutting issues that are highly relevant to the quality of ML algorithms. Furthermore, we contribute a list of quality guidelines for each of these dimensions. Subsequent work should aim to devise quality metrics and standardised quality documentations.

4.5 Quality Dimensions and Guidelines

As part of the evaluation of ML methods for use in official statistics, colleagues from statistical offices have already attended to the task of deriving a set of suitable quality dimensions (Step 1 of the list above). Building on de Broe et al. (2021)—perhaps the first such contribution—a group of national experts from UNECE member states and Australia under the umbrella of the UNECE High-Level Group for the Modernisation of Official Statistics developed a 'Quality Framework for Statistical Algorithms' (Yung et al. 2022). In it, they take up many of the quality aspects elaborated by de Broe et al. (2021), further flesh them out with regard to the methodological peculiarities of 'statistical algorithms' (including ML algorithms) in statistics production and suggest five quality dimensions: explainability, accuracy, reproducibility, timeliness and cost-effectiveness. While these dimensions cover many of challenges and advantages associated with ML, they omit at least one

[5] Note that the quality dimensions and quality guidelines in the following section may appear trivial at first sight. The real challenge lies not in theoretical musings but rather in integrating these quality dimensions and quality guidelines into practical work, ensuring compliance with them and making the degree of fulfilment transparent.

important topic: How reliable is a model, once trained, in production, when it is confronted with previously unanticipated conditions? For instance, relationships that a model has learned from the training data may change in the real world over time. This and related problems are covered by the term 'robustness' (also: 'stability'), which is essential for an adequate quality assessment of machine learning methods. Once robustness is added to the above list, the six quality dimensions cover all indicators from the QAF that are relevant for the use of ML: Indicators associated with sound methodology, appropriate statistical procedures and non-excessive burden on respondents can be assigned to the dimensions accuracy, robustness, explainability and reproducibility. Relevant indicators from accuracy and reliability are split up into the two dimensions, i.e. accuracy and robustness, the latter of which also covers coherence and comparability. Accessibility and clarity relate to the dimensions reproducibility and explainability. The principles timeliness and punctuality, as well as cost-effectiveness, remain as dimensions with the same name.

Table 4.2 summarises the six quality dimensions and offers a very brief description. Furthermore, it arranges the dimensions in increasing order of abstraction: While accuracy and (partly) robustness concern themselves with individual predictions, cost-effectiveness is assessed at the level of business processes. In the following, each quality dimension is briefly discussed in more detail, accompanied by a list of quality guidelines (Step 2 of the above-mentioned list). While the quality dimensions represented a natural extension of the dimensions from Yung et al. (2022), the quality guidelines were created from scratch, in a collaborative process that involved both subject matter and methodological experts. They were published in a first version in Saidani et al. (2023) but further developed since. While compiling quality guidelines, the authors faced the following competing objectives: On the one hand, comprehensive quality assurance is best achieved through diverse and detailed quality guidelines. On the other hand, an excessive number of guidelines would limit practicability and possibly also acceptance. This conflict was eventually resolved by a moderate number of quality guidelines that are limited to the essentials. Before publication, the revised quality guidelines were also tested by subject matter experts for their suitability. In German official statistics, the

Table 4.2 Quality dimensions for machine learning (summarised)

Dimension	Description	Level of abstraction
Accuracy	Phenomenon is described correctly	Predictions
Robustness	Stable but useful results despite small perturbations	Predictions, model
Explainability	Understand how results are generated	Predictions, model
Reproducibility	Reproduce results identically	IT infrastructure
Timeliness and punctuality	Deliver up-to-date results punctually	IT infrastructure, business processes
Cost-effectiveness	Appropriate costs	Business processes

Table 4.3 Quality guidelines for the dimension *accuracy*

A1	Machine learning methods follow standard scientific methodology in relation to statistics and ML. The methodology used was documented (including reasons for selecting a particular method) and communicated with the target audience
A2	The suitability and quality of the training data were analysed and evaluated during development
A3	Relevant accuracy metrics to be used in the optimisation and selection of ML models were determined by subject matter experts
A4	All relevant metrics were reported along with suitable uncertainty estimates
A5	Relevant subgroups were specified by subject matter experts; accuracy metrics were evaluated for those subgroups and checked for systematic prediction errors
A6	During the model selection process, the performance of the shortlisted models was communicated transparently

resulting guidelines have already been established as binding in principle: Whenever machine learning is to be used in the production of official statistics, the respective subject matter department decides which of these guidelines are to be fulfilled.

4.5.1 Accuracy

Accuracy is the degree to which a statistical output is able to correctly describe the phenomenon being measured, i.e. to minimise the suitably measured distance between the estimate and the true value. Accuracy is thus not a binary criterion; how much accuracy suffices depends on the specific use case. In supervised learning, the metrics used to evaluate accuracy depend on the phenomenon under observation, in particular whether the goal is classification or regression. Which metric is deemed most relevant for model selection depends, again, on the use case and is a decision to be made by subject matter experts. In any case, accuracy should not just be reported as a point estimate, but also be accompanied by a measure of uncertainty such as a confidence interval.[6] Furthermore, model accuracy should also consider the uncertainty or bias in the training, validation and test datasets. Thus, ensuring high-quality training data is essential for accurate ML models. Last, beyond the average result, certain subgroups might be of particular interest (e.g. firms of different sizes, economic sectors, influential data points, entities of public or political interest, administrative regions and age groups). In this case, accuracy should not just be reported on an aggregate level, but also for those relevant subgroups (Table 4.3).

[6] For a discussion on this approach, see Schulz-Kümpel et al. (2025) in this book.

Table 4.4 Quality guidelines for the dimension *robustness*

R1	The desired object of robustness analyses was defined: specific predictions, model coefficients, accuracy metrics or aggregate values of the target variable
R2	Robust ML methods were considered as candidate models
R3	The effect of deliberate manipulation or the addition of grossly incorrect data points was analysed; a suitable resampling procedure was used
R4	The effect of violating model assumptions and different hyperparameter choices were analysed
R5	The risk of concept drift was assessed; if relevant, a procedure for detecting concept drift was implemented

4.5.2 Robustness

Robustness is the degree to which a model produces stable (but useful) results given small perturbations in the environment—which may be outliers in the data, changes in its distribution, violations of model assumptions, structural changes in the observed phenomenon over time (concept drift) or different choices of hyperparameters. Concept drift in particular is almost certainly an issue if a model is employed for multiple years and can be dealt with by regular retraining or by implementing a mechanism for drift detection. But which 'results' (i.e. which quantities) should be stable given small perturbations? Plausible candidates are specific predictions (e.g. for influential data points), model coefficients (e.g. of a linear model), accuracy metrics (like an F1 score for all or certain classes in a classification problem) or aggregates that are produced downstream in the statistical production process (e.g. total revenue by industry, export volume by enterprise type). The latter seems most relevant in almost all cases, yet the large number of processing steps—often conducted using separate tools that may or may not be connected by automated interfaces—makes assessing robustness on this level very difficult. In practice, the most feasible and useful approach is to evaluate the stability of accuracy metrics under simulated, adverse scenarios (Table 4.4).

4.5.3 Explainability

Explainability is the ability to understand which relationships the algorithm uses to make predictions, i.e. to be able to demonstrate the (possibly local) relationship between input and output variables—in other words, on the basis of which relationship outputs are generated. Defined in this way, explainability as post hoc interpretability[7] is becoming more and more relevant in the face of new laws

[7] See also Dandl et al. (2025) in this book.

Table 4.5 Quality guidelines for the dimension *explainability*

E1	The desired level of explainability was determined—in particular, whether decomposability, algorithmic transparency or interpretability of the results are required, and whether additional requirements arise from the legal situation for the given statistics
E2	After the preselection of suitable ML methods, these were evaluated in terms of their explainability. Only those ML methods that fulfil the above-defined requirements were then used. If two models yield (approximately) similar results, the more explainable model was used
E3	Post hoc explainability analyses were carried out, if necessary

and regulations on the use of artificial intelligence systems (European AI Act).[8] Independently thereof, explainable models are preferable because they generally increase trust among users and allow developers to spot specification mistakes more easily. In addition to post hoc interpretability, Lipton (2018) widens the concept of explainability to include decomposability and algorithmic transparency. Decomposability is present when the model components have intuitive meanings. This applies in particular to the features and their (estimated) parameters, provided that such a relation exists in a procedure at all. Algorithmic transparency, on the other hand, is present when the behaviour of the algorithm can be described well not in terms of content but rather in technical terms, for example, on the basis of convergence or optimality criteria (Table 4.5).

4.5.4 Reproducibility

Reproducibility is the ability to achieve identical results when using the same data and the same code for model training, selection and application. In practice, this is achieved by versioning, documenting and archiving data, codes and libraries for a specific application or—better—as a standard procedure for all use cases. Reproducibility ensures trust in official statistics and guarantees the auditability of administrative activities. The reproducibility of results is perhaps the most natural of the quality dimensions mentioned here: What is a statistician to do with a procedure that does not allow him or her to reproduce previous results with the exact same previous data using the same code—clearly, such a procedure would be arbitrary. The practical challenges, however, often lie in being able to recreate those exact conditions at any given time (Table 4.6).

[8] See also Krög (2025) in this book.

Table 4.6 Quality guidelines for the dimension *reproducibility*

P1	All data and code files used in the analysis were stored in a retrievable format. If deletion periods restrict archiving, this was documented
P2	It was ensured that the archived data remains unchanged
P3	Future access to the data and code files was clarified and documented, and suitable access permissions were set up (taking into account all necessary legal requirements and security levels)
P4	A dataset description has been made available in which, among other things, the scope of the data is precisely defined and documented. A codebook, which describes the information and variables in the dataset, has been made available. Potentially necessary additional information (e.g. peculiar variable definitions, temporary sample characteristics) have also been written down
P5	The software used was documented in writing, and information on the packages, modules or libraries used was stored. Versions were documented. If those differ from the latest version, this choice was justified and documented. The code files contain information on the time of creation and the authors
P6	All code blocks were sufficiently commented in order to be able to trace all steps—i.e. input, output and modification of the data by the code. If functions or classes have special characteristics, this was sufficiently documented

Table 4.7 Quality guidelines for the dimension *timeliness and punctuality*

T1	Sufficient time was scheduled for the design, development and implementation of ML methods
T2	Sufficient time was scheduled for data procurement and data preparation
T3	The (possibly longer) lead time of the developed ML application in production was measured and scheduled

4.5.5 *Timeliness and Punctuality*

Timeliness and punctuality describe the ability to design, train and apply the ML algorithm within the required time frame and to publish up-to-date results. Obviously, if a machine learning procedure is not available in time, it will not fulfil its purpose (at least in the short term) and thus offer no added value. At the same time, it is important to ensure that sufficient time is available for (re)training (including evaluating the quality of the training data; preparing the training, validation and test data material; selecting variables and models; estimating the final quality measures; etc.). Quick ad hoc 'solutions' are usually not suitable for production. Finally, it must be possible to apply the produced ML model, i.e. to generate the predictions, within the time allotted for this purpose—computing processes that take several days can be problematic if, for example, short-term statistics have to be produced on a monthly basis (Table 4.7).

Table 4.8 Quality guidelines for the dimension *cost-effectiveness*

C1	Requirements for the ML application were systematically collected and documented
C2	The feasibility of the project was assessed before the start. If the prospects of fulfilling the requirements are low, or if an assessment is difficult to make, a feasibility study (e.g. use case, proof of concept) was carried out in which the key features of the ML application were tested
C3	A profitability forecast was carried out. Expected expenses (time, personnel, other resources) were estimated and compared with the expected benefits
C4	A post hoc cost-benefit analysis of the ML application was carried out after its implementation. Actual costs were determined and compared with the realised benefits

4.5.6 Cost-Effectiveness

Cost-effectiveness describes the implementation costs relative to the above-mentioned quality dimensions. Thus, lower costs given fixed accuracy, robustness, explainability, reproducibility, timeliness and punctuality imply better cost-effectiveness (Table 4.8).

4.6 Cross-Cutting Issues

Last but not least, there are two cross-cutting topics that are not quality dimensions themselves, but are nonetheless closely related to the six dimensions discussed above: fairness—whether an ML model produces 'fair' results for subgroups of interest—and machine learning operations (MLOps), a set of technical standards, which in practice are necessary for a time- and cost-efficient implementation of many quality requirements.

4.6.1 Fairness

The fairness of statistical procedures refers to the effects that algorithmic decisions or classifications can have on the individuals or administrative units surveyed. In the context of official statistics, such effects are usually indirect, for example, through political decisions based on the published data.[9] In addition to general considerations such as the quality of the training data—a machine learning model can learn false correlations regardless of the estimation accuracy if the training data already contain structural biases (Mehrabi et al. 2022)—the accuracy of an ML procedure can have implications for fairness if statistical aggregates for certain

[9] For fairness of ML in official statistics, see also Schenk et al. (2025) in this book.

subgroups are systematically over- or underestimated. This is particularly relevant for subgroups that are less represented in the data or tend to be at the edge of the distribution. Strategies to increase the accuracy of the model for small subgroups and thus avoid bias include up- and downsampling methods, hybrid forms such as SMOTE and ROSE, active learning using expert feedback, weighting observations or the use of adapted ML models that are specially optimised for unbalanced data. In practice, these methods may not be absolutely necessary if simpler models provide satisfactory estimation accuracy for subgroups. This can be evaluated by calculating whether simple aggregates for subgroups of interest are under- or overestimated (to a relevant extent) in the trained model. Besides accuracy, fairness also relates to explainability in so far as understanding the effect of subgroup characteristics on the target variable allows ML developers to draw valuable conclusions about the inner workings and reliability of the model. For instance, if a model has learned multivariate relationships that are supported by empirical research or common-sense, this might inspire confidence in its ability to deal with unseen data.

4.6.2 MLOps

Standardised data processing and data management are essential for quality: The evaluation of accuracy is facilitated immensely by tools that output relevant quality metrics and their variance during model training by default. Ensuring robustness and explainability can be extremely time-consuming if specific tests and evaluation routines are not pre-programmed and easily accessible. Standardised processes, once developed, also allow for better timeliness and higher efficiency. In order to ensure that data, codes and environments are reproducible, certain procedures and technical tools must be established and used across the whole organisation. Practices and processes that aim to reliably and efficiently develop, productively deploy, manage, monitor and maintain machine learning models are summarised under the term 'machine learning operations' (MLOps) (Kreuzberger et al. 2023). To the extent that such processes, tools and practices are necessary to fulfil the quality dimensions specified in the quality frameworks of official statistics, MLOps are also a basic prerequisite for machine learning in official statistics.[10] Consequently, it is essential that MLOps are given ample consideration during the development of IT platforms and data management systems. As part of the ONS-UNECE-ML-2022 project, requirements for MLOp systems and possible system architectures have been gathered and evaluated (Engdahl et al. 2022).

[10] For MLOps, see also Sect. 8.4 of Avouac et al. (2025) in this book.

4.7 Conclusion

Machine learning has great potential to replace manual, labour-intense tasks such as classification. As a result, ML (or statistical) models are now being introduced into the production of official statistics, often in areas where hard-coded rules have previously enabled the partial automation of processes. However, ML is no panacea and comes with its own particular challenges, types of uncertainty and sources of inaccuracy. Therefore, existing quality frameworks cannot be applied without further specification. This chapter has aimed to contribute to the development of a tailor-made, comprehensive guidance for the use of ML, thus paving the way for the widespread usage of such methods in official statistics. Further work is required on quality indicators and standards for quality documentation. Subsequently, theoretical standards need to be implemented in day-to-day statistics production.

Acknowledgments This chapter is in parts based on previous work in German, conducted by the authors in collaboration with Christian Borgs, Alexander Brand, Andreas Nickl, Alexandra Rittmann, Johannes Rohde, Christian Salwiczek, Nina Storfinger and Selina Straub and published as Saidani et al. (2023).

References

G. Ascari, K. Blix, G. Brancato, T. Burg, A. McCourt, A. van Delden, D. Krapavickaitę, N. Ploug, S. Scholtus, P. Stoltze, T. de Waal, L.-C. Zhang, Quality of multisource statistics – the KOMUSO project. Survey Stat. **81**, 36–51 (2020)

R. Avouac, T. Faria, F. Comte, A cloud-native data science platform for official statistics, in *Foundations and Advances of Machine Learning in Official Statistics*, ed. by F. Dumpert, Chap. 8 (Springer, Berlin, 2025)

S. Barragán, A. Pérez-Bote, C. Sáez, D. Salgado, L. Sanguiao-Sande, Streamlining business functions in official statistical production with machine learning, in *Foundations and Advances of Machine Learning in Official Statistics*, ed. by F. Dumpert, Chap. 13 (Springer, Berlin, 2025)

S. Dandl, B. Bischl, L. Bothmann, Interpretable machine learning for official statistics, in *Foundations and Advances of Machine Learning in Official Statistics*, ed. by F. Dumpert, Chap. 5 (Springer, Berlin, 2025)

S. de Broe, P. Struijs, P. Daas, A. van Delden, J. Burger, J. van den Brakel, O. ten Bosch, K. Zeelenberg, W. Ypma, Updating the paradigm of official statistics: new quality criteria for integrating new data and methods in official statistics. Stat. J. IAOS **37**(1), 343–360 (2021)

F. Dumpert, Machine Learning in der amtlichen Statistik – Ergebnisse und Bewertung eines internationalen Projekts. WISTA Wirtschaft und Statistik **73**(4), 53–63 (2021)

F. Dumpert, J. Reichel, E. Oertel, H. Leerhoff, C. Salwiczek, Ethische Fragen beim Einsatz von KI/ML in der Produktion amtlicher Statistiken – Teil 1: Identifikation. WISTA Wirtschaft und Statistik **77**(1), 15–24 (2025a)

F. Dumpert, J. Reichel, E. Oertel, H. Leerhoff, C. Salwiczek, Ethische Fragen beim Einsatz von KI/ML in der Produktion amtlicher Statistiken – Teil 2: Auseinandersetzung. WISTA Wirtschaft und Statistik **77**(1), 25–36 (2025b)

J. Engdahl, I. Choi, E. Deeben, J. Karanka, A. Karlsson, M. Meszaros, J. Pocknee, P. Holroyd, A. Baily, Building an ML Ecosystem in Statistical Organisations. https://statswiki.unece.org/display/ML/Machine+Learning+Group+2022 (2022)

European Statistical System Committee, European Statistics Code of Practice – revised edition 2017. https://ec.europa.eu/eurostat/documents/4031688/8971242/KS-02-18-142-EN-N.pdf (2017)

European Statistical System Committee, Quality Assurance Framework of the European Statistical System, version 2.0. https://ec.europa.eu/eurostat/documents/64157/4392716/ESS-QAF-V2.0-final.pdf (2022)

Y.A. Gootzen, P. Daas, A. van Delden, Quality framework for combining survey, administrative and big data for official statistics. Stat. J. IAOS **392**, 439–446 (2023)

C. Julien, UNECE – HLG-MOS Machine Learning Project: Project report (2020). https://statswiki.unece.org/display/ML/Machine+Learning+Project+Report

A. Kowarik, M. Six, Quality guidelines for the acquisition and usage of big data with additional insights on web data, in *4th International Conference on Advanced Research Methods and Analytics (CARMA 2022)*. (Universitat Politècnica de València, Valencia, 2022)

D. Kreuzberger, N. Kühl, S. Hirschl, Machine learning operations (MLOps): overview, definition, and architecture. IEEE Access **11**, 31866–31879 (2023)

L. Krög, Legal implications for the use of machine learning in official statistics, in F. Dumpert (ed.) *Foundations and Advances of Machine Learning in Official Statistics*, Chap. 7 (Springer, Berlin, 2025)

Z.C. Lipton, The mythos of model interpretability. Commun. ACM **61**(10), 36–43 (2018)

N. Mehrabi, F. Morstatter, N. Saxena, K. Lerman, A. Galstyan, A survey on bias and fairness in machine learning. ACM Comput. Surveys **54**(6), 1–35 (2022)

M. Puts, P. Daas, Machine learning from the perspective of official statistics. Survey Stat. **84**, 12–17 (2021)

Y. Saidani, F. Dumpert, C. Borgs, A. Brand, A. Nickl, A. Rittmann, J. Rohde, C. Salwiczek, N. Storfinger, S. Straub, Qualitätsdimensionen maschinellen Lernens in der amtlichen Statistik. AStA Wirtschafts- und Sozialstatistisches Archiv **17**(3), 253–303 (2023)

D. Salgado, S. Barragán, E. Rosa-Pérez, Timeliness and accuracy with machine learning algorithms: early estimates of the industrial turnover index (2023). https://unece.org/statistics/documents/2023/05/ml2023s1spainsalgadopaperpdf

P.O. Schenk, C. Kern, F. Kreuter, Fairness in machine learning for national statistical organizations, in *Foundations and Advances of Machine Learning in Official Statistics*, ed. by F. Dumpert, Chap. 6 (Springer, Berlin, 2025)

H. Schulz-Kümpel, A.-L. Boulesteix, S. Fischer, R. Hornung, Challenges in resampling based performance estimation, in *Foundations and Advances of Machine Learning in Official Statistics*, ed. by F. Dumpert, Chap. 3 (Springer, Berlin, 2025)

W. Yung, S.-M. Tam, B. Buelens, H. Chipman, F. Dumpert, G. Ascari, F. Rocci, J. Burger, I. Choi, A quality framework for statistical algorithms. Stat. J. IAOS **38**(1), 291–308 (2022)

Chapter 5
Interpretable Machine Learning for Official Statistics

Susanne Dandl, Bernd Bischl, and Ludwig Bothmann

5.1 Introduction

Machine learning (ML) has huge potential to aid decision-making processes due to its predictive performance. Unfortunately, these ML models are often black boxes and are too complex to be understood by humans, for example, a random forest that consists of multiple decision trees or a deep neural network with multiple layers of neurons. The lack of explanation can be a barrier to applying ML models, especially in critical domains like official statistics, where predictions can have detrimental effects on certain groups. In the last decade, a whole research field emerged around the interpretation of machine learning models, known as interpretable machine learning (IML) or explainable artificial intelligence (XAI). The field can broadly be divided into two research areas: The first one is concerned with the development of (high-performance) interpretable models, e.g., by distilling a neural network into a simple decision tree (Frosst and Hinton 2017); the second one is concerned with the development of methods to interpret complex models post hoc, i.e., after model fitting. In this book chapter, we address the latter field of post hoc interpretation methods. We summarize its purposes (Sect. 5.2) and provide an overview of methods (Sect. 5.3). We also highlight some of the latest advancements in the field, which can especially be applied in official statistics (Sects. 5.3.1–5.3.3), and discuss open research questions (Sect. 5.4).

S. Dandl · B. Bischl · L. Bothmann (✉)
Department of Statistics, LMU Munich, Munich, Germany

Munich Center for Machine Learning (MCML), Munich, Germany
e-mail: bernd.bischl@stat.uni-muenchen.de; ludwig.bothmann@lmu.de

F. Dumpert (ed.), *Foundations and Advances of Machine Learning in Official Statistics*, Society, Environment and Statistics,
https://doi.org/10.1007/978-3-032-10004-7_5

Throughout this chapter, we use the task of predicting whether or not a potential error (soft edit rule, “Kann-Fehler”) in an input requires a correction to illustrate how official statistics can benefit from the insights of the interpretation methods presented. Potential errors are implausibilities indicated by hard-coded rules. An example of a potential error is the entry of a 16-year-old male who is married with an annual income of € 120,000. This entry seems implausible, but because it is not necessarily incorrect, a review is required. To reduce the human burden of manual review, an ML model can be trained on previous review data to predict whether a potential error should result in a change or not.

5.2 Interpretation Goals

The interpretability of ML models is important from multiple perspectives. First, interpretation methods can help to discover and gain *global* insights into a model. The user can learn something about which features affect a prediction the most and how these features affect the prediction on average. Interpretation methods can also help to understand and control *individual* decisions. The right to an explanation for individual algorithmic decisions is also grounded in the General Data Protection Regulation (Recital 71, GDPR):[1]

> In any case, such processing should be subject to suitable safeguards, which should include specific information to the data subject and the right to obtain human intervention, to express his or her point of view, to obtain an explanation of the decision reached after such assessment, and to challenge the decision.

These insights can be the basis for identifying flaws in the model or data. Therefore, they can also help with model auditing, for which the gained insights must be compared to domain knowledge. The inaccuracies can then be corrected for a subsequent model. This is especially relevant if the model reveals discriminatory behavior—for example, when the prediction for individuals depends on their gender or ethnicity. This brings about ethical issues if the prediction is used as a basis for further actions or decisions in the real world. Bothmann et al. (2024), for example, present an approach of translating philosophical questions into an actionable framework of fairness-aware machine learning, and Caton and Haas (2024) discuss current research on algorithmic fairness. The following section presents an overview of post hoc interpretation methods that aim for a single or subset of interpretation goals.

---

[1] See https://gdpr.eu/.

5.3 Overview of Post Hoc Interpretability Methods

Post hoc interpretation methods can be divided into classes based on certain properties of those methods. The following differentiation is based on Molnar (2022) and Molnar et al. (2022).

First, we can distinguish between model-specific or model-agnostic methods. Model-specific methods are only applicable to specific ML models, e.g., the Gini importance method by Breiman (2001) for random forests or saliency maps for neural-network-based image classifiers by Simonyan et al. (2013). These methods make use of the model structure, i.e., the tree structure of forests or access to the gradients of the neural network. Therefore, they cannot be applied to other types of models. In contrast, model-agnostic methods do not exploit the model structure and can be applied to any type of model.[2] Model-agnostic methods are especially useful if diverse models should be compared against each other with respect to their explanations, e.g., to assess whether the model corresponds to the user's own domain knowledge.[3] For the remainder of this chapter, we focus only on model-agnostic methods.

A second differentiation of methods is based on whether the methods actually aim to explain the model behavior in general—these methods are called global—or whether the methods aim to explain the behavior of the model for a single data point of interest and its close surroundings; these methods are called local.

Third, we can divide methods into feature effect and feature importance methods. The goal of feature effect methods is to assess the direction and size of a change in the outcomes due to changes in the feature values. Feature importance methods assess a feature's contribution to model performance (e.g., through a loss function) or to the variance of the prediction function.

The following two subsections focus on two subclasses of feature importance and feature effect methods: loss-based feature importance methods (Sect. 5.3.1) and semi-factual explanations (Sect. 5.3.2).

5.3.1 Spotlight: Loss-Based Feature Importance

This section provides a short, high-level overview of global, loss-based, model-agnostic feature importance (FI) methods. It is heavily based on the work by Ewald et al. (2024), a comprehensive guide aiming to equip researchers with the tools to

[2] This does not necessarily mean that methods can be applied to binary, multi-class, or regression models. For example, counterfactual explanation methods are often tailored to binary classifiers; however, it does not matter if the model is a random forest, linear model, or neural network.

[3] For example, if an expert in a certain field is certain that a feature X has a huge impact on an outcome Y, it would worry this person if the feature X is not one of the most important features for the given prediction models.

make informed decisions in selecting the most appropriate method for their specific analytical needs.

FI methods serve as a bridge between the complex predictions generated by machine learning models and the need for interpretable insights into the underlying data-generating process. By quantifying the contribution of each feature to the prediction process, these methods offer valuable insights into the relative importance of different features in driving model outcomes. This understanding is essential not only for model interpretation but also for hypothesis generation, feature selection, and domain-specific insights. In the example of potential errors above, the most important features can be an indicator for what feature might require changes.

Ewald et al. (2024) differentiate between three classes of FI techniques: univariate perturbations, marginalization, and model refitting. The following briefly introduces three of the most prominent loss-based feature importance methods and discusses their differences. For further methods and in-depth insight, we refer to Ewald et al. (2024).

Permutation Feature Importance (PFI) (Breiman 2001; Fisher et al. 2019) is a univariate perturbation FI method. To calculate the PFI for a feature of interest (FOI) X_j, the respective observations are permuted such that the dependencies between the FOI and the target, as well as between the FOI and all other features, are broken. The discrepancy between the expected loss of the model using the perturbed feature and the model with the original feature yields the PFI for feature X_j. Ewald et al. (2024) state that PFI can be used to assess unconditional feature importance. However, PFI requires assumptions about feature independence, which are typically unrealistic in real-world datasets. This can limit their practical applicability despite their theoretical soundness.

The next two methods are especially suitable if conditional feature importances are of interest, i.e., a feature's importance conditional on the remaining features.

Conditional Feature Importance (CFI) (Strobl et al. 2008) is also a method that is based on univariate permutation. It is similar to the PFI, the only difference being that the permutation of observations of the feature of interest is performed such that the dependencies with the other features are preserved while the relationship between the target and X_j is broken. CFI requires accurate models of the univariate conditional distribution, which might not always be readily available or complex to derive.

Leave-One-Covariate-Out (LOCO) (Lei et al. 2018) is a model refitting FI method, in contrast to PFI and CFI. As its name suggests, LOCO determines the importance of a feature by removing it from the data and refitting the model (using the same learner) without it. The discrepancy in risk of the model without the feature and the full model quantifies the LOCO importance measure. Due to multiple model refits, the method is computationally expensive.

Overall, selecting the most appropriate feature importance method requires careful consideration of several factors. Researchers must assess the nature of the data, the complexity of the model, and the specific questions they seek to address. While each FI technique has its strengths, its practicality depends on the specific application and computational resources available. For example, in the potential error example, the input data might be implausible by (hard-coded) definitions such that the independence assumption of PFI is less harmful than for other use cases. In general, practitioners should choose methods that balance accuracy, assumptions, and computational demands based on their specific needs.

To illustrate the practical application of these methods, Ewald et al. (2024) use the well-known “bike sharing” dataset. The dataset includes 731 observations and 12 features related to weather, temperature, wind speed, season, and day of the week. By applying both PFI and LOCO to this dataset, they highlight how different methods can yield varying results and discuss the implications of these differences for understanding the data-generating process. An example of applying feature importances in survey statistics is given by Felderer et al. (2024) who use machine learning methods to predict reliability and validity of survey questions and explain the trained models via PFI.

Uncertainty estimation is a critical aspect of interpreting feature importance measures. Ewald et al. (2024) discuss various techniques for estimating the uncertainty of these measures, such as resampling methods and statistical inference techniques. They emphasize the importance of using independent test data to avoid biased estimates and highlight best practices for ensuring reliable results.

Finally, the chapter identifies several open challenges and areas for future research, including (i) developing methods that can accurately estimate feature importance in the presence of complex interactions between features, (ii) creating benchmarks and empirical studies to compare the performance of different feature importance methods in various scenarios, and (iii) investigating the causal relationships between features and the target variable, moving beyond mere associations to understand the underlying mechanisms.

By addressing these challenges, future research can enhance the reliability and applicability of feature importance methods, making them more useful for scientific inference and practical applications alike.

5.3.2 *Spotlight: Counterfactual and Semi-factual Explanations*

5.3.2.1 Motivation

Counterfactual explanations (CFEs) and semi-factual explanations (SFEs) are local interpretation methods aimed at explaining a model’s behavior for individual observations (Doshi-Velez and Kim 2017). CFEs highlight minimal changes in features required to alter a prediction, while SFEs show the maximum possible changes to keep a prediction the same. For instance, in the potential error scenario, a CFE might suggest that increasing the age leads to a lower change probability. Conversely, an SFE might indicate that even with changes in the marital status to single, there is a high change probability. Understanding why a certain event happened involves identifying its causes, a concept rooted in counterfactual reasoning. As Hume (1748) and Lewis (1973) suggested, this involves considering what would have happened if the circumstances were different.

While for the generation of CFEs many approaches are available (see Dandl et al. 2023b, for an overview of counterfactual explanation methods and related software packages), research on the generation of SFEs is rather limited. The following subsection describes a specific method of SFEs, namely, "interpretable regional descriptors". This method was proposed by Dandl et al. (2023a), and we refer the reader for details of the method to the original source.

5.3.2.2 Interpretable Regional Descriptors

Interpretable regional descriptors (IRDs) represent a novel technique for generating local, model-agnostic interpretations. This method employs hyperboxes to delineate the manner in which feature values of an observation can be altered without affecting its prediction. This approach facilitates an understanding of the robustness of predictions and provides SFEs, which indicate the range of feature values that will maintain the prediction constant. Such insights are of value to both model developers and individuals affected by the decisions made by machine learning models.

IRDs address the question of how much a feature value can be varied while maintaining the same prediction. They produce a set of "even if" explanations, which help justify a decision by showing that even with different feature values, the outcome would remain unchanged. For example, in the potential error scenario above, an IRD might indicate that even if the individual's marital status is single instead of married, and even if the income is reduced to €50,000, the probability of required changes would still be high (e.g., above 60%).

The process of generating IRDs involves formulating an optimization problem where the objective is to find the largest hyperbox (i.e., one that covers much of the feature space) around a point of interest. This hyperbox must contain feature values that yield predictions within a user-defined closeness region. The optimization seeks to find a hyperbox with maximal *coverage* of the feature space while ensuring that it maintains a high *precision*, meaning the points within the hyperbox should result in predictions within the specified range.

The process of generating an IRD involves several steps:

1. **Restriction of the search space**: The initial search space is restricted to the largest local hyperbox around the point of interest. For numerical features, this involves varying the feature values on a grid until the prediction falls outside the desired range. For categorical features, all categories that still yield predictions within the desired range are included.
2. **Selection of the dataset**: The dataset used to generate the IRD can be either the training data or data generated to uniformly cover the feature space of interest. The selection affects the empirical measures of coverage and precision.

3. **Initialization**: Depending on the approach, the initial hyperbox can be the largest local box covering all data points (inside the search space) or the smallest possible box that only includes the point of interest.
4. **Optimization**: The boundaries of the hyperbox are iteratively adjusted to maximize coverage while maintaining precision. Top-down methods shrink the largest box while ensuring the point of interest remains within the hyperbox, whereas bottom-up methods expand the smallest box around the point of interest.
5. **Post-processing**: To refine the hyperbox boundaries, additional data points are sampled from the feature space, and the box is adjusted to improve precision and coverage. This step helps in addressing any regions in the hyperbox that might have suboptimal coverage or precision.

By following these steps, IRDs provide a comprehensive and interpretable method to understand the stability of predictions made by complex machine learning models. They are particularly useful for justifying decisions and identifying features that are not locally influential on the prediction.

5.3.3 Spotlight: Model Summaries in R

The R package `mlr3summary` by Dandl et al. (2024) is designed to generate concise and interpretable summaries for machine learning models. Inspired by the `summary` function for generalized linear models (GLMs) in R, `mlr3summary` extends its functionality to be model-agnostic, thereby providing unified summary outputs for both parametric and nonparametric machine learning models (Dandl et al. 2024). This package enables information on dataset characteristics, model performance, complexity, estimated feature importances, feature effects, and fairness metrics,[4] all evaluated using resampling strategies for unbiased performance estimates.

As machine learning becomes integral in decision-making across various fields, the need for interpretable models is paramount. Traditional summary functions in R, such as those for GLMs, offer insights into model parameters and fit but are limited to specific model types and do not generalize to more complex machine learning models. The `mlr3summary` package addresses this gap by providing a standardized diagnostic output for a diverse set of models, facilitating easier model comparison and selection.

`mlr3summary` incorporates several key features to enhance the interpretability and evaluation of machine learning models:[5]

[4] All fairness metrics from the `mlr3fairness` package (Pfisterer et al. 2023) are available; this includes common measures like the demographic parity or equalized odds.

[5] See also Dandl et al. (2024) for a thorough applied example.

- **Model-agnostic summaries**: Provides a consistent summary format across various types of models, including both linear and complex models like random forests and gradient boosted trees.
- **Resampling-based evaluation**: Uses techniques such as cross-validation to provide unbiased estimates of model performance and feature importance.
- **Detailed metrics**: Includes performance metrics (e.g., AUC, F1-score), model complexity measures (e.g., sparsity, interaction strength), feature importance (e.g., partial dependence plots, permutation feature importance), and fairness assessments.
- **Customizable output**: Users can customize the summary output using the `summary_control` function, allowing adaptation to specific needs and preferences.

The core function of `mlr3summary` is its S3-based summary function for `mlr3` (Lang et al. 2019) `Learner` objects. The typical workflow involves initializing a task, selecting a learner, training the model, and applying a resampling strategy to evaluate the model. The summary function then provides a comprehensive overview of the model and its performance, including sections on general model information, residuals, performance metrics, model complexity, feature importance, and effect plots. Each section is derived from the resampling results to ensure unbiased evaluation.

The summary output can be tailored to specific needs using the `summary_control` function. Users can specify which measures to include, the number of important features to display, and which sections to hide. Additionally, fairness metrics can be incorporated by specifying a protected attribute. The package includes features to handle large datasets and complex models efficiently.

`mlr3summary` provides a robust tool for summarizing machine learning models in a consistent and interpretable manner. This is especially valuable for official statistics for which transparency and full reporting are of high importance. Future developments may include extended support for additional interpretability methods, enhanced visualization capabilities, and integration with other model diagnostic tools.

For more information and access to the package, visit the GitHub repository at https://github.com/mlr-org/mlr3summary and the CRAN page at https://cran.r-project.org/package=mlr3summary.

5.4 Discussion

The work by Ewald et al. (2024) and Dandl et al. (2023a, 2024) featured above highlights several advancements in the field of interpretable machine learning. Each contribution enhances the possibilities of using machine learning techniques in official statistics: While Ewald et al. (2024) describe how to choose between different feature importance methods, Dandl et al. (2023a) show how to create interpretable semi-factual explanations—answering "even if" questions—and Dandl et al. (2024) provide a user-friendly tool for practitioners allowing to summarize black-box machine learning models and thereby improving possibilities for model diagnosis and model selection. However, additionally to advanced interpretability methods, ethical issues are of importance in official statistics. As described by Bothmann et al. (2023) and applied to the question of "fair" rental prices by Bothmann and Peters (2024), methods for accounting for historical biases in real-world data, using methods of causal inference, should come more into focus in the future—especially in official statistics. As Leininger et al. (2025) have shown, different fairness metrics can be simultaneously achieved with high predictive performance when causal preprocessing methods are used to account for historical biases. Finally, the method "privilege scores" by Bothmann et al. (2025) can be used for individual and global quantification of privilege, which can be used in official statistics, e.g., for the identification of unwanted discrimination with respect to protected attributes such as gender or race.

References

L. Bothmann, K. Peters, Fairness als Qualitätskriterium im Maschinellen Lernen – Rekonstruktion des philosophischen Konzepts und Implikationen für die Nutzung außergesetzlicher Merkmale bei qualifizierten Mietspiegeln. AStA Wirtschafts- und Sozialstatistisches Archiv **18**(2), 185–201 (2024)

L. Bothmann, S. Dandl, M. Schomaker, Causal fair machine learning via rank-preserving interventional distributions, in *Proceedings of the 1st Workshop on Fairness and Bias in AI co-located with 26th European Conference on Artificial Intelligence (ECAI 2023)*. CEUR Workshop Proceedings (2023). https://ceur-ws.org/Vol-3523/

L. Bothmann, K. Peters, B. Bischl, What is fairness? On the role of protected attributes and fictitious worlds (2024). https://arxiv.org/abs/2205.09622

L. Bothmann, P.A. Boustani, J.M. Alvarez, G. Casalicchio, B. Bischl, S. Dandl, Privilege scores (2025). https://arxiv.org/abs/2502.01211

L. Breiman, Random forests. Mach. Learn. **45**(1), 5–32 (2001)

S. Caton, C. Haas, Fairness in machine learning: a survey. ACM Comput. Surveys **56**(7), 1–38 (2024)

S. Dandl, G. Casalicchio, B. Bischl, L. Bothmann, Interpretable regional descriptors: hyperbox-based local explanations, in *Machine Learning and Knowledge Discovery in Databases: Research Track*, ed. by D. Koutra, C. Plant, M. Gomez Rodriguez, E. Baralis, F. Bonchi (Springer, Berlin, 2023a), pp. 479–495

S. Dandl, A. Hofheinz, M. Binder, B. Bischl, G. Casalicchio, counterfactuals: an R package for counterfactual explanation methods (2023b). https://arxiv.org/abs/2304.06569

S. Dandl, M. Becker, B. Bischl, G. Casalicchio, L. Bothmann, mlr3summary: concise and interpretable summaries for machine learning models, in *Joint Proceedings of the xAI 2024 Late-breaking Work, Demos and Doctoral Consortium co-located with the 2nd World Conference on eXplainable Artificial Intelligence (xAI-2024), Valletta, July 17–19, 2024*, ed. by L. Longo, W. Liu, G. Montavon, vol. 3793. CEUR-WS.org (2024)

F. Doshi-Velez, B. Kim, Towards a rigorous science of interpretable machine learning (2017). https://arxiv.org/abs/1702.08608

F.K. Ewald, L. Bothmann, M.N. Wright, B. Bischl, G. Casalicchio, G. König, A guide to feature importance methods for scientific inference, in *Explainable Artificial Intelligence*, ed. by L. Longo, S. Lapuschkin, C. Seifert (Springer, Berlin, 2024), pp. 440–464

B. Felderer, L. Repke, W. Weber, J. Schweisthal, L. Bothmann, Predicting the validity and reliability of survey questions (2024). https://doi.org/10.31219/osf.io/hkngd

A. Fisher, C. Rudin, F. Dominici, All models are wrong, but many are useful: Learning a variable's importance by studying an entire class of prediction models simultaneously. J. Mach. Learn. Res. **20**(177), 1–81 (2019)

N. Frosst, G.E. Hinton, Distilling a neural network into a soft decision tree, in *Proceedings of the First International Workshop on Comprehensibility and Explanation in AI and ML 2017 co-located with 16th International Conference of the Italian Association for Artificial Intelligence (AI*IA 2017), Bari, November 16th and 17th, 2017*, vol. 2071. *CEUR Workshop Proceedings*, ed. by T.R. Besold, O. Kutz (eds.) CEUR-WS.org (2017)

D. Hume, An enquiry concerning human understanding (1748)

M. Lang, M. Binder, J. Richter, P. Schratz, F. Pfisterer, S. Coors, Q. Au, G. Casalicchio, L. Kotthoff, B. Bischl, mlr3: a modern object-oriented machine learning framework in R. J. Open Source Software **4**(44), 1903 (2019)

J. Lei, M. G'Sell, A. Rinaldo, R.J. Tibshirani, L. Wasserman, Distribution-free predictive inference for regression. J. Am. Stat. Assoc. **113**(523), 1094–1111 (2018)

C. Leininger, S. Rittel, L. Bothmann, Overcoming fairness trade-offs via pre-processing: a causal perspective (2025). https://arxiv.org/abs/2501.14710

D.K. Lewis, *Counterfactuals* (Blackwell, Hoboken, 1973)

C. Molnar, *Interpretable Machine Learning*, 2nd edn. (2022). https://christophm.github.io/interpretable-ml-book

C. Molnar, G. König, J. Herbinger, T. Freiesleben, S. Dandl, C.A. Scholbeck, G. Casalicchio, M. Grosse-Wentrup, B. Bischl, General pitfalls of model-agnostic interpretation methods for machine learning models, in *xxAI - Beyond Explainable AI: International Workshop*, ed. by A. Holzinger, R. Goebel, R. Fong, T. Moon, K.-R. Müller, W. Samek (Springer, Berlin, 2022), pp. 39–68

F. Pfisterer, W. Siyi, M. Lang, *mlr3fairness: fairness auditing and debiasing for 'mlr3'* (2023). https://CRAN.R-project.org/package=mlr3fairness. R package version 0.3.2

K. Simonyan, A. Vedaldi, A. Zisserman, Deep inside convolutional networks: visualising image classification models and saliency maps. CoRR, abs/1312.6034 (2013). https://api.semanticscholar.org/CorpusID:1450294

C. Strobl, A.-L. Boulesteix, T. Kneib, T. Augustin, A. Zeileis, Conditional variable importance for random forests. BMC Bioinform **9**(1), 307 (2008)

Chapter 6
Fairness in Machine Learning for National Statistical Organizations

Patrick Oliver Schenk, Christoph Kern, and Frauke Kreuter

6.1 Introduction

Machine learning (ML) applications are widely used at national statistical organizations (NSOs) to improve the timeliness and cost-effectiveness of data production processes. As NSOs offer new or refine existing products with the use of ML algorithms, they need to ensure that high-quality standards are upheld. Such standards may be codified in frameworks such as the "Quality Framework for Statistical Algorithms" (QF4SA) by Yung et al. (2022), which includes five dimensions: accuracy, timeliness, cost-effectiveness, explainability, and reproducibility.

We argue that fairness should be factored into quality assessments at NSOs, both in interaction with current quality concepts and as its own quality dimension. Specifically, we posit that the rich research on algorithmic fairness has developed concepts and methodology that align with the goal of NSOs to ensure a safe and reliable deployment of ML. At the same time, we emphasize the crucial role of data as a result of a series of design decisions, collection, and processing steps that jointly affect fairness errors downstream. The experiences, practices, and standards at NSOs can enrich the fairness in ML literature given its focus on developing high-

P. O. Schenk (✉) · C. Kern
Department of Statistics, LMU Munich, Munich, Germany

Munich Center for Machine Learning (MCML), Munich, Germany
e-mail: p.o.s.on.stats@gmail.com; christoph.kern@stat.uni-muenchen.de

F. Kreuter
Department of Statistics, LMU Munich, Munich, Germany

Munich Center for Machine Learning (MCML), Munich, Germany

University of Maryland, College Park, MD, USA
e-mail: frauke.kreuter@stat.uni-muenchen.de

F. Dumpert (ed.), *Foundations and Advances of Machine Learning in Official Statistics*, Society, Environment and Statistics,
https://doi.org/10.1007/978-3-032-10004-7_6

quality data products—with respect to both measurement and representation—that may be used for a wide range of downstream tasks.

In response to the widespread application of ML in various critical domains, research on algorithmic fairness has developed into a multifaceted and cross-disciplinary research field, with strong roots in the computer science and machine learning community. Fair ML research focuses on the impacts of algorithms that may be deployed to inform decisions or to allocate resources in social contexts (Mitchell et al. 2021; Mehrabi et al. 2021; Barocas et al. 2019). In such settings, fairness concerns arise when model outputs depend on personal ("protected") attributes beyond an individual's agency. However, as models learn from historical data, social disparities based on such protected attributes (e.g., gender, age, or ethnic origin; Barocas and Selbst 2016) may be incorporated into model structures and fed forward through their predictions. The fair ML literature has developed a plethora of fairness notions and metrics to quantify such model-based disparities, with a strong focus on group-specific evaluations of model outputs and error. This has given rise to methodology that algorithmically assesses fairness properties of a given prediction model (Kim et al. 2019; von Zahn et al. 2023) and various techniques for "de-biasing" machine learning models at different stages of the modeling pipeline (i.e., pre-, in-, and post-processing methods; Carton et al. 2016). Given its focus on prediction modeling and algorithmic decision-making, however, the fairness literature focuses on ML for data analysis and not on ML for data collection, processing, and production, i.e., the main work of NSOs.

In the following, we outline how fairness considerations expand beyond the algorithmic decision-making setting and interact with data production processes at NSOs (Sect. 6.2), draw connections between fairness and the QF4SA (Sect. 6.3), and critically discuss the implications for fairness-impacting decisions at NSOs (Sect. 6.4). Section 6.5 concludes.

6.2 The Role and Importance of Fairness

Data and Decision-Making

In our view, the main reasons why humans collect, produce, and analyze data include describing, learning about, and predicting the world. Results acquired in this manner are arguably the most consequential when they inform decisions. First, this may occur rather indirectly, e.g., when these results enter public and political discourse or advance scientific knowledge, or directly, e.g., when laws mandate specific changes based on census statistics. This clearly falls into the domain of official statistics. Second, for an example of data-driven decision-making on a lower level, consider a public employment agency training a ML model to predict the likelihood of long-term employment for each individual they serve (Desiere et al. 2019; Körtner and Bonoli 2023). For anyone for whom this probability exceeds a certain threshold, the agency would offer a job training program, and everyone below this threshold would be excluded from this training program. This is an example of *automated decision-making* (ADM).[1] Although NSOs do not engage in such data-driven decision-making themselves, we emphasize decision-making for two reasons. (i) NSOs' data products may be used by downstream users to train models for ADM. As such models are only as good as the data they are trained on—and the documentation that is available for these data—NSOs play a crucial upstream role for achieving fairness. (ii) Decision-making, and in particular ADM, is central to the fair ML literature, and thus this background is necessary to understand much of the literature's perspective.

Of course, fairness is neither a new concept nor one that has been hitherto unimportant. Yet, in a world with ML-based predictions and ADM, two important aspects have changed: agency and scale. (i) As model predictions become more and more important in decision-making, with ADM at the end of the spectrum, human agents lose their discretionary power in decision-making. This is crucial insofar as we as humans expect other humans (e.g., the employment officers in our example above) to care about justice and fairness in their decisions even without being explicitly told or incentivized to do so. If the human is taken increasingly out of the loop and without carefully designing the whole decision-making pipeline, and in particular the trained model it employs, one cannot expect decisions to be just. (ii) Investing the resources to train a model for ADM is especially worthwhile for large-scale deployment. Organizations that engage in high-stake, large-scale (automated) decision-making, such as a country's employment agency, may also be among the most likely users of NSOs' data products for such purposes,[2] highlighting the potential impact of the fairness of NSOs' data products.

Justice and Fairness

With this background on decision-making, we can clarify the terms *justice* and *fairness*, as they are often used somewhat haphazardly, even within the fair ML literature. We base our presentation on Kuppler et al.'s (2022) more thorough discussion. Global *justice* concerns properties of a society as a whole, whereas local justice operates on a lower level: as in our example of the public employment agency, it typically involves the allocation of scarce resources by institutions to individuals. Commonly accepted justice principles include equality (of treatment or of outcome), "desert" (claims to the benefits or goods according to one's productive contribution, effort, or costs; Moriarty 2018), need (for survival, for a human life, or for functioning in one's role in society), and efficiency (the society-wide outcome is maximized by allocating the resources to those who benefit from it the most). These principles can be and often are in conflict with each other and with other goals. *Fairness* is less fundamental than justice: from a mechanistic perspective, fairness can be seen as a means to achieve justice (in society/in decision-making). Although

---

[1] In contrast, semiautomated decision-making denotes setups in which a model's prediction—here, the probability of long-term unemployment—is only one of the factors that a human decision-maker considers.

[2] Or they may serve a dual role, with a decision-making branch and a data-producing branch that is very similar to NSOs.

which justice principle is most appropriate and the trade-off between justice and other desirable properties may vary by situation, it has long been recognized across several fields that, in general, justice is an attractive quality.[3] As justice is desirable to humans and fairness represents a or the tool to attain justice, fairness is generally desirable too.

Kuppler et al. (2022) and Scantamburlo et al. (2024) recognize that it is helpful in data-driven decision-making to distinguish between, first, the prediction step (training and applying the ML model) and, then, the decision step. They both share the abovementioned illuminating definition of fairness as a tool: a fair predictive ML model (Kuppler et al. 2022) or a fair whole system (i.e., both steps together, Scantamburlo et al. 2024) contribute to just decision-making and a just society.

Often in the literature, the prediction step is viewed from the perspective of a data analyst who is given (fixed) data that they then feed into model training, perhaps after a little bit of pre-processing. We suggest to broaden this point of view: everything before the decision step should be considered part of the prediction step. Thus, not just the eventual model training but all steps along the chain contribute to the (un)fairness of the eventual outcome, from designing a data collection through conducting it and into all the changes made to the data during (pre-)processing. Particularly in NSOs, this is a multistep process, as described, e.g., by the total survey error (TSE) framework (see Groves et al. 2009, Chapter 2.3) or by some of its offspring frameworks that are more tailored to ML data (e.g., Amaya et al. 2020, West et al. 2023, and Puts et al. 2025, in this book). Along this chain, fairness errors compound, which has several implications. First, it is important to consider the cumulative effect on fairness along the chain and not just the additional unfairness introduced in one's own particular step along the chain, so as to avoid the boiling frog problem. Second, fairness errors may be more than additive: e.g., in ongoing research, we find that effects of misrepresenting the population in the training data can be worse when there are also important features missing. The earlier in the chain one's work is, the higher the leverage of one's fairness-impacting choices. In addition to the two just-mentioned reasons: if a certain group is (largely) removed from the data in a step, then in subsequent steps, it is no longer possible to evaluate fairness (reliably) for this group, or the group may become completely forgotten.

The prediction step provides input for the decision step. Likewise, a step in the chain of creating the data product provides input for the next step. While NSOs may not engage in the very last steps (training the model and decision) of a long series of steps, they provide crucial input, rendering the fairness of their important data products vital.[4]

[3] Perhaps the veil of ignorance is the most well-known argument: not knowing into which position in society one would be born, one would prefer a just over an unjust society.

[4] In addition, the role of NSOs is not fundamentally different from many real-world situations, as the decision-making institutions such as employment agencies (see Allhutter et al. 2020) may often outsource steps, e.g., the model training.

Multiple Objectives, Quantification, and Optimization

In contrast to traditional statistical models, which aim to estimate some parameters of interest, supervised ML models are not trained for their own sake but to be deployed (on some target population). In that, however, only trained models that are deemed to work well enough are actual candidates for deployment. First and foremost, the general model performance, as measured by the metric to evaluate and select ML models, must be high enough. Among the quality dimensions guiding NSOs (see QF4SA, Yung et al. 2022, and Section 6.3), this corresponds to *accuracy*. However, there are additional objectives to consider: in particular, the model shall be interpretable, fair, robust, and reproducible. For some quality dimensions such as interpretability, quantifying directly how well a model does on them is very hard (Molnar 2020, Chapter 3.4), so that auxiliary constructs such as model simplicity and model sparsity are used (e.g., Rudin 2019). Fortunately, there do exist concrete metrics for fairness. This, together with the existence of measures to improve fairness (see below), means that fairness is not a qualitative, intangible, or merely theoretical and academic concept, but one that can be put into practice.

As it is generally very unlikely that one happens to be in a situation in which there exists exactly one model that is optimal (i.e., better than every other model) in every dimension, satisfying multiple objectives at once is problematic. Unless one is willing to drop all objectives but one, there are three main approaches. First, the multiple objectives can be combined into a single objective, e.g., as a weighted sum. The organization must then decide on the relative weight of each objective in that sum and, if the dimensions have different units, also how to convert units. Second, multi-objective optimization methods aim to find the set of Pareto-optimal models, that is, the set of all models for which there exists no other model that is strictly better in at least one dimension and no worse in all the other dimensions. Aside from being computationally complex, this produces a (potentially very large) set of models: hence, in order to choose which of these models to deploy, an organization would still need to find decision rules about the trade-off between the multiple objectives. Third, one can choose only one primary objective to optimize—typically, performance—and impose the other objectives as mathematical constraints, e.g., specifying the maximum level of unfairness that is tolerated. As this level must be stated precisely and with legal requirements of this kind not (yet) in place, organizations may have to self-impose fairness constraints on their data analyses and data products. Thus, all three approaches require organizations to deliberate and make decisions, and the decisions may depend on context (i.e., the specific data product).

How to Handle (Un)fairness

In practice, fairness evaluation is often only an ex post consideration, that is, after a model has been trained and selected. If the model fails the fairness check, there are several strategies, none of which is ideal. First, it is possible to simply not deploy a trained model that fails quality requirements and to stop all work there. For NSOs, this may be an option for a new, optional data product or experimental procedure that fails, but it is not a viable general strategy. Without any tangible output, this also has the worst cost-effectiveness possible. Second, an unfair data product may still be shipped. With adequate warning and proper documentation, potential downstream data users may then decide whether to use this data product or a different, fairer product, and for what purposes it is fit, and for which groups in the population it can be used. Good documentation of the limitations is key, as i) downstream data users may be less able to evaluate fairness than NSOs as producers of the product, and ii) such information may also be useful in the mitigation of unfairness of models trained on such data. In any case, this puts the onus on the users. Except for experimental, optional data products, shipping an unfair data product may run counter to the quality dimension of accuracy and, in the bigger picture, might also not be the best strategy to foster the perceived trustworthiness of NSOs. Third, one may push the project back up the pipeline, either all the way or, if feasible, to the step in the chain at which the fairness problem started. We touch on the ways to get to a fairer data product below.

A final, sometimes forgotten, strategy comes from *model multiplicity*: i.e., the frequent existence of multiple models that all perform nearly equally as well as the best-performing model, which has been observed empirically (Breiman 2001, may be the first, or at least first prominent, observer) and discussed theoretically (Semenova et al. 2022). From this so-called Rashomon set of models that have functionally the same predictive performance, one then selects the one that is the best on another objective, such as interpretability (Rudin 2019) or fairness (Rodolfa et al. 2020, Chapter 11.4.2), or several of these objectives. Typically, this approach is not very resource-intensive, if the models are trained anyway (to find the one with the highest performance) and since fairness evaluations are generally comparatively low in computational costs. However, there is no guarantee that model multiplicity will likely produce fair candidate models. In fact, if the only difference between models is their model class, then only this source of unfairness might be mitigated.[5] There are some other strategies such as changing the set of features (e.g., exclude features with high measurement error, missingness, or little variability in some protected groups). Yet, any root causes of unfairness that are upstream from these steps (inclusion of model classes, engineering and selection of features, etc.) generally will not be fixed by model multiplicity.

In sum, considering fairness only after training can waste the limited resources of NSOs and the energy required to compute elaborate models, and it may deliver the data product late, with unfairness, or not at all. It is therefore prudent to mind fairness throughout the whole chain. The strategies for fair ML that are discussed in the literature can be generally categorized into pre-, in-, and post-processing methods (Mehrabi et al. 2021, Chapter 5).

Post-processing, i.e., putting some layer on top of a trained model to alter its predictive behavior, makes the most sense when data and model training are a black box (ibid). This should typically not be the case for NSOs unless parts of the chain

[5] Such as when a data-hungry model class performs very well on average but generally poorly for small minorities.

are not done in-house. *In-processing* methods involve changes to the optimization, such as incorporating fairness directly in the loss function or as a constraint. These options should always receive strong consideration. *Pre-processing* "tr[ies] to transform the data so the underlying discrimination is removed" (Mehrabi et al. 2021, Chapter 5). Similar to the abovementioned limited perspective of the prediction step, the literature is mainly focused on the fixed or found data perspective. NSOs (and other data producers) are engaged in the whole pipeline and have a well-developed toolbox including the design of data collection instruments (e.g., surveys), sampling schemes, and so on. There is much room for designing and producing fair data or to bring new or additional data where necessary. Similarly, continuous, near real-time monitoring of data quality is already common when, i.e., conducting surveys.

6.3 Fairness and Quality Dimensions for Official Statistics

Assessing and documenting the fitness of an ML model along multiple dimensions is a key prerequisite for their deployment in high-stake settings at NSOs. Concepts such as the Quality Framework for Statistical Algorithms (QF4SA; Yung et al. 2022)—as well as recent extensions that consider fairness aspects (Saidani et al. 2023; Saidani and Dumpert 2025)—are tailored to typical uses of ML in official statistics and provide invaluable guidelines for NSOs to continue to uphold high-quality standards. We argue that systematically mapping fairness considerations to the QF4SA enriches and expands current quality dimensions, sharpens existing requirements, and enables a comprehensive documentation of potential model limitations. Re-evaluating quality concepts through the fairness lens can point to blind spots and introduce additional safeguards with a focus on the various consequential downstream uses of data products of NSOs. For a detailed discussion of the interactions between fairness and the QF4SA, we point the reader to Schenk and Kern (2024).

Multigroup Quality

Ingesting fairness considerations into the QF4SA's quality dimensions starts by recognizing the central role of protected attributes in the fairness literature. In algorithmic fairness, models are assessed with respect to their potential of disparate impact on social groups, most commonly defined by personal characteristics beyond individuals' agency such as gender, age, or ethnic origin (Mehrabi et al. 2021; Makhlouf et al. 2021; Mitchell et al. 2021). Multigroup fairness takes an intersectional perspective by expanding beyond the simple group-based focus and considers (complex) interactions between (both protected and non-protected) attributes (Hebert-Johnson et al. 2018; Kim et al. 2019). Both group and multigroup fairness connect to key quality dimensions of the QF4SA: *accuracy* of predictions in supervised learning contexts, for example, can be required not only overall but also for subgroups that may be viewed as vulnerable in a given application context. Groups may be defined not only by protected attributes but also by other measures that define meaningful categories for the use case of interest, including spatial or temporal units. The underlying conception is that multi-accurate models do not only provide more reliable (or honest) predictions even for small subsets of the target population but also offer some protection against practical issues that may arise in downstream applications such as distribution shift (Kim et al. 2022). The *robustness* dimension (Saidani et al. 2023; Saidani and Dumpert 2025) can be similarly connected to the (multi)group perspective by expanding the monitoring of models over time to include measures of group-specific performance that may serve as early warning signs of instability. The *interpretability* dimension strongly connects to fairness considerations, which includes the use of interpretable machine learning techniques (Molnar 2020) to assess whether a model's functioning differs between vulnerable subpopulations. Insufficient *reproducibility* can have considerable fairness implications, particularly as manifold design decisions along the modeling pipeline can have subgroup-specific impacts on model performance and predictions (Simson et al. 2024). Fairness concerns further link to the degree of human oversight that may be needed in the construction of data products and thus should also be factored into assessments of *cost-effectiveness* and *timeliness*.

Fairness Beyond the QF4SA

While the quality dimensions of QF4SA are predominantly focused on the (algorithmic) production of NSOs' outputs, the data products themselves can also be discussed through the lens of algorithmic fairness. Data produced by NSOs may serve as training data or for evaluation and benchmarking purposes for third parties such as external agencies. As biases in data is one of the root causes of fairness issues downstream (Mehrabi et al. 2021), the importance of high-quality and inclusive data products of NSOs is reinforced. While the QF4SA undoubtedly caters to this goal, data outputs of NSOs may be specifically evaluated from a fairness perspective. For individual-level data, this includes adequate representation of minorities and group-specific considerations of measurement error (Rodolfa et al. 2020). It is important to note that not only the data collection but also the data (post-)processing process can lead to group-specific data deficits, e.g., when small groups are disproportionately affected by record linkage errors. Once a data product is ready for release, we further note that common fairness metrics (e.g., (conditional) statistical parity of an outcome variable; Makhlouf et al. 2021) can be applied prior to any model training to assess the degree as to which social disparities are reflected in the data itself.

6.4 Discussion

The mechanistic connection of fairness to justice, the feeding function of the prediction step to the decision step, and the input mindset we discussed in Sect. 6.2 all imply: each step in the chain can produce fairer outcomes when it is informed by what comes after it—how and for what purposes its outputs are used and which justice principles and other goals will eventually be pursued. The first implication is that seeking such information can help to produce fairer products. This is most feasible in case of important, repeated, known downstream users of a data product. Similarly, within NSOs, stakeholders along the chain should not operate in isolation. Second, some data products may have many or a priori unknown users and thus cannot be tailored to the (fairness) needs of a specific downstream use. In this case, we suggest that the input mindset makes group and multigroup performance attractive fairness measures of a ML model. Third, to the extent that NSOs also engage in data analyses, they must deliberate over which criteria are relevant and what their order or trade-off between them is. These decisions may vary by data product and should be documented and communicated both within NSOs and to outside data users.

Regarding data, sources of unfairness may fall into the three categories: (mis-)-representation, errors of measurement, and missing variables. Considerations can be both absolute (e.g., Is each group sufficiently represented in the data? Is the accuracy of a classifier for each group above a minimum acceptable threshold?) and relative (e.g., Are there more measurement errors for some group? Is a classifier's accuracy much worse for some group?). Testing, documenting, and communicating (multi)group fairness in interaction with key dimensions of data quality is vital. A data-centric approach to fairness also points to blind spots in the current fair ML literature: its strong focus on data-driven (automated) decision-making and binary classification tends to divert the attention away from upstream causes of unfairness and the cumulative effect of errors along the data processing pipeline as well as considerations of valid inference and measures of uncertainty.

6.5 Conclusion

NSOs' data products inform public and political discourse, help to advance scientific knowledge, and are employed by various downstream data users. Therefore, the fairness of these data products is important for a productive and just society, and it will only become more important as data-driven decision-making continues to scale up. Note that this perspective is not at odds with the primary role of NSOs to *describe* social realities—in line with the distinction between the prediction and decision step in ADM (Kuppler et al. 2022), fairness in data products rather serves as a pre-condition to identify disparities and as an input for downstream decision-making that may eventually act on them.

ML models are trained to be deployed and only models that apparently work well enough will be shipped. In addition to a typical measure of performance such as accuracy, this evaluation must consider further quality dimensions such as fairness. As quality dimensions interact with each other, it is generally not possible to optimize a model with respect to all dimensions simultaneously. Hence, decisions about the relative importance need to be made by NSOs, either together with crucial downstream data users or on their own.

Considering fairness up front and during all the steps of creating a data product, instead of only after model training, can save resources (time, energy, money, etc.) invested in a product that ends up subpar, perhaps even too unfair to deploy.

We emphasized the role of data. First, data quality is arguably the biggest single contributor to the fairness of ML models. Second, while much of the fair ML literature views data as fixed, data producers such as NSOs are in a unique position as they have much discretion regarding the design of data (collection) and have amassed vast knowledge and a rich toolkit for how to do so. This is a much-needed broadening of perspective.

In our view, what To et al. (2023) write about human-computer interaction also applies to fair ML: the effect of current research and design practices is to consider protected or potentially disadvantaged groups "predominantly from the lens of deficit and damage." Trying to keep the unfairness in a ML model regarding protected groups just within acceptable levels may make fairness seem like a tedious chore. Instead, fairness can also be understood as shining a light on opportunities: NSOs can create new data products for affected groups, improve existing products particularly for these groups, and use fair ML machinery to find (sub)groups that are actually disadvantaged by a particular data product beyond those protected by law. Fairness can stimulate and guide innovation.

References

D. Allhutter, F. Cech, F. Fischer, G. Grill, A. Mager, Algorithmic profiling of job seekers in Austria: how austerity politics are made effective. Front. Big Data **3**, 1–17 (2020)

A. Amaya, P.P. Biemer, D. Kinyon, Total error in a big data world: adapting the tse framework to big data. J. Survey Stat. Methodol. **8**(1), 89–119 (2020)

S. Barocas, A.D. Selbst, Big data's disparate impact. California Law Rev. **104**(3), 671–732 (2016)

S. Barocas, M. Hardt, A. Narayanan, Fairness and Machine Learning (2019). https://fairmlbook.org

L. Breiman, Statistical modeling: the two cultures. Stat. Sci. **16**(3), 199–231 (2001)

S. Carton, J. Helsby, K. Joseph, A. Mahmud, Y. Park, J. Walsh, C. Cody, C.E. Patterson, L. Haynes, R. Ghani, Identifying police officers at risk of adverse events, in *Proceedings of the 22nd ACM SIGKDD International Conference on Knowledge Discovery and Data Mining* (Association for Computing Machinery, New York City, 2016), pp. 67–76

S. Desiere, K. Langenbucher, L. Struyven, Statistical profiling in public employment services. OECD Social, Employment and Migration Working Papers, No. 224 (2019). https://doi.org/10.1787/b5e5f16e-en

R.M. Groves, F.J. Fowler Jr, M.P. Couper, J.M. Lepkowski, E. Singer, R. Tourangeau, *Survey Methodology*, 2nd edn. (John Wiley & Sons, Hoboken, 2009)

U. Hebert-Johnson, M. Kim, O. Reingold, G. Rothblum, Multicalibration: calibration for the (computationally-identifiable) masses, in *Proceedings of the 35th International Conference on Machine Learning*, ed. by J. Dy, A. Krause, vol. 80. *Proceedings of Machine Learning Research* (2018), pp. 1939–1948

M.P. Kim, A. Ghorbani, J. Zou, Multiaccuracy: black-box post-processing for fairness in classification, in *Proceedings of the 2019 AAAI/ACM Conference on AI, Ethics, and Society* (Association for Computing Machinery, New York City, 2019), pp. 247–254

M.P. Kim, C. Kern, S. Goldwasser, F. Kreuter, O. Reingold, Universal adaptability: target-independent inference that competes with propensity scoring. Proc. Natl. Acad. Sci. **119**(4), e2108097119 (2022)

J. Körtner, G. Bonoli, Predictive algorithms in the delivery of public employment services, in *Handbook of Labour Market Policy in Advanced Democracies*, ed. by D. Clegg, N. Durazzi (Edward Elgar Publishing, Cheltenham, 2023), pp. 387–398

M. Kuppler, C. Kern, R. Bach, F. Kreuter, From fair predictions to just decisions? Conceptualizing algorithmic fairness and distributive justice in the context of data-driven decision-making. Front. Sociol. **7**, 1–18 (2022)

K. Makhlouf, S. Zhioua, C. Palamidessi, On the applicability of machine learning fairness notions. ACM SIGKDD Explorations Newsl. **23**(1), 14–23 (2021)

N. Mehrabi, F. Morstatter, N. Saxena, K. Lerman, A. Galstyan, A survey on bias and fairness in machine learning. ACM Comput. Surveys **54**(6), 1–38 (2021)

S. Mitchell, E. Potash, S. Barocas, A. D'Amour, K. Lum, Algorithmic fairness: choices, assumptions, and definitions. Ann. Rev. Stat. Its Appl. **8**(1), 141–163 (2021)

C. Molnar, *Interpretable Machine Learning. A Guide for Making Black Box Models Explainable*, 2nd edn. (Leanpub, Victoria, 2020). https://christophm.github.io/interpretable-ml-book

J. Moriarty, Desert-based justice, in *The Oxford Handbook of Distributive Justice*, ed. by S. Olsaretti (Oxford University Press, Oxford, 2018), pp. 152–174

M. Puts, D. Salgado, P. Daas, Leveraging machine learning for official statistics. in *Foundations and Advances of Machine Learning in Official Statistics*, ed. by F. Dumpert, Chap. 2 (Springer, Cham, 2025)

K.T. Rodolfa, P. Saleiro, R. Ghani, Bias and fairness, in *Big Data and Social Science*, ed. by I. Foster, R. Ghani, R.S. Jarmin, F. Kreuter, J. Lane, Chapter 11, 2nd edn. (CRC Press, Boca Raton, 2020)

C. Rudin, Stop explaining black box machine learning models for high stakes decisions and use interpretable models instead. Nat. Mach. Intell. **1**(5), 206–215 (2019)

Y. Saidani, F. Dumpert, Quality dimensions and quality guidelines for machine learning in official statistics, in *Foundations and Advances of Machine Learning in Official Statistics*, ed. by F. Dumpert, Chap. 4 (Cham, Berlin, 2025)

Y. Saidani, F. Dumpert, C. Borgs, A. Brand, A. Nickl, A. Rittmann, J. Rohde, C. Salwiczek, N. Storfinger, S. Straub, Qualitätsdimensionen maschinellen Lernens in der amtlichen Statistik. AStA Wirtschafts- und Sozialstatistisches Archiv **17**(3), 253–303 (2023)

T. Scantamburlo, J. Baumann, C. Heitz, On prediction-modelers and decision-makers: why fairness requires more than a fair prediction model. AI & Society (2024). https://doi.org/10.1007/s00146-024-01886-3

P.O. Schenk, C. Kern, Connecting algorithmic fairness to quality dimensions in machine learning in official statistics and survey production. AStA Wirtschafts- und Sozialstatistisches Archiv **18**(2), 131–184 (2024)

L. Semenova, C. Rudin, R. Parr, On the existence of simpler machine learning models, in *Proceedings of the 2022 ACM Conference on Fairness, Accountability, and Transparency* (Association for Computing Machinery, New York City, 2022), pp. 1827–1858

J. Simson, F. Pfisterer, C. Kern, One model many scores: using multiverse analysis to prevent fairness hacking and evaluate the influence of model design decisions, in *Proceedings of the 2024 ACM Conference on Fairness, Accountability, and Transparency* (Association for Computing Machinery, New York City, 2024), pp. 1305–1320

A. To, A.D.R. Smith, D. Showkat, A. Adjagbodjou, C. Harrington, Flourishing in the everyday: moving beyond damage-centered design in HCI for BIPOC communities, in *Proceedings of the 2023 ACM Designing Interactive Systems Conference* (Association for Computing Machinery, New York Ciry, 2023), pp. 917–933

M. von Zahn, O. Hinz, S. Feuerriegel, Locating disparities in machine learning, in *2023 IEEE International Conference on Big Data* (IEEE, Piscataway, 2023), pp. 1883–1894

B.T. West, J. Wagner, J. Kim, T.D. Buskirk, The Total Data Quality Framework. https://www.coursera.org/specializations/total-data-quality (2023). Online; Accessed Feb 05 2025

W. Yung, S.-M. Tam, B. Buelens, H. Chipman, F. Dumpert, G. Ascari, F. Rocci, J. Burger, I. Choi, A quality framework for statistical algorithms. Stat. J. IAOS **38**(1), 291–308 (2022)

Chapter 7
Legal Implications for the Use of Machine Learning in Official Statistics

Leon Krög

7.1 Relevance and Purpose of Official Statistics in Germany

Statistics form a necessary infrastructure for any modern and efficient society. Looking at statistics from a legal[1] perspective, there is no way around the census ruling of the German Federal Constitutional Court (Bundesverfassungsgericht, BVerfG) from 1983, which dealt with the violation of fundamental rights through statistical surveys (Kühling 2023, Introduction, para 122m). In this ruling, the court stated that a state policy based on the social state principle must not simply accept economic, social, and ecological changes as unalterable fate.[2] As an information tool and source, official statistics provide the necessary informational basis for political decision-making. They provide an informational basis both for government action and for forming individual opinions, as they "in principle, enable everyone to observe and assess social, economic, and ecological phenomena" (Radermacher 2017).

However, statistics are bound by the principle that a statistical survey must result in the reproduction of data in a structured and anonymous form.[3] Official statistics provide results on mass phenomena and do not serve to present information directly relating to individuals or institutions. Consequently, decisions made based on these representations are not directed at individual cases but observe and address problems

[1] BT-Drs. 10/5345, p. 13.

[2] BVerfGE 65, 1 (47).

[3] BT-Drs. 10/5345, p. 13; BVerfGE 65, 1 (53, 54).

L. Krög (✉)
Universität Mannheim, Mannheim, Germany
e-mail: leon.kroeg@uni-mannheim.de

F. Dumpert (ed.), *Foundations and Advances of Machine Learning in Official Statistics*, Society, Environment and Statistics,
https://doi.org/10.1007/978-3-032-10004-7_7

affecting society as a whole (Kühling and Schmid 2023, § 1 BStatG, para 7, 11). At the same time, statistics should be in line with the current state of the methodological discussion, i.e., use modern methods of data processing and exploit new sources of knowledge.[4]

In its decision, the BVerfG also dealt with the principles of data collection conducted by public authorities. For example, it is found that when data are collected for statistical purposes, special rules apply to purpose limitation, as it is in the nature of statistics that the data are used for different purposes after they have been processed.[5] The court has confirmed this relaxation of purpose limitation in recent case law on the 2011 census.[6] The 1983 census ruling is regarded a landmark ruling in the field of German data protection law.

The BVerfG defines statistics as the methodical collection, compilation, presentation, and evaluation of data and facts for state purposes.[7] A reference to internal facts and events, as well as political evaluations and opinions, can also be part of statistics as long as they are of a purely functional nature.[8] The purpose of official statistics is to collect factual material and not to bring about political actions or activities.[9] They are limited to the abstract presentation of general developments and phenomena (Kühling and Schmid 2023, § 1 BStatG, para 7).

According to § 1 Federal Statistics Act (Bundesstatistikgesetz, BStatG), official German statistics are subject to the principles of neutrality, objectivity, and professional independence in the performance of their tasks.[10] Neutrality requires that no preference be given to individual third-party interests. Objectivity requires the development and dissemination of federal statistics in a systematic, reliable, and unbiased manner. Professional independence applies above all to the procedures, definitions, methods, and sources, as well as the timing and content of all forms of dissemination (Kühling and Schmid 2023, § 1 BStatG, para 19 et seq.).

7.2 Importance of Data Protection in Official Statistics

One of the basic ideas behind the census ruling was that population statistics should be as little burdensome as possible for individual citizens. The fundamental right to informational self-determination, which the court derived from Art. 2 (1) and

---

[4] BVerfGE 65, 1 (55, 56); BVerfGE 150, 1 (110).

[5] BVerfGE 65, 1 (47).

[6] BVerfGE 150, 1 (108).

[7] BVerfGE 150, 1 (79).

[8] BVerfGE 8, 104 (111).

[9] BVerfGE 8, 104 (111).

[10] These principles are in line with those of regulation (EC) No. 223/2009 (regulation about European statistics). The regulation also extends the principles to include reliability, confidentiality, and cost-effectiveness.

Art. 1 (1) of the German Basic Law (Grundgesetz, GG), protects against information processing by official authorities (Schantz and Wolff 2017, para 153). It leaves it up to the individual to decide what is disclosed about them.[11] However, the fundamental right does not provide unlimited protection against interference,[12] as individual reality is also partly a reflection of society and therefore serves the general public.[13] Suitable measures should be taken to ensure that the level of interference is as low as possible.

However, according to the Federal Constitutional Court, official statistics are prohibited from creating comprehensive personality profiles of individuals.[14] If this was possible, intimidation effects could arise because citizens would no longer know "who knows what about them, when and on what occasion,"[15] which could restrict them in the exercise of individual fundamental rights. Statistical confidentiality can therefore be described as the "foundation of official statistics" (Dorer et al. 1988, § 16 BStatG, para 1). The task of official statistics is to depict and statistically represent mass phenomena that are no longer related to individuals. The abstract-general presentation must not serve the purpose of administrative enforcement, as this is designed to address the individual person (Kühling and Schmid 2023, § 1 BStatG, para 11).

As an expression of the abovementioned fundamental right to informational self-determination, § 16 BStatG requires the confidentiality of individual statistical data for official federal statistics (Kühling and Sauerborn 2023, § 16 BStatG, para 1). It represents a counterbalance to the respondents' obligation to provide information and the fact that the exact purposes of the statistical survey cannot be precisely defined in advance (Kühling 2023, Introduction, para 43).[16]

The confidentiality of individual statistical data is also part of the Code of Practice for European Statistics.[17] Article 5 of this code requires data protection and the confidentiality of individual data through measures on the part of the statistical offices, legal obligations, and voluntary commitments on the part of the employees of the statistical offices. Although this code is not a binding legal act, it is mentioned in Arts. 1, 2 (1) and 11 of Regulation (EC) 223/2009, which is also referred to as the "statistical framework regulation" (Kingreen 2022, Art. 338 TFEU, para 3; Kühling 2023, Introduction BStatG, para 77).

[11] BVerfGE 65, 1 (43).

[12] The first section of the German Constitution protects citizens against actions of public authorities that interfere with those rights.

[13] BVerfGE 65, 1 (43, 44).

[14] BVerfGE 65, 1 (53).

[15] BVerfGE 65, 1 (42, 43); BVerfGE 115, 166 (188).

[16] BVerfGE 65, 1 (48).

[17] Publications Office of the European Union (2018): European Statistics Code of Practice for National Statistical Institutes and Eurostat (EU statistical authority), adopted by the European Statistical System Committee on 16 November 2017; available at https://ec.europa.eu/eurostat/documents/4031688/8971242/KS-02-18-142-EN-N.pdf/e7f85f07-91db-4312-8118-f729c75878c7?t=1528447068000, as of 24 January 2025.

7.2.1 General Data Protection Regulation

Since the production of statistics requires collection and storage—and therefore processing—of personal data (e.g., in the census), the General Data Protection Regulation[18] (GDPR) is applicable. The regulation establishes a uniform legal framework in the EU member states, which is intended to guarantee the European fundamental right to data protection under Art. 16 TFEU[19] through the rights of the data subjects and the obligations of those who process the data. The GDPR applies to both automated and manual processing of personal data.

For the field of official statistics, however, the GDPR provides for specific privileges as long as the interests of the data subjects are protected. The implementation of these is partly left to the member states through opening clauses. The term *statistics* in the sense of the regulation is to be understood as the methodical handling of empirical data (Buchner and Tinnefeld 2024, Art. 89, para 15).

Article 89 (1) GDPR requires appropriate safeguards when handling personal data for statistical purposes that protect the rights and freedoms of the data subject. These safeguards[20] include the extensive anonymization[21] and, alternatively, pseudonymization[22] of personal data (Pauly 2021, Art. 89 GDPR, para 13). In particular, Art. 89 (2) GDPR permits the restriction of data subjects' rights, insofar as these significantly impair or render impossible the statistical purposes. This is particularly conceivable if these rights are used excessively, resulting in a high administrative burden for the data controllers concerned or depriving them of their data basis (Pauly 2021, Art. 89 GDPR, para 14). Whether the specific purposes are actually restricted or even rendered impossible depends on a prognosis in the individual case (Pauly 2021, Art. 89 GDPR, para 14).

Another exception concerns the purpose limitation of data processing. In principle, Art. 5 (1) b GDPR requires specified, explicit, and legitimate purposes for the use of personal data. If further processing is intended, this usually requires a compatibility test in accordance with Art. 6 (4) GDPR. However, further processing

---

[18] Regulation (EU) 2016/679 of the European Parliament and of the Council of 27 April 2016 on the protection of natural persons with regard to the processing of personal data and on the free movement of such data and repealing Directive 95/46/EC (General Data Protection Regulation), OJ L 119/1, 4.5.2016.

[19] TFEU means Treaty on the Functioning of the European Union, one of the sources of EU primary law.

[20] Concerning the scope and extent of these guarantees, see also Weichert (2020).

[21] Recital 26 of the GDPR defines anonymous data as "information which does not relate to an identified or identifiable natural person or to personal data rendered anonymous in such a manner that the data subject is not or no longer identifiable."

[22] This is defined in Art. 4 Nr. 5 GDPR which states "'pseudonomyzation' means the processing of personal data in such a manner that the personal data can no longer be attributed to a specific data subject without the use of additional information, provided that such additional information is kept separately and is subject to technical and organisational measures to ensure that the personal data are not attributed to an identified or identifiable natural person."

for statistical purposes is considered compatible—subject to the guarantees of Art. 89 GDPR—unless the purposes can also be achieved with anonymous data (Buchner and Tinnefeld 2024, Art. 89 GDPR, para 21).

When the GDPR refers to data for statistical purposes, it assumes that the data are no longer personal when they are published (see Recital 162, sentences 3–5). Hence, the privilege only applies in this case. A personal reference, which may still exist during collection and processing, is eliminated through the separation of auxiliary information[23] for internal purposes—at German level required by § 12 BStatG (Buchner and Tinnefeld 2024, Art. 89 GDPR, para 15; Weichert 2020, pp. 22, 23). If auxiliary information is still present in aggregated statistical results, these are no longer privileged statistical purposes (Buchner and Tinnefeld 2024, Art. 89 GDPR, para 15, 15a).

Data protection in statistics thus serves to maintain this privilege in relation to the provisions of the GDPR in order to ensure efficient, functioning statistics. In addition, the function of statistics would also be limited by the fact that, with a low level of confidentiality, the participation of respondents would not be equally accurate, and false information could be expected (Couper et al. 2008). If there is no strict confidentiality, this could lead to a decreasing willingness to cooperate on the part of the citizens surveyed.[24]

7.2.2 Advantages of Machine Learning for Official Statistics

Technology journalist Thomas Ramge summarizes the essence of ML as follows:[25] “In machine learning, computer systems recognize patterns in examples and can transfer their ‘findings’ to other examples. In this way, they learn to draw ever more precise conclusions from data and derive decisions” (Ramge 2018, p. 16).

The use of ML in official statistics is conceivable, for example, for classification procedures that would otherwise be difficult or impossible to carry out manually.[26] This includes the *coding* of collected microdata,[27] whereby data are assigned to an

---

[23] Auxiliary data are used for the technical implementation of statistics and enables the identification of individual persons, primarily for the initial verification of the collected data and to allow for queries. Auxiliary characteristics are, for example, name, surname, or address. See Dorer et al. (1988, § 10 BStatG, para 4).

[24] Cf. BVerfGE 65, 1 (50); cf. also BT-Drs. 10/534, p. 20 on § 16 of the draft of the Federal Statistics Act (BStatG); Dorer et al. (1988, § 16 BStatG, para 4).

[25] Translated from German.

[26] https://www.destatis.de/DE/Service/Hauptstadtkommunikation/Zukunft/maschinelles-lernen.html, as of 24 January 2025.

[27] Microdata, also known as individual data, is information about the personal and factual circumstances of an individual, e.g., about a specific person (their occupation, health, grades, etc.). Individual data can be found in official statistics for persons, households, or companies, for example. Individual data contain the maximum amount of information of a statistic which makes

official systematic numerical code. This is done by recognizing the text of the survey data, which is then assigned using an algorithm. The classified survey variables can be different characteristics such as age, salary, a job description, or the description of an injury.[28] One example is the assignment of occupations according to an Occupational Classification System.[29]

Plausibility checks offer another possible application. This process is also known as *editing* and aims to identify and flag unusual values in the data.[30] In addition, it can help to identify rules for plausibility checks that were previously based on the intuition of the employees responsible; it is also possible to find hidden structures in data.[31] This process could be accelerated through the use of ML.[32] Closely linked to this is *imputation*, in which (previously recognized) missing or incorrect values are added or replaced (Preising et al. 2021).

When considering the legal implications for the use of ML in official statistics, the potential uses and advantages offered by the technology must be considered. An initial indication of the need for an expansion of methods in official statistics is the fact that the Federal Constitutional Court has called for a continuous discussion of methods in the census ruling.[33] The purpose of this is to assess whether technical progress is producing or has already produced survey methods that reduce the depth of interference with the fundamental right to informational self-determination.[34] The BVerfG warned that future developments in methods should be "monitored to see whether they enable procedures that are more respectful of fundamental rights."[35] The quality manual for official statistics also contains the principle that the methodology of statistical processes must correspond to the current state of scientific research (Statistische Ämter des Bundes und der Länder 2021; Dumpert 2021). Full surveys can lead to the disclosure of data that are not part of the actual

---

them the raw material of the statistician (https://www.destatis.de/DE/Service/Statistik-Campus/ESC/mikrodaten.html, as of 24 January 2025).

[28] UNECE ML C&C Theme Report, p. 3 (https://statswiki.unece.org/display/ML/WP1+-+Theme+1+Coding+and+Classification+Report, as of 24 January 2025).

[29] UNECE ML C&C Theme Report, p. 2. This is a classification system for the occupational affiliation of people.

[30] UNECE HLG-MOS Machine Learning Project Theme report of the editing and imputation group (https://statswiki.unece.org/display/ML/WP1+-+Theme+2+Edit+and+Imputation+Report, as of 24 January 2025).

[31] UNECE HLG-MOS Machine Learning Project Theme report of the editing and imputation group.

[32] UNECE ML WP1 Executive Summary (https://statswiki.unece.org/spaces/ML/pages/293536330/WP1+-+Executive+Summary?preview=/293536330/293536328/ML_WP1_ExecutiveSummary.pdf, as of 24 January 2025).

[33] BVerfGE 65, 1 (55); see also BT-Drs. 10/5345, p. 13.

[34] BVerfGE 150, 1 (110, 113, 114).

[35] BVerfGE 150, 1 (133); cf. also BVerfGE 65, 1 (55, 56).

survey.[36] If necessary, ML can be used to check plausibility in such a way that fewer queries need to be made to the respondents, which would also reduce the severity of interference with their rights.

Another argument is that a more effective use of human resources can be expected if the employees of the statistical offices were supported by ML (Dumpert 2021). This would take account of the general interest in cost-efficient administration. It may also result in a higher level of data protection because fewer people have contact with the collected data. On the one hand, this offers advantages in the event of an increased collection frequency, which is to be expected for social statistics at European level as a result of a new statistical regulation (Söllner and Körner 2022; COM(2023) 31 final). On the other hand, the past few years have presented politics and society with difficult tasks, which are not exhaustively mentioned with the corona pandemic, the energy crisis, and a growing housing shortage in German cities. Reducing the time required for the production of statistics through the use of ML could lead to greater flexibility and adaptability (Dumpert and Beck 2017). This strenghtens the role of statistics as a foundation for making smart and well-founded decisions on crucial policy questions.

7.2.3 Data Protection Risks from the Use of Machine Learning

However, the use of ML also poses risks. The past has shown in several cases that it can be possible to identify people based on just a few characteristics. Latanya Sweeney, for example, proved during her time as a Ph.D. student that a large part of the US population could be clearly identified on the basis of three characteristics alone. She demonstrated this impressively by comparing health data from an insurance company with cheaply available data from a voter registration list. The assignment of the data enabled the unambiguous classification of individuals, including the governor of Massachusetts (Sweeney 2002).

Another case deals with the publication of movie ratings by Netflix with the aim of determining the preferences of users based on their history and ratings. Arvind Narayanan and Vitaly Shmatikov from the University of Texas showed that these data were sufficient to uniquely identify at least some of the users, provided they were linked to other data about films the people had already watched (Narayanan and Shmatikov 2008).

Examples such as these make clear that a specific personal reference can be established based on a small amount of data with supposedly little informative value. As explained above, ML has the ability to link large volumes of data and recognize patterns in them. However, this ability to better link data could pose a risk to data security and make statistical data assignable at an individual level. Hence, it

[36] BVerfGE 150, 1 (134). This is possible in surveys with an on-site interviewer who could receive information that is not part of the actual survey.

is relevant to ask whether such cases could also occur using modern data processing in the form of ML in official statistics.

The aforementioned processes of coding, plausibility checks, and imputation of survey data are used for data preparation, i.e., for the most part they take place before the separation and deletion of the auxiliary characteristics, so that the data are still personal anyway.[37] In this respect, the methods do not differ from their analog counterparts. Here too, the auxiliary characteristics are only separated and, if necessary, deleted once the plausibility of the values has been checked. The possibility of quickly assigning the data to the auxiliary characteristics in order to enable any queries to be made must be available until the plausibility check has been carried out (Dorer et al. 1988, § 12 BStatG, para 3). In this respect, it is therefore not evident that modern, ML-supported data processing is disadvantageous.

However, it is conceivable that an ML model used to check the data could store data that makes it possible to reestablish a personal reference at a later point in time (Song et al. 2017). In order to prevent the deleted auxiliary characteristics from being restored via the ML algorithm after deletion, it would be advisable for the algorithm to only come into contact with the survey features and not with personal auxiliary information. To ensure a high level of data protection, training with already anonymous or synthetic data could also be considered (Raji 2021).

A model may also require auxiliary features to be used either during training or in application. In this case, a later deletion of the model data would be required if auxiliary characteristics are separated and deleted to prevent reconstruction via the model after successful plausibility checks or coding. This is simply because § 12 BStatG requires irreversible deletion of the data (Isfort et al. 2023, § 12 BStatG, para 11). In addition, a restoration of the personal reference would also be excluded by § 21 BStatG, which prohibits reidentification.

Nevertheless, it is still conceivable that the analysis with an ML algorithm could reveal a pattern that makes it possible to assign the data to a single person again. In particular, the linking of several datasets could result in unique combinations of characteristics. If a model offered the technical feasibility of reestablishing a personal reference at a later point in time, this would not be permitted due to the aforementioned prohibition on reidentification. That poses the question: whether a mere prohibition is sufficient as a safeguard.

The means to be considered in order to determine whether a personal reference can be established and, thus, whether personal data can be assumed are controversial (Karg 2025, Art. 4, No. 1, para 58; Klar and Kühling 2024, Art. 4 GDPR, para 25). In some cases, an absolute understanding is assumed, according to which information is considered personal if the controller or any third party has the knowledge to establish the personal reference (Karg 2025, Art. 4, No. 1, para 59; Klar and Kühling 2024, Art. 4 GDPR, para 25). According to the relative understanding of personal reference, on the other hand, only information that is likely to be used—considering factors like cost and time—and that is actually available in the specific individual

[37] Cf. Isfort et al. (2023, § 12 BStatG, para 15); BT-Drs. 10/5348, 18.

case is to be taken into account for the question of whether the personal reference can be established (Karg 2025, Art. 4, No. 1, para 60). This (relative) view is also shared by the European Court of Justice (ECJ)[38] and, in more recent case law, by the European General Court (EGC).[39] Therefore, even in the case of a technical possibility, a reference to a person cannot be assumed, since in cases in which a reference to a person could theoretically be established, it is the relative criterion that is important.[40] The use of unlawful means must be ruled out in accordance with the case law of the ECJ (Karg 2025, Art. 4, No. 1, para 66; Klar and Kühling 2024, Art. 4 GDPR, para 29). In any case, this can be assumed for official statistics, as there is a ban on reidentification under § 21 BStatG.

To sum up: Although the use of intelligent data processing poses an abstract threat to anonymity, this is put into perspective on closer inspection. The use of ML is limited to the area within official statistics, so it has no influence on statistical confidentiality to the outside world when the data are published. Reidentification is also prohibited when using modern data technology. Therefore, when training the models, care must be taken to ensure that as little personal data as possible be used. The strict legal requirements of confidentiality for employees of official statistics also rule out reidentification by them. Overall, the appropriate guarantees can be ensured when applying ML.

7.3 AI Act

Another regulatory approach that affects ML is the European Union's Artificial Intelligence (AI) Act.[41] The EU Commission's original proposal[42] has undergone various changes as a result of the opinions of the Council[43] and the European Parliament.[44] The regulation was published on 12 June 2024 and, according to its Art. 113, entered into force (as is usual) 20 days after its publication—on 1 August 2024—and in general applies from 2 August 2026. As is typical of

[38] ECJ, 19 October 2016—C-582/14, ECLI:EU:C:2016:779 (para 31–49).

[39] EGC, 26 April 2023—T-557/20, ECLI:EU:T:2023:219 (para 94–105).

[40] ECJ, 19 October 2016—C-582/14, ECLI:EU:C:2016:779 (para 46–49).

[41] Regulation (EU) 2024/1689 of the European Parliament and of the Council of 13 June 2024 laying down harmonized rules on artificial intelligence and amending Regulations (EC) No. 300/2008, (EU) No. 167/2013, (EU) No. 168/2013, (EU) 2018/858, (EU) 2018/1139 and (EU) 2019/2144 and Directives 2014/90/EU, (EU) 2016/797 and (EU) 2020/1828 (Artificial Intelligence Act), OJ L, 2024/1689, 12.7.2024.

[42] COM(2021) 206 final. The proposal was adopted on 21 April 2021.

[43] Available at https://www.consilium.europa.eu/en/press/press-releases/2022/12/06/artificial-intelligence-act-council-calls-for-promoting-safe-ai-that-respects-fundamental-rights/, as of 24 January 2025.

[44] P9_TA(2023)0236; available at https://www.europarl.europa.eu/RegData/seance_pleniere/textes_adoptes/definitif/2023/06-14/0236/P9_TA(2023)0236_EN.pdf, as of 24 January 2025.

secondary legislation, the regulation is divided into recitals, which help to interpret and understand the text of the legislative act, and legally binding articles. Being a regulation, the AI Act is directly applicable in the individual member states and does not have to be transposed first by the member states to become national law (Chalmers et al. 2024, p. 107).

The aim of the regulation is to establish uniform rules for the development, use, and marketing of AI in line with the values of the European Union.[45] At the same time, it intends to provide protection against the risks—both material and immaterial—that artificial intelligence can pose.[46] On the one hand, the AI Act is seen as a necessary regulatory measure aiming to contain the risks of a technology area that is growing (too) fast. On the other hand, it is seen as an obstacle to innovation, and there are fears that possible overregulation will lead to the migration of technology companies (Bomhard and Siglmüller 2024; Hacker 2023).

7.3.1 Scope and Structure of the AI Act

The personal scope of application is defined in Art. 2 and includes not only providers of AI systems but also operators and importers. Based on the assumption that an ML model is developed in-house and for its own purposes, the Federal Statistical Office would be regarded as a provider of an AI system (Art. 3, No. 3) and would also fall under the regulation in geographical terms, as a planned AI system would be used within the EU (Art. 1 (1) a).

Article 3 then provides definitions for the terms the regulation uses and in its No. 1 first and foremost defines the term *AI system*. While the Commission draft was still working with a reference to Annex I, the final legislative act defines an AI system conclusively: The definition is intended to differentiate it from traditional software and especially algorithms.[47] The decisive factor is that an AI system is machine-based, can act with a certain degree of autonomy, and has the ability to infer[48]—based on its input values—how to generate output that can influence the environment. ML is a subarea of artificial intelligence (Baum 2021, Part 9.1, para 8). Therefore, ML models used in official statistics would generally be subject to the AI Act.

AI systems that are used for statistical purposes might then fall under one of the exceptions that Art. 2 provides. Article 2 (6) exempts AI systems that are used for scientific research. Although the term *scientific research* is to be understood broadly (Wendehorst 2024, Art. 2 AI Act, para 84), European legislation has explicitly

---

[45] Recital 1.

[46] Recitals 4 and 5.

[47] Recital 12.

[48] Steen (2024) explains how inference works and why it differentiates an AI system from a mere algorithm.

included statistical purposes in other legislative acts like the GDPR. Thus, it can be concluded that statistical purposes have been left out on purpose, and Art. 2 (6) does not apply.

Article 2 (8) exempts AI systems from the scope of application before they are placed on the market or put into service, during research, testing, or development activities regarding those systems, only postponing applicability of the regulation. As a result, the AI Act would not be applicable while an AI system has not been put into service by a statistical organization. Besides that other exceptions are not relevant for AI used for statistical purposes.

The regulation takes a risk-based approach[49] and imposes stricter restrictions on AI systems the greater their potential threat to fundamental rights.[50] Accordingly, it distinguishes between four risk levels: unacceptable risk, high risk, limited risk, and low risk.[51] The higher the risk level when using an AI system, the more far-reaching the regulation. At which risk level ML for statistical purposes is to be classified will be explained below, based on the abovementioned methods of coding, plausibility checks, and imputation.

These methods undoubtedly do not classify as prohibited AI practices under Art. 5. Such practices are, in particular, methods of cognitive behavior control, social scoring, biometric identification and categorization, or biometric real-time remote identification (von Welser 2024). However, none of these fields of application correspond to the use cases of ML for statistical purposes.

The second risk class (high risk) includes systems that pose a high potential threat to fundamental rights.[52] As the main addressee of the regulation, providers of high-risk AI are subject to numerous restrictions in accordance with Art. 6 et seq. Article 6 (1) classifies AI systems as high risk that are used in products that fall under EU product safety regulations, such as aviation, vehicles, medical devices, and toys (Bomhard and Siglmüller 2024). Article 6 (2) extends the number of systems that are considered high risk with specific areas listed in Annex III of the regulation. The final version of the AI Act in Art. 6 (3) then provides for exceptions from this classification for systems that are listed in Annex III but comply with one of the following requirements (Bomhard and Siglmüller 2024):

- AI systems that have a very narrow range of tasks within a process
- Systems that aim to improve a result that was previously achieved purely by humans

[49] Recital 14.

[50] Recitals 6 and 48.

[51] COM(2021) 206 final, p. 15; Recital 26 of the final regulation. Although the European Commission used four risk categories in its proposal, the regulation does not use the term "limited risk." See also Ebert and Spiecker gen. Döhmann (2021). Limited risk refers to special uses of AI, such as chatbots and deepfakes, whose providers/deployers have to comply with transparency obligations according to Art. 50. See https://digital-strategy.ec.europa.eu/en/policies/regulatory-framework-ai, as of 24 January 2025.

[52] Recital 48.

- Systems that are intended to recognize decision-making patterns or deviations from them but are not intended to influence or replace human decisions or
- The AI system is intended to perform a preparatory task to an assessment relevant for the purposes of the use cases listed in Annex III.

The use cases of ML in statistics are support methods for human evaluation of statistical survey data. It is not clear to what extent they could be subsumed under one of the listed categories of high-risk AI. Additionally, it could be argued that classification or plausibility checks for data processing only represent a narrow field of application anyway or serve to improve procedures previously carried out purely by humans, which would justify an exception. Thus, classifying as neither unacceptable nor high risk, the application of ML in official statistics would have to be considered as low risk.

The European Parliament suggested to make ethical principles mandatory for the developers of AI systems; this, however, has not been included in the final version (Bomhard and Siglmüller 2024). Instead, it has been included in the recitals,[53] which are not legally binding.[54] The consideration of ethical guidelines is also included in the voluntary commitments of Art. 95. Ethical principles within the meaning of this regulation refer to the principles established by the High-Level Expert Group (HLEG) on AI in 2019.[55] These principles include humane conduct and oversight, technical robustness and security, data protection and data governance, transparency, diversity, nondiscrimination and fairness, social and environmental wellbeing, and accountability. Some of these principles already overlap with the principles of the European Statistics Code of Practice, for example, in terms of data protection and transparency. There are also parallels between these principles and those of the Quality Framework for Statistical ML (Saidani et al. 2023; Saidani and Dumpert 2025), particularly between explainability and transparency.[56]

7.3.2 Obligations for National Statistical Organizations

A new aspect that has been included in the regulation since the parliamentary proposal is the requirement for AI literacy (Art. 4) for employees and third parties who work with the AI system or are used to operate it. AI literacy refers to knowledge of the opportunities and risks and possible harm that the use of AI entails

[53] Recitals 8, 25, and 27.

[54] ECJ, 19 June 2014—C-345/13, ECLI:EU:C:2014:2013 (para 31).

[55] Recitals 7, 27. Ethics Guidelines for Trustworthy AI. Available at https://digital-strategy.ec.europa.eu/en/library/ethics-guidelines-trustworthy-ai, as of 24 January 2025.

[56] Recital 27.

in order to be able to make informed decisions when dealing with an AI system.[57] According to Art. 113, since it is part of Chapter 1, it applies from 2 February 2025 regardless of the risk category the AI system falls under (Wendehorst 2024, Art. 113 AI Act).

The regulation also aims to incentivize organizations to implement voluntary commitments. Article 95 thus requires that codes of conduct be made attractive for organizations as an optional measure. These are to be promoted by the AI Office[58] and the Member States and include the rules that also apply to high-risk AI. Although these are voluntary commitments, one incentive could be the increased trust of people whose data are used by an AI system. According to Art. 95 (2), the voluntary commitments through codes of conduct should have clearly measurable objectives such as

- Applicable elements provided for in Union ethical guidelines for trustworthy AI
- Sustainability of AI systems in terms of their energy consumption during development, training, and use
- The promotion of AI literacy
- Inclusive and diverse design of AI systems
- Risk prevention in the use of AI systems, in particular the nondiscriminatory design of AI systems

Overall, it can be stated that ML in official statistics is subject to the AI Act. However, the already existing and planned applications are low-risk AI methods. Consequently, they are not prohibited by the regulation and the obligations that the regulation places on high-risk AI do not apply to them either. Consequently, the requirements for the Federal Statistical Office as a provider of an AI system are primarily training of the responsible employees in the area of AI literacy. This needs to be complied with starting 02 February 2025, as it is part of the general provisions. In addition, the establishment of a code of conduct based on the HLEG's Ethics Guidelines should be considered as a voluntary commitment under Art. 95. However, this is not obligatory, and there are already some overlaps with other quality standards in official statistics.

References

L. Baum, Teil 9 – Künstliche Intelligenz, in *IT-Recht*, ed. by A. Leupold, A. Wiebe, S. Glossner, 4th edn. (C.H. Beck, Munich, 2021)

D. Bomhard, J. Siglmüller, AI Act – Das Trilogergebnis, in *Recht Digital (RDi)* (2024), pp. 45–96

[57] Article 3, No. 56.

[58] The AI Office is a function of the European Commission contributing to the implementation, monitoring, and supervision of AI systems and general-purpose AI models, and AI governance (Article 3, No. 47).

B. Buchner,d M.-T. Tinnefeld, Art. 89 GDPR, in *Datenschutzgrundverordnung/BDSG – Kommentar*, ed. by J. Kühling, B. Buchner, 4th edn. (C.H. Beck, Munich, 2024)

D. Chalmers, G. Davies, G. Monti, V. Heyvaert, *European Union Law: Text and Materials* (Cambridge University Press, Cambridge, 2024)

M.P. Couper, E. Singer, F.G. Conrad, R.M. Groves, Risk of disclosure, perceptions of risk, and concerns about privacy and confidentiality as factors in survey participation. J. Official Stat. **24**(2), 255–275 (2008)

P. Dorer, H. Mainusch, H. Tubies, *Bundesstatistikgesetz – Kommentar* (C.H. Beck, Munich, 1988)

F. Dumpert, Machine Learning in der amtlichen Statistik – Ergebnisse und Bewertung eines internationalen Projekts. WISTA Wirtschaft und Statistik **73**(4), 53–63 (2021)

F. Dumpert, M. Beck, Einsatz von Machine-Learning-Verfahren in amtlichen Unternehmensstatistiken. AStA Wirtschafts- und Sozialstatistisches Archiv **2**(11), 83–106 (2017)

A. Ebert, I. Spiecker gen. Döhmann, Der Kommissionsentwurf für eine KI-Verordnung der EU. NVwZ Neue Zeitschrift für Verwaltungsrecht (2021), pp. 1188–1193

P. Hacker, Die Regulierung von ChatGPT et al. – ein europäisches Trauerspiel, in *GRUR-Gewerblicher Rechtsschutz und Urheberrecht* (2023), pp. 289–290

C. Isfort, J. Kühling, C. Schmid, Introduction, in *Bundestatistikgesetz – Kommentar*, ed. by J. Kühling (C.H. Beck, Munich, 2023)

M. Karg, Art. 4 No. 1 GDPR, in *Datenschutzrecht – Kommentar*, ed. by S. Simitis, G. Hornung, and I. Spiecker gen. Döhmann, 2nd edn. (Nomos, Glashütte, 2025)

T. Kingreen, Art. 338 TFEU, in *EUV/AEUV – Kommentar*, ed. by C. Callies, M. Ruffert, 6th edn. (C.H. Beck, Munich, 2022)

M. Klar, J. Kühling, Art. 4 GDPR, in *Datenschutzgrundverordnung/BDSG – Kommentar*, ed. by J. Kühling, B. Buchner, 4th edn. (C.H. Beck, Munich, 2024)

J. Kühling, Introduction, in *Bundestatistikgesetz – Kommentar*, ed. by J. Kühling (C.H. Beck, Munich, 2023)

J. Kühling, C. Sauerborn, §16 BStatG, in *Bundestatistikgesetz – Kommentar*, ed. by J. Kühling (C.H. Beck, Munich, 2023)

J. Kühling, C. Schmid, §1 BStatG, in *Bundestatistikgesetz – Kommentar*, ed. by J. Kühling (C.H. Beck, Munich, 2023)

A. Narayanan, V. Shmatikov, Robust de-anonymization of large sparse datasets, in *2008 IEEE Symposium on Security and Privacy (sp 2008)*. IEEE (2008), pp. 111–125

D. Pauly, Art. 89 GDPR, in *Datenschutz-Grundverordnung, Bundesdatenschutzgesetz – Kommentar*, ed. by B. Paal, D. Pauly, 3rd edn. (C.H. Beck, Munich, 2021)

M. Preising, K. Lange, F. Dumpert, Imputation zur maschinellen Behandlung fehlender und unplausibler Werte in der amtlichen Statistik. WISTA Wirtschaft und Statistik **73**(5), 40–52 (2021)

W.J. Radermacher, Governance in der amtlichen Statistik. AStA Wirtschafts- und Sozialstatistisches Archiv **11**(2), 65–81 (2017)

B. Raji, Rechtliche Bewertung synthetischer Daten für KI-Systeme. DuD Datenschutz und Datensicherheit **45**(5), 303–309 (2021)

T. Ramge, Mensch fragt, Maschine antwortet. APuZ Aus Politik und Zeitgeschichte **S**, 15–21 (2018)

Y. Saidani, F. Dumpert, Quality dimensions and quality guidelines for machine learning in official statistics, in *Foundations and Advances of Machine Learning in Official Statistics*, ed. by F. Dumpert, Chap. 4 (Springer, Berlin, 2025)

Y. Saidani, F. Dumpert, C. Borgs, A. Brand, A. Nickl, A. Rittmann, J. Rohde, C. Salwiczek, N. Storfinger, S. Straub, Qualitätsdimensionen maschinellen Lernens in der amtlichen Statistik. AStA Wirtschafts- und Sozialstatistisches Archiv **17**(3), 253–303 (2023)

P. Schantz, H.A. Wolff, *Das neue Datenschutzrecht – Datenschutz-Grundverordnung und Bundesdatenschutzgesetz in der Praxis* (C.H. Beck, Munich, 2017)

R. Söllner, T. Körner, Der Registerzensus: Ziele, Anforderungen und Umsetzungsansätze. WISTA Wirtschaft und Statistik **74**(4), 13–24 (2022)

C. Song, T. Ristenpart, V. Shmatikov, Machine learning models that remember too much, in *Proceedings of the 2017 ACM SIGSAC Conference on Computer and Communications Security* (2017), pp. 587–601

Statistische Ämter des Bundes und der Länder, *Qualitätshandbuch der Statistischen Ämter des Bundes und der Länder*. Statistische Ämter des Bundes und der Länder (2021)

A. Steen, Ableitungen als wesentliche Fähigkeit von KI-Systemen nach der KI-VO – Begriffsbestimmung und Darstellung der verschiedenen Ableitungsprinzipien. *KIR Künstliche Intelligenz und Recht* (2024), pp. 7–11

L. Sweeney, k-anonymity: a model for protecting privacy. Int. J. Uncertain. Fuzziness Knowl.-Based Syst. **10**, 557–570 (2002)

M. von Welser, Die KI-Verordnung – ein Überblick über das weltweit erste Regelwerk für künstliche Intelligenz. *GRUR-Prax Gewerblicher Rechtsschutz und Urheberrecht in der Praxis* (2024), pp. 485–489

T. Weichert, Die Forschungsprivilegierung in der DS-GVO. Gesetzlicher Änderungsbedarf bei der Verarbeitung personenbezogener Daten für Forschungszwecke. ZD Zeitschrift für Datenschutz **10**(1), 18–23 (2020)

C. Wendehorst, Art. 113, in *KI-VO: Verordnung über Künstliche Intelligenz – Kommentar*, ed. by C. Wendehorst, M. Martini (C.H. Beck, Munich, 2024)

C. Wendehorst, Art. 2, in *KI-VO: Verordnung über Künstliche Intelligenz – Kommentar*, ed.by C. Wendehorst, M. Martini (C.H. Beck, Munich, 2024)

Part III
Technological Aspects

Chapter 8
A Cloud-Native Data Science Platform for Official Statistics

Romain Avouac, Thomas Faria, and Frédéric Comte

8.1 Introduction

In recent years, the European Statistical System (ESS) has committed to leverage nontraditional data sources in order to improve the process of statistical production, an evolution that is encapsulated by the concept of Trusted Smart Statistics (Ricciato et al. 2019). This dynamic is accompanied by innovations in the statistical processes, so as to be able to take advantage of the great potential of these sources (greater timeliness, increased spatio-temporal resolution, etc.), but also to cope with their complexity or imperfections. At the forefront of these innovations are machine learning methods and their promising uses in the coding and classification fields, data editing and imputation (Gjaltema 2022). The multiple challenges faced by statistical offices because of this evolution are addressed in the Bucharest Memorandum on Official Statistics in a Datafied Society (Trusted Smart Statistics), which predicts that "the variety of new data sources, computational paradigms and tools will require amendments to the statistical business architecture, processes, production models, IT infrastructures, methodological and quality frameworks, and the corresponding governance structures," and consequently invites the ESS to assess the required adaptations and prioritize them (DGINS 2018).

In line with these recommendations, much work has been done in the context of successive projects at the European level in order to operationalize the use of nontraditional data sources in the production of official statistics. Within the scope of the ESSnet Big Data II project (2018–2020), National Statistical Offices (NSOs) have been working across a wide range of themes (online job vacancies, smart energy, tracking ships, etc.) in order to put together the building blocks for

---

R. Avouac (✉) · T. Faria · F. Comte
INSEE (National Institute of Statistics and Economic Studies), Montrouge, France
e-mail: romain.avouac@insee.fr; thomas.faria@insee.fr; frederic.comte@insee.fr

F. Dumpert (ed.), *Foundations and Advances of Machine Learning in Official Statistics*, Society, Environment and Statistics,
https://doi.org/10.1007/978-3-032-10004-7_8

using these sources in actual production processes and identify their limitations (EUROSTAT 2021). However, while a substantial amount of work has been devoted to developing methodological frameworks (Descy et al. 2019; Salgado et al. 2020), quality guidelines (Kowarik and Six 2022), as well as devising business architectures that make third-party data acquisition more secure (Ricciato et al. 2018), not much has been said about the IT infrastructures and skills needed to properly deal with these new objects.

The characteristics of big data sources make them particularly complex to process, be it their volume, their velocity (speed of creation or renewal), or their variety (structured but also unstructured data, such as text and images). Besides, the "skills and competencies to automate, analyze, and optimize such complex systems are often not part of the traditional skill set of most National Statistical Offices" (Ashofteh and Bravo 2021). Not incidentally, an increasing number of public statisticians trained as data scientists have joined NSOs in recent years. Within its multiple meanings, the term "data scientist" reflects the increased involvement of statisticians in the IT development and orchestration of their data processing operations, beyond merely the design or validation phases (Davenport and Patil 2012). However, based on our observations at INSEE and other French statistical offices, the ability of these new data professionals to derive value from big data sources and machine learning methods is limited by several challenges.

A first challenge is related to the lack of proper IT infrastructures to tackle the new data sources that NSOs now have access to as well as the accompanying need for new statistical methods. Big data sources, including both nontraditional data sources and "traditional big data" sources (e.g., tax data, exhaustive census data, etc.), require huge storage capacities and often rely on distributed computing frameworks to be processed (Liu 2013). Similarly, the adoption of new statistical methods based on machine learning algorithms often requires IT capacities—in particular, GPUs (graphical processing units)—to massively parallelize computations (Saiyeda and Mir 2017). Such resources are not readily available in traditional IT infrastructures. Furthermore, these new infrastructures generally require specific skills—especially to build and maintain them—that are not easily found in NSOs.

Another major challenge lies in equipping statisticians with development environments that allow them to experiment more freely. The essence of innovation in statistical work lies in the ability to swiftly adapt to and incorporate new tools and methodologies. This agility is hampered when statisticians depend excessively on IT departments to provision resources or install new software packages. In traditional setups—personal computers or virtual desktops on centralized architectures—IT departments generally prioritize security and system stability over the provision of new services, which limits the innovation potential. In addition, these rigid environments make it harder to implement and develop best practices, such as collaborative work—which requires environments where experiments can be easily shared with peers—and reproducibility.

A third challenge is related to the difficulty of transitioning from innovative experiments to production-grade solutions. Even when statisticians have access to development environments in which they can readily experiment, the step toward deploying an application or a model is generally very large. Production environments often differ from development environments in such a way that the additional development costs needed to go from a proof of concept to an industrialized solution that actually serves users can limit the feasibility of this transition. Furthermore, in the case of machine learning projects, models that have been deployed require proper monitoring to ensure that they maintain their accuracy and utility over time, and generally require periodic or continuous improvements. Again, this pleads for more flexible environments that enable statisticians to manage the complete life cycle of their data science projects in a more continuous way.

We argue that these various challenges have an underlying common theme: the need for more autonomy. The ability of data science methods to improve and potentially transform the production of official statistics crucially depends on the ability of statisticians to carry out innovative experiments more freely. To do so, they need to have access to substantial and diverse computing resources that enable them to tackle the volume and diversity of big data sources and leverage machine learning methods. Such experimental projects require, in turn, flexible development environments that foster collaborative work in order to capitalize the diversity of profiles and skills that compose project teams. Finally, to derive value from these experiments, statisticians require tools to deploy applications as proof of concepts and orchestrate their statistical operations autonomously.

Against this background, INSEE developed Onyxia: an open-source project that enables organizations to deploy data science platforms that foster innovation by giving statisticians more autonomy.[1] This chapter aims at describing the full thought process that led to this project and at exemplifying how it empowers statisticians at INSEE, thus becoming a cornerstone of our innovation strategy. Section 8.2 provides an in-depth analysis of the data ecosystem's latest developments, casting light on the technological choices that have shaped the development of a modern data science environment tailored to the specific needs of statisticians. In particular, we show how cloud-native technologies—particularly containers and object storage—are key to building scalable and flexible environments that can enhance autonomy while promoting reproducibility in the production of official statistics. However, despite their appealing attributes for modern data science applications, the complexity of configuring and utilizing cloud technologies often poses barriers to their broad adoption. In Sect. 8.3, we detail the core of the Onyxia project: How we made cloud technologies accessible to statisticians through a user-friendly interface and an extensive catalogue of ready-to-use data science environments, while circumventing potential vendor lock-in effects for both the institution and their users. We also show how providing an open-innovation instance of Onyxia, the SSP Cloud, greatly facilitated the adoption of these technologies and fostered improved development practices. Finally, through the case study of the classification of French companies' activity (NACE), Sect. 8.4 illustrates how leveraging these

[1] https://github.com/InseeFrLab/onyxia

technologies greatly facilitated the deployment of machine learning models at INSEE in alignment with the industry best practices—namely, MLOps principles.

8.2 Principles for Building a Modern and Flexible Data Architecture for Official Statistics

With the emergence of big data sources and new methodologies offering significant promise to improve the production process of official statistics, statisticians trained in data science techniques are eager to innovate. However, their ability to do so is limited by several challenges. Central among these challenges is the need for greater autonomy—be it in scaling resources to match statistical workloads, deploying proofs of concept with agility and in a collaborative manner, etc. Against this background, our aim was to design a data science platform that not only manages big data efficiently but also empowers statisticians by enhancing their autonomy. To achieve this, we looked at the evolution of the data ecosystem to identify significant trends that could help overcome the above-mentioned limitations.[2] Our findings indicate that leveraging cloud-native technologies, particularly containers and object storage, is key to building infrastructures capable of handling large and varied datasets in a flexible and cost-effective manner. Furthermore, these technologies significantly enhance autonomy, facilitating innovation and promoting reproducibility in the production of official statistics.

8.2.1 Limitations of Traditional Big Data Architectures

Over the last decade, the landscape of big data has dramatically transformed. Following the publication of Google's seminal papers that introduced the MapReduce paradigm (Ghemawat et al. 2003; Dean and Ghemawat 2008), Hadoop-based systems rapidly became the reference architecture of the big data ecosystem, celebrated for their capability to manage extensive datasets through the use of distributed computing. The introduction of Hadoop marked a revolutionary step, enabling organizations to process and analyze data at an unprecedented scale. Basically, Hadoop provided companies with all-rounded capabilities for big data analytics: tools for ingestion, data storage (HDFS), and computing capacities (Spark, among

---

[2] As a preamble to this review, we should note that, although we did our best to ground our insights in the academic literature, a lot of it stems from informal knowledge gathered through diligent and ongoing technology watch. In the rapidly evolving data ecosystem, traditional research papers are increasingly giving way to blog posts as the primary references for cutting-edge developments. This shift is largely due to the swift pace at which big data technologies and methodologies such as machine learning are advancing, making the lengthy publication process of formal research often not the preferred way of disseminating timely insights and innovations.

others) (Dhyani and Barthwal 2014), thus explaining its rapid adoption across industries.

In the late 2010s, Hadoop-based architectures have experienced a clear decline in popularity. In traditional Hadoop environments, storage and compute were co-localized by design: If the source data is distributed across multiple servers (horizontal scaling), each section of the data is directly processed on the machine hosting that section, so as to avoid network transitions between servers. In this paradigm, scaling the architecture often meant a linear increase in both compute and storage, regardless of the actual demand. In a recent article provocatively titled "Big Data is Dead,"[3] Jordan Tigani, one of the founding engineers behind Google BigQuery, explains why this model does not fit the reality of most data-centric organizations anymore. First, because "in practice data sizes increase much faster than compute sizes." While the amount of data generated and thus needing to be stored may grow linearly over time, it is generally the case that we only need to query the most recent portions of it, or only some columns and/or groups of rows. Besides, Tigani points out that "the big data frontier keeps receding": Advancements in server computing capabilities and declining hardware costs mean that the number of workloads that do not fit on a single machine—a simple yet effective definition of big data—has been continually decreasing. As a result, by properly separating storage and compute functions, even substantial data processing jobs may end up using "far less compute than anticipated [...] and might not even need to use distributed processing at all."

These insights strongly align with our own observations at INSEE in recent years. For instance, an INSEE team set up a Hadoop cluster as an alternative architecture to the one already in use to process sales receipt data in the context of computing the consumer price index. An acceleration of data processing operations by up to a factor of 10 was achieved, for operations that previously took several hours to perform (Leclair et al. 2019). Despite this increase in performance, this type of architectures was not reused later for other projects, mainly because the architecture proved to be expensive and complex to maintain, necessitating specialized technical expertise rarely found within NSOs (Vale 2015). Although these new projects could still involve substantial data volumes, we observed that effective processing could be achieved using conventional software tools (R, Python) on single-node systems by leveraging recent important innovations from the data ecosystem. First, by using efficient formats to store the data such as Apache Parquet (Foundation 2013), which properties—columnar storage (Abadi et al. 2013), see Fig. 8.1, optimization for "write once, read many" analytics, ability to partition data, etc.—make it particularly suited to analytical tasks such as those generally performed in official statistics (Abdelaziz et al. 2023). Second, by performing computations using optimized in-memory computation frameworks such as Apache Arrow (Foundation 2016) or DuckDB (Raasveldt and Mühleisen 2019). Also based on columnar representation—thus working in synergy with

[3] https://motherduck.com/blog/big-data-is-dead/, assessed 21 January 2025.

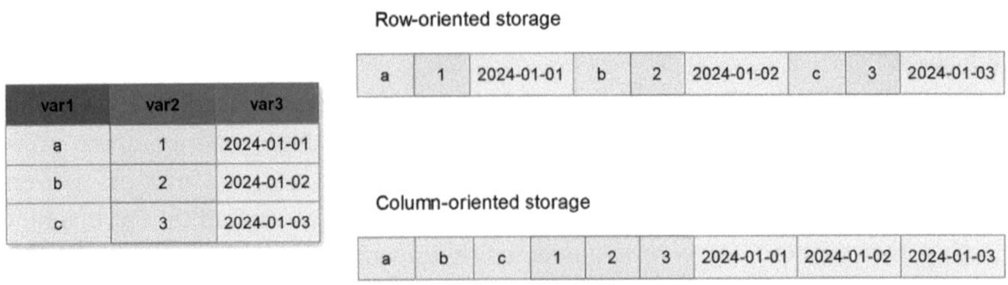

Fig. 8.1 Row-oriented and column-oriented representation of a same dataset. Note: Many statistical operations are analytical (OLAP) in nature: They involve selecting specific columns, computing new variables, performing group-based aggregations, etc. Row-oriented storage is not well-suited to analytical operations as it requires the full dataset to be read in memory to query it. Conversely, column-based storage allows only relevant data columns to be queried, significantly reducing read and processing times for analytical workloads. In practice, popular columnar formats such as Parquet use a hybrid representation: They are primarily column-oriented but also implement clever row-based grouping to optimize filtering queries

Parquet files—both of these frameworks greatly improve data queries performance through the use of "lazy evaluation": Instead of doing lots of separate operations (e.g., selecting columns and/or filtering rows, then computing new columns, then performing aggregations, etc.), they process them all at once in a more optimized way. As a result, computations are limited to the data effectively needed by the queries, enabling much larger-than-memory data processing on usual single-node machines.

8.2.2 *Embracing Cloud-Native Technologies*

In light of this evolution of the big data ecosystem, there has been a notable shift in recent years within the industry toward more flexible and loosely coupled architectures. The advent of cloud technologies has been instrumental in facilitating this shift. Unlike the era where Hadoop was prominent, network latency has become much less of a concern, making the traditional model of on-premise and co-located storage and compute solutions less relevant. Regarding the nature of the data that needs to be processed, we observe an evolution that some have described as moving "from big data to flexible data." Modern data infrastructures are required not only to process large volumes but also to be adaptable in multiple dimensions. They must accommodate various data structures (ranging from structured, tabular formats to unstructured formats like text and images), ensure data portability across multicloud and hybrid cloud environments, and support a diverse range of computational workloads (from parallel computations to deep learning models necessitating GPUs, as well as the deployment and management of applications) (Li et al. 2020). In recent years, two technologies have emerged in the data ecosystem as foundational technologies to achieve such flexibility in cloud-based environments: containerization and object storage.

In a cloud environment, the computer of the user becomes a simple access point to perform computations on a central infrastructure. This enables both ubiquitous access to and scalability of the services, as it is easier to scale a central infrastructure—usually horizontally, i.e., by adding more servers. However, such centralized infrastructures have two well-identified limitations that need to be dealt with: the competition between users in access to physical resources and the need to properly isolate deployed applications. The choice of containerization is fundamental as it tackles these two issues (Bentaleb et al. 2022). By creating "bubbles" specific to each service, containers guarantee application isolation while remaining lightweight, as they share the support operating system with the host machine (see Fig. 8.2). In order to manage multiple containerized applications in a systematic way, containerized infrastructures generally rely on an orchestrator software—the most prominent one being Kubernetes, an open-source project initially developed by Google to manage its numerous containerized workloads in production (Vaño et al. 2023). Orchestrators automate the process of deploying, scaling, and managing

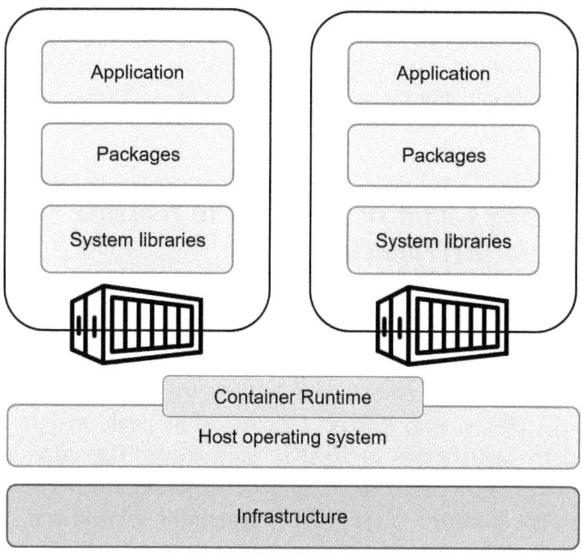

Fig. 8.2 Architecture of a containerized environment. Note: A container is a logical grouping of resources that makes it possible to encapsulate an application (e.g., Python code), the packages used (e.g., Pandas, NumPy), and system libraries (the Python interpreter, other OS-dependent libraries, etc.), in a single package. Containerized applications are isolated from one another through virtualization, which makes it possible to attribute specific physical resources to each application while guaranteeing complete independence between them. But contrary to virtual machines which also virtualize the operating system (OS), containers rely on a lightweight form of virtualization: The container shares the OS of the host infrastructure through the container runtime (e.g., Docker). As a result, containers are much more portable and can be readily deployed and redistributed

containerized applications, coordinating their execution across various servers. Interestingly, this property makes it possible to handle very large volumes of data in a distributed way: containers break down big data processing operations into a multitude of small tasks, organized by the orchestrator. This minimizes the required resources while providing more flexibility than Hadoop-based architectures (Zhang et al. 2018).

The other fundamental choice in a data architecture is the nature of data storage. In the cloud ecosystem, the so-called object storage has become the de facto reference (Samundiswary and Dongre 2017).[4] In this paradigm, files are stored as "objects" consisting of data, an identifier, and metadata. This type of storage is optimized for scalability, as objects are not limited in size and the underlying technology enables cost-effective storage of (potentially very) large files. It is also instrumental in building a decoupled infrastructure such as discussed before: The data repositories—referred to as "buckets"—are directly searchable using standard HTTP requests through a standardized REST API. In a world where network latency is not the main bottleneck anymore, this means that storage and compute do not have to be on the same machines or even in the same location and can thus scale independently according to specific organization demands. Finally, object storage is a natural complement to architectures based on containerized environments for which it provides a persistence layer—containers are stateless by design—and easy connectivity without compromising security, or even with increased security compared to a traditional storage system (Mesnier et al. 2003).

8.2.3 Leveraging Cloud Technologies to Increase Autonomy and Foster Reproducibility

Understanding how the technological choices described in the technical discussion above are relevant in the context of official statistics requires an in-depth review of the professional practices of statisticians in their use of computing environments. At the end of the 2000s, with microcomputing at its peak, many of the technical resources used by statisticians at INSEE were local: The code and processing software were located on personal computers, while data was accessed through a file-sharing system. Because of the limited scalability of personal computers, this setup greatly limited the ability of statisticians to experiment with big data sources or computationally intensive statistical methods and involved security risks because of the widespread data dissemination within the organization. In order to overcome these limitations, a transition was made toward centralized IT infrastructures, concentrating all—and thus overall much more—resources on central servers. Such infrastructures, made available to statisticians through a shared, virtual desktop

[4] Mainly because of Amazon's "S3" (Simple Storage Service) implementation.

environment for ease of use, remain the dominant method for conducting statistical computations at INSEE at the time of writing these lines.

Through our observations and discussions with fellow statisticians, it became obvious that although the current IT infrastructure adequately supported the core activities of statistical production, it noticeably restricted statisticians' capacity to experiment freely and innovate. The primary bottleneck in this organization is the dependency of statistical projects on centralized IT decision-making, such as the allocation of computing resources, access to shared data storage, the use of preconfigured programming languages and packaging environments, etc. Besides, such dependencies often lead to a well-known phenomenon within the software development community, where the priorities of developers—iterate rapidly to improve functionality in a continuous manner—often clash with IT's focus on security and process stability. On the contrary, it is our understanding that modern data science practices reflect an increased involvement of statisticians in the IT development and orchestration of their data processing operations, beyond merely the design or validation phases. New data science infrastructures must take this expanded role of their users into account, giving them more autonomy than conventional infrastructures.

We argue that cloud technologies stand out as a powerful solution to give statisticians much more autonomy in their daily work, enabling a culture of innovation. Through object storage, users gain control over the storage layer, allowing them to experiment with diverse datasets without being constrained by the limited storage spaces typically allocated by IT departments. Containerization empowers users to customize their working environments to their specific needs—be it programming languages, system libraries, or package versions—while also providing the flexibility to scale their applications according to the required computing power and storage capacities. By design, containers also foster the development of portable applications, which enables smoother transitions between environments (development, testing, staging, production), ensuring that applications can be moved seamlessly without the hurdles of environmental inconsistencies. Finally, with orchestration tools like Kubernetes, statisticians can deploy applications and APIs more easily and automate the whole building process. This capability aligns with the DevOps approach, which advocates building proofs of concept in an iterative manner, rather than building the optimal (but time-consuming) solution for a predefined objective (Leite et al. 2019).

Besides scalability and autonomy, these architectural choices also foster reproducibility of statistical computations. The concept of reproducibility—namely the ability to reproduce the result of an experiment by applying the same methodology to the same data—is a fundamental criterion of scientific validity (McNutt 2014). It is also highly relevant in official statistics, as it serves as a foundation for transparency, which in turn is crucial for building and maintaining public trust (European Commission 2018; Yung et al. 2022; Saidani and Dumpert 2025). Fostering reproducibility in statistical production involves devising processing solutions that can produce reproducible statistics on the one hand, and that can be shared with peers on the other hand (Luhmann et al. 2019). Traditional IT infrastructures—either a personal computer or a shared infrastructure with remote desktop access—fall short in this regard, as building a project or just computing a statistical indicator there generally involves a series of manual steps (installing system libraries, the programming language binary, projects packages, dealing with potentially conflicting versions, etc.) that cannot be fully reproduced across projects. In comparison, containers are reproducible by design, as their build process involves defining precisely all the needed resources as a set of processing operations in a standardized manner, from the "bare machine" to the running application (Moreau et al. 2023) (Fig. 8.3). Furthermore, these reproducible environments can be easily shared to peers as they can be readily published on open registries (e.g., a container registry such as DockerHub) along with the source code of the application (e.g., on a public software forge like GitHub or GitLab). This approach significantly enhances the reusability of code projects, fostering a community-driven model of development and innovation.

8.3 Onyxia: An Open-Source Project to Build Cloud-Native Data Science Platforms

This section explores how Onyxia, an open-source project initiated at INSEE, democratizes access to cloud technologies for statisticians by providing modern data science environments that foster autonomy. We discuss how this initiative fits in with the general aim of creating "knowledge commons" by promoting and building software that can be easily reused in the field of official statistics and beyond.

8.3.1 Making Cloud Technologies Accessible to Statisticians

Our technology watch and literature review highlighted cloud-native technologies, in particular containerization and object storage, as instrumental in building a data science platform that is both scalable and flexible. Building on these insights, we established our initial on-premise Kubernetes cluster in 2020, integrating it with MinIO, an open-source object storage system designed to work seamlessly with Kubernetes. Yet, our first experiments highlighted a significant barrier to the widespread adoption of cloud-native technologies: the complexity of their integration. This is an important consideration when building data architectures that prioritize modularity—an essential feature for the flexibility we aim to achieve.[5]

[5] A telling example of the importance of building a modular architecture is the ability to switch between storage sources (on-premise, public cloud provider, etc.). The storage solution we chose, MinIO, is compatible with Amazon's S3 API, which has become a de facto standard in the cloud ecosystem due to the success of Amazon's AWS S3 storage solution. As a result, organizations that

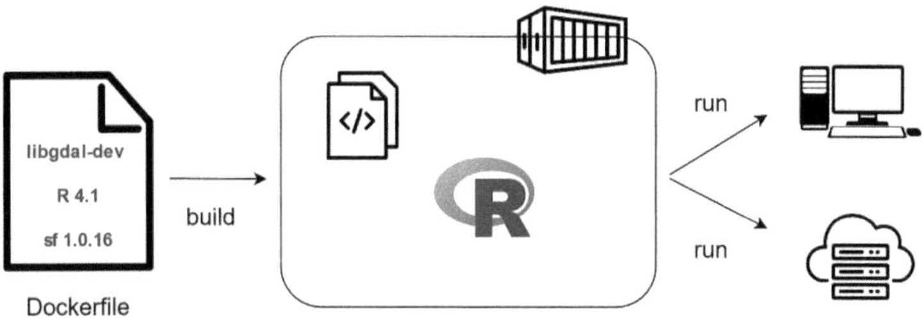

Note: In a containerized environment, applications are created through script specifications – a paradigm known as "infrastructure as code". In a text file conventionally named "Dockerfile", data scientists can specify the working environment of their application: the application code, the software to be included (e.g. R), the packages used for their processing operations (e.g. the R package for geospatial computation *sf*), and the OS-dependent system libraries that are called by these packages (e.g. GDAL, the translator library for raster geospatial data formats used by most packages dealing with geospatial data). Importantly, the versions of the software and packages that were used to develop the application can be precisely specified, which guarantees reproducibility of the computations performed. A build step then creates an image associated to the Dockerfile, i.e. a packaged and compressed form of the working environment of the application. Images created this way are portable: they can be readily distributed – usually through a container registry – and executed in a reproducible manner on any infrastructure that has a container runtime. — The R logo, © 2016 The R Foundation, is used without any changes according to the Creative Commons Attribution-ShareAlike 4.0 International license (CC-BY-SA 4.0), `https://creativecommons.org/licenses/by-sa/4.0/`.

Fig. 8.3 Containers foster reproducibility and portability by design. Note: In a containerized environment, applications are created through script specifications—a paradigm known as "infrastructure as code." In a text file conventionally named "Dockerfile," data scientists can specify the working environment of their application: the application code, the software to be included (e.g., R), the packages used for their processing operations (e.g., the R package for geospatial computation *sf*), and the OS-dependent system libraries that are called by these packages (e.g., GDAL, the translator library for raster geospatial data formats used by most packages dealing with geospatial data). Importantly, the versions of the software and packages that were used to develop the application can be precisely specified, which guarantees reproducibility of the computations performed. A build step then creates an image associated to the Dockerfile, i.e., a packaged and compressed form of the working environment of the application. Images created this way are portable: They can be readily distributed—usually through a container registry—and executed in a reproducible manner on any infrastructure that has a container runtime—The R logo, © 2016. The R Foundation is used without any changes according to the Creative Commons Attribution-ShareAlike 4.0 International license (CC-BY-SA 4.0), https://creativecommons.org/licenses/by-sa/4.0/

However, modularity of the architecture components also entails that any data application launched on the cluster must be configured so as to communicate with all the components. For instance, in a big data setup, configuring Spark to operate on

choose to use Onyxia are not tied to a specific storage solution: They can choose any solution that complies with the standards defined by the S3 API.

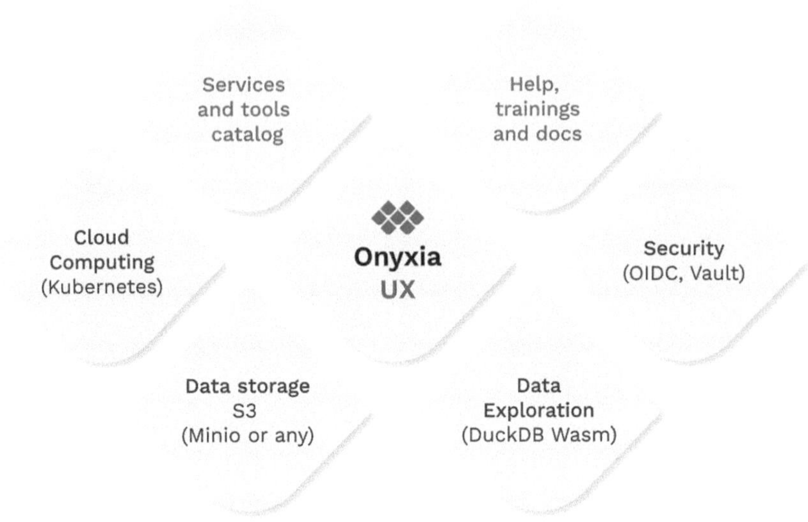

Fig. 8.4 Onyxia is the technical binder between cloud-native modular components

Kubernetes while interacting with datasets stored in MinIO requires an intricate set of configurations (specifying endpoints, access tokens, etc.), a skill set that typically lies beyond the expertise of statisticians.

For instance, due to MinIO's compatibility with the Amazon S3 API, the storage source could easily be switched to one managed by another public cloud provider, without requiring substantial modifications.

This insight is really the base of the Onyxia project: Choosing technologies that foster autonomy will not actually foster autonomy if their complexity acts as a barrier from widespread adoption in the organization. In recent years, statisticians at INSEE already needed to adapt to a changing environment in terms of their everyday tools: transitioning from proprietary software (SAS®) to open-source ones (R, Python), acculturating to technologies that improve reproducibility (version control with Git), consuming and developing APIs, etc. These changes, making their job more and more akin to the one of software developers, already imply significant training and changes in daily work practices. Against this background, adoption of cloud technologies was utterly dependent on making them readily accessible.

To bridge this gap, we developed Onyxia, an application that essentially acts as interface between the modular components that compose the architecture (see Fig. 8.4). The main entry point of the user is a user-friendly web application[6] that enables users to launch services from a data science catalog (see Sect. 8.3.3) as running containers on the underlying Kubernetes cluster. The interface between the

[6] https://github.com/InseeFrLab/onyxia-ui

UI and Kubernetes is done by a lightweight custom API[7] that essentially transforms the application request of the user into a set of manifests to deploy Kubernetes resources. For a given application, these resources are packaged under the form of Helm charts, a popular way of packaging potentially complex applications on Kubernetes (Gokhale et al. 2021). Although users can configure a service to tailor it to their needs, they will most of the time just launch an out-of-the-box service with default settings and start developing straight away. This point really illustrates the added value of Onyxia in facilitating the adoption of cloud technologies. By injecting authentication information and configuration into the containers at the initialization, we ensure that users can launch and manage data science services in which they can interact seamlessly with the data from their bucket on MinIO, their sensitive information (tokens, passwords) in a secret management tool such as Vault, etc. This automatic injection, coupled with the preconfiguration of data science environments in Onyxia's catalogs of images[8] and associated Helm charts,[9] makes it possible for users to execute potentially complex workloads—such as running distributed computations with Spark on Kubernetes using data stored in S3, or training deep learning models using a GPU—without getting bogged down by the technicalities of configuration.

8.3.2 Architectural Choices Aimed at Fostering Autonomy

The Onyxia project is based on a few structuring principles, with a central theme: fostering autonomy, at both the organizational and individual levels. First, at the level of the organization by preventing vendor lock-in. In order to get a competitive edge, many commercial cloud providers develop applications and protocols that customers need to use to access cloud resources, but that are not interoperable, greatly complexifying potential migrations to another cloud platform (Opara-Martins et al. 2016). Recognizing these challenges, there is a trend toward endorsing cloud-neutral strategies (Opara-Martins et al. 2017) in order to reduce reliance on a single vendor's specific solutions. In contrast, the use of Onyxia is inherently not restrictive: When an organization chooses to use it, it chooses the underlying technologies—containerization and object storage—but not the solution. The platform can be deployed on any Kubernetes cluster, either on-premise or in public clouds. Similarly, Onyxia was designed to be used with MinIO because it is an open-source object storage solution, but is also compatible with object storage solutions from various cloud providers (AWS, GCP).

Onyxia also fosters autonomy at the level of users. Proprietary software that have been used intensively in official statistics—such as SAS or STATA—also produce

[7] https://github.com/InseeFrLab/onyxia-api

[8] https://github.com/InseeFrLab/images-datascience

[9] https://github.com/InseeFrLab/helm-charts-interactive-services

a vendor lock-in phenomenon. The costs of licensing are high and can evolve quickly, and users are tied in certain ways of performing computations, preventing progressive upskilling. On the contrary, Onyxia aspires to be removable; we want to enhance users' familiarity and comfort with the underlying cloud technologies rather than act as a permanent fixture in their workflow. An illustrative example of this philosophy is the platform's approach to user actions: For tasks performed through the UI, such as launching a service or managing data, we provide users with the equivalent terminal commands, promoting a deeper understanding of what actually happens on the infrastructure when triggering something. Furthermore, all the services offered through Onyxia's catalog are open-source.

Naturally, the way Onyxia makes statisticians more autonomous in their work depends on their needs and familiarity with IT skills. Statisticians that just want to have access to extensive computational resources to experiment with new data sources or statistical methods will have access in a few clicks to easy-to-use, preconfigured data science environments, so that they can directly start to experiment. However, many users want to go deeper and build actual prototypes of production applications for their projects: configuring initialization scripts to tailor the environments to their needs, deploying an interactive app that delivers data visualization to users of their choice, deploying other services than those available in our catalogs, etc. For these advanced users to continue to push the boundaries of innovation, Onyxia gives them access to the underlying Kubernetes cluster. This means that users can freely open a terminal on an interactive service and interact with the cluster—within the boundaries of their namespace—in order to apply custom resources and deploy custom applications or services.

Besides autonomy and scalability, the architectural choices of Onyxia also foster reproducibility of statistical computations. In the paradigm of containers, the user must learn to deal with resources which are by nature ephemeral, since they only exist at the time of their actual mobilization. This fosters the adoption of development best practices, notably the separation of the code—put on an internal or open-source forge such as GitLab or GitHub—the data—stored on a specific storage solution, such as MinIO—and the computing environment. While this requires an entry cost for users, it also helps them to conceive their projects as pipelines, i.e., a series of sequential steps with well-defined inputs and outputs (akin to directed acyclic graph (DAG)). The projects developed in that manner are usually more reproducible and portable—they can work seamlessly on different computing environments—and thus also more readily shareable with peers.

8.3.3 An Extensive Catalog of Services to Cover the Entire Life Cycle of Data Science Projects

In developing the Onyxia platform, our intention was to provide statisticians with a comprehensive environment designed to support end-to-end development of data

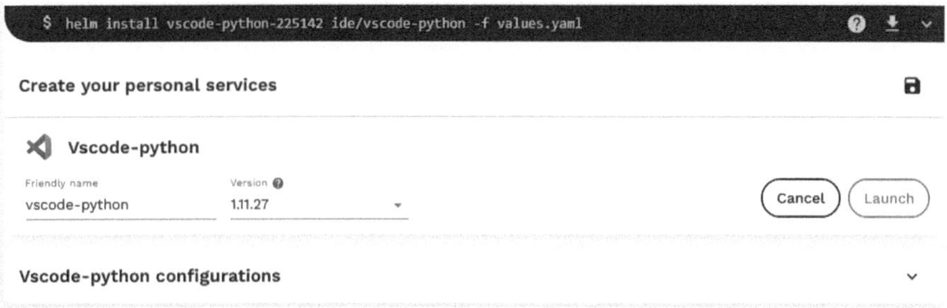

Fig. 8.5 Launching a service through Onyxia's UI. Note: Services from Onyxia's catalog can either be used vanilla or configured by the users to tailor them to their specific needs. In order to limit the dependence of users on Onyxia, each action performed by the user on the UI is accompanied by the actual command that is executed on the Kubernetes cluster

science projects. The platform offers a vast array of services that span the complete life cycle of a data science project.

The primary usage of the platform is the deployment of interactive development environments (IDE), such as RStudio, Jupyter, or VSCode (Fig. 8.5). These IDEs come equipped with the latest kernels of major open-source programming languages commonly employed by public statisticians (R, Python, Julia), as well as an extensive collection of packages commonly used in data science for each language. In order to ensure that services remain up-to-date and consistent between them, we maintain our own stack of underlying Docker images and rebuild it weekly. The stack of images is fully open-source[10] and can thus be reused outside Onyxia.

As discussed in previous sections, the persistence layer of these interactive environments is mainly carried out by MinIO, Onyxia's default object storage solution. As it is based on a standardized REST API, files can be easily queried directly from R or Python using high-level packages. This in itself is an important step of ensuring reproducibility: The input files of a project are not mounted manually and then specified via paths adherent to a specific infrastructure and filesystem. Rather, files are specified as HTTP queries, making the overall structure of projects much more extendable. In our experience, the object storage paradigm covers very well the needs of most statistical projects we accompany. However, additional database services such as PostgreSQL and MongoDB are available for applications with specific needs, such as those requiring online transaction processing (OLTP) capabilities or document-oriented storage.

As Onyxia was developed to allow experimentation with big data sources and machine learning methods, we also provide services optimized for scalability. For instance, frameworks like Spark and Trino that enable to perform distributed computations within Kubernetes. These services come preconfigured to integrate seamlessly with S3 storage, thus facilitating building integrated and efficient data pipelines.

---

[10] https://github.com/InseeFrLab/images-datascience

Beyond mere experimentation, our goal is to empower statisticians to transition from trial phases to production-grade projects. In line with principles from the DevOps approach, this involves facilitating the deployment of prototypes and their continuous improvement over time. To this end, we provide a set of open-source tools aimed at automatizing and industrializing the process of deploying data-intensive applications (ArgoCD, Argo-Workflows, MLflow). For projects leveraging machine learning models, statisticians can serve their models through APIs, deploy them using the aforementioned tools, and manage their life cycle using an API manager (e.g., Gravitee). Section 8.4 illustrates how these tools, particularly MLflow, have been central in putting machine learning models in production at INSEE, in accordance with MLOps principles.

In Sect. 8.3.2, we stressed that one of Onyxia's fundamental design principles was to avoid vendor lock-in. In line with this idea, organizations that implement Onyxia are free to customize catalogs to suit their specific requirements, or even opt to construct their own catalogs independent of Onyxia's default offerings. This flexibility ensures that organizations are not confined to a single solution or provider and can adapt the platform to their evolving needs.

8.3.4 Building Commons: An Open-Source Project and an Open-Innovation Platform

As a fully open-source initiative, the Onyxia project aims at building "knowledge commons" by promoting and building software that can be easily reused in official statistics and beyond (Schweik 2006). This concerns, first of all, the components on which Onyxia are based: both its constitutive technological bricks (Kubernetes, MinIO, Vault) and all the services from the catalog are open-source. But more crucially, all the code of the project is available openly on GitHub.[11] Alongside an in-depth documentation,[12] this greatly facilitates the potential for other organizations to create instances of data science platforms built upon the Onyxia software and tailor it to their respective needs (see Fig. 8.6). This enabled the project to attract a growing community of contributors from official statistics (Statistics Norway), NGOs (Mercator Ocean), research centers, and even industry, thus transitioning progressively toward a more decentralized governance of the project. In the next years, the involvement of NSIs from the European Statistical System is expected to increase as Onyxia was chosen as the reference data science platform in the context of the AIML4OS project, a "One-Stop-Shop" for Artificial Intelligence/Machine Learning for Official Statistics in the European Statistical System.[13]

[11] https://github.com/InseeFrLab/onyxia

[12] https://docs.onyxia.sh/

[13] More information on this project available at https://cros.ec.europa.eu/dashboard/aiml4os.

Fig. 8.6 One project, multiple instances: The UI is adaptable to the graphic identity of the organization

Another major way in which we try to build commons is by developing and maintaining a showcase instance of the Onyxia project, the SSP Cloud (Comte et al. 2022). This platform, equipped with extensive and scalable computational resources,[14] is designed to be a sandbox for experimenting with cloud technologies and new data science methods. The full catalog of services of Onyxia is available on the platform, enabling motivated users to go beyond mere experimentation by producing “proof of concepts,” with full autonomy regarding the configuration and orchestration of their services.

Beyond its technical capabilities, the SSP Cloud is an endeavor at embodying the principles of open innovation (Chesbrough 2003). Deployed on Internet,[15] it is open not only to INSEE employees, but also more broadly to French governmental agencies, French Universities, and other European NSIs, and is dedicated to experimenting with data science methods using open data (Aymon et al. 2024). Thus, the projects carried out on this platform showcase the growing abundance of datasets published openly by organizations. The fundamentally collaborative nature of the SSP Cloud has proven especially beneficial for organizing innovative events such as hackathons—both at the national and international levels—and in the academic sphere. It has become an integral resource for several universities and Grandes Ecoles in France, fostering the use of cloud-native and reproducible environments, and preventing vendor lock-in effect due to the over-reliance of educational organizations on proprietary cloud solutions. As a result, the platform is now widely used in the French National Statistical System and beyond, with about 1,000 unique users per month in 2024. These users form a dynamic community, thanks to a centralized discussion canal; they help improve the user experience by reporting bugs, suggesting new features, and thus contribute directly to the project.

[14] On the physical side, the SSP Cloud consists in a Kubernetes cluster of about 20 servers, for a total capacity of 10 TB of RAM, 1100 CPUs, 34 GPUs, and 150 TB of storage.

[15] https://datalab.sspcloud.fr/

8.4 Case Study: Using MLOps to Improve NACE Classification

This chapter aims, through a concrete example, to illustrate how INSEE managed to deploy its first machine learning (ML) model into production. It provides an in-depth description of the MLOps approach that this project strived to adhere to, focusing on the various technologies that were employed. In particular, we highlight how cloud technologies were instrumental in building a solution iteratively and how Onyxia greatly facilitated this process by providing flexible development environments as well as tools to deploy and monitor models, promoting a continuous improvement loop. The entire project is available in open-source[16] and remains under active development.

8.4.1 Improving the NACE Classification Process Using ML Methods

8.4.1.1 Motivation

Coding tasks are common operations for NSOs and can sometimes be challenging due to the size of statistical classifications. At INSEE, a sophisticated coding tool called Sicore was developed in the 1990s to perform various classification tasks (Meyer and Rivière 1997). It consists in a coding engine containing numerous deterministic rules which identify ground-truth labels. Each input label goes through these rules, and when a ground-truth label is recognized, the associated code is assigned. When the label is not recognized, it must be manually classified by an INSEE agent.

Two main reasons drove the experimentation of new coding methods.

Firstly, there was an internal change with the redesign of the French statistical business register, which lists all companies in France and assigns them a unique identifier used across public administrations. The main goals of this revamping were to improve the daily management of the registry for INSEE agents and to reduce waiting times for companies. Additionally, at the national level, the French government launched a one-stop shop (the "guichet unique") for business formalities, allowing more flexibility for business owners in describing their main activities. Initial testing exercises revealed that Sicore was no longer the suitable tool for performing NACE classification, as only 30% of the input data were being automatically coded.

Three stakeholders were involved in this project: the business team responsible for managing the French statistical business register, the IT team developing

[16] https://github.com/orgs/InseeFrLab/teams/codification-ape/repositories

Table 8.1 NACE nomenclature

Level	NACE	Title	Size
Section	H	Transportation and storage	21
Division	52	Warehousing and support activities for transportation	88
Group	522	Support activities for transportation	272
Class	5224	Cargo handling	615
Subclass	**5224A**	**Harbor handling**	**732**

software related to the register's operation, and the innovation team responsible for implementing the new coding tool. The latter team is the SSP Lab, which was created in 2018 with the objective of providing support to other teams on innovation topics to streamline their various projects.

8.4.1.2 Classification Task

The project we describe consists in a standard NLP classification problem. Starting from a textual description of the activity, we want to predict the associated class in the NACE Rev. 2 statistical classification. Comparable projects involving automated German NACE coding are described in Beuter et al. (2025), where the focus is on presenting the methodological investigations. This classification has the particularity of being hierarchical and contains five different levels:[17] section, division, group, class, and subclass. In total, 732 subclasses are included in the classification, which is the level at which we aim to perform the classification. Table 8.1 provides an example of this hierarchical structure.

With the establishment of the "guichet unique," business owners now describe their activity description with a free-text field. As a result, the new labels are very different from the harmonized labels that were previously received. Therefore, it was decided to work with ML models that are known to be effective for supervised text classification tasks (Li et al. 2022). This represents a significant paradigm shift from INSEE's perspective, as ML was not traditionally used in the actual production of official statistics. Besides, the perspective of putting the new model in production was considered from the outset, guiding numerous methodological and technical choices. As such, several strategic choices had to be made from the outset, including the methodology, the choice of a development environment consistent with the target production environment, and the adoption of collaborative work methods.

[17] Actually, there are five different levels in France but only four at the European level.

8.4.1.3 Methodology

Text classification from the free-text field provided by business owners is a complex task: The activity descriptions are relatively short and thus contain limited statistical information, can contain spelling mistakes, and often require domain knowledge to be properly classified. On such task, traditional text analysis methods such as count vectorization or TF-IDF often fall short, whereas neural-network-based embedding methods tend to perform better (Li et al. 2022). However, such architectures often impose greater computational demands, as they are much larger and might require specific hardware such as GPUs to perform inference with acceptable latency. These constraints led us away from the most powerful language models, such as Transformer models, and instead directed us toward the fastText model (Joulin et al. 2016), a simpler embedding-based classifier. The fastText model is extremely fast to train, even from scratch, and inference does not require a GPU to achieve low latency time. Besides, the model yielded excellent performance results in our use case that, considering the time and human resource constraints, were more than sufficient to enhance the existing process. Finally, the model's architecture is relatively simple, simplifying communication and adoption within the various INSEE teams.

The fastText model relies on a bag-of-words model to obtain embeddings and a classification layer based on logistic regression. The bag-of-words approach involves representing a text as the set of vector representations of each of its constituent words. The specificity of the fastText model compared to other embeddings-based approaches is that embeddings are not only computed on words but also on word n-grams and character n-grams, providing more context and reducing biases due to spelling mistakes. Then, the embedding of a sentence is computed as a function of the individual token embeddings, typically the average. In the case of supervised text classification, the embedding matrix and the classifier's parameters are learned simultaneously during training by gradient descent, minimizing the cross-entropy loss function. Figure 8.7 represents the full pipeline of operations performed by fastText on an example text input.

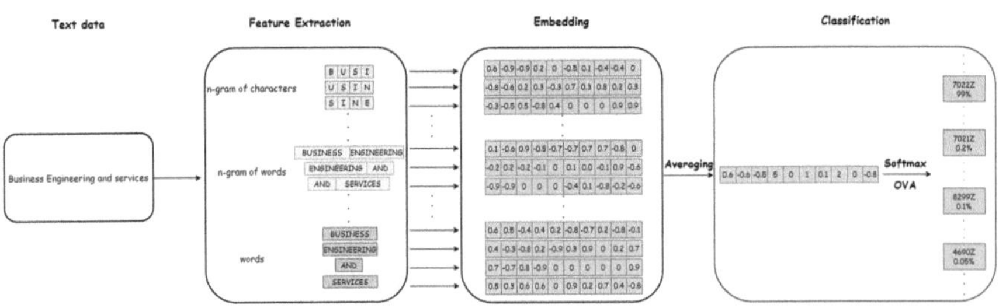

Fig. 8.7 Overview of the simplified process behind fastText classifications

8.4.2 A Production-First Approach with MLOps

From the very onset of this project, the target was to go beyond mere experimentation and put the model in production. Besides, the goal with this pilot project was also to build a template for future ML projects at INSEE. We thus strove to enforce best development practices from the very beginning of the project: following community standards for code quality, using scripts-based development over notebooks, building a modular package-like structure, etc. However, compared to traditional development projects, machine learning projects have specific features that make it necessary to apply a complementary set of best practices, gathered under the name of MLOps.

8.4.2.1 From DevOps to MLOps

DevOps is a set of practices designed to foster collaboration between development (Dev) and operations (Ops) teams. The fundamental idea is to integrate the full life cycle of a project in a single automated continuum. An important tool to achieve this continuity is CI/CD pipelines. With continuous integration (CI), each commit of new source code will trigger a pipeline of standardized operations, such as building the application, testing it, and making it available as a release. Then, continuous deployment (CD) consists of tools to automate the deployment of the new code and limit manual intervention, while ensuring proper monitoring to guarantee process stability and security. This approach promotes a faster, continual release of necessary feature changes or additions. Furthermore, by encouraging collaboration between teams, DevOps also promotes a quicker cycle of innovation, allowing teams to address issues as they arise and incorporate feedback effectively throughout the project life cycle.

The MLOps approach can be seen as an extension of DevOps, developed to address the specific challenges related to managing the life cycle of ML models. Fundamentally, both DevOps and MLOps aim at building software in a more automated and robust manner. The main difference is that in MLOps, this software also has a machine learning component. Consequently, the life cycle of the project gets more complex. The underlying ML model needs to be retrained regularly, to avoid any loss of performance over time. Data ingestion must also be included in the pipeline, as new data may be used to improve performance. Figure 8.8 presents the steps of an ML project using the continuous representation traditionally seen in DevOps. This illustrates a fundamental principle of MLOps, the need for continuous improvement, described in more detail in Sect. 8.4.2.2.

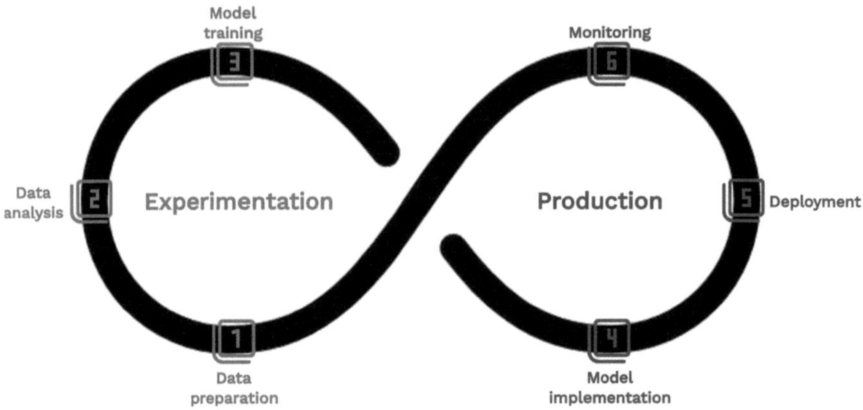

Fig. 8.8 The MLOps approach promotes a continuous management of ML projects' life cycle

8.4.2.2 Principles of MLOps

MLOps is defined by a few core principles that are crucial for building production-grade and scalable ML applications. These principles are designed to address the specific challenges associated with ML workflows.

The most fundamental principle of MLOps is continuous improvement, reflecting the iterative nature of ML projects. In the experimentation phase, the model is developed using a training dataset that generally differs from the target data in some respect. When a model is deployed in production, the new data that the model needs to perform prediction on can reveal insights about the model's performance and potential shortcomings. These insights necessitate a return to the experimentation phase, where data scientists adjust or redesign their models to address any discovered issues or to improve accuracy. This principle thus emphasizes the importance of building a feedback loop that enables ongoing enhancements throughout the life cycle of a model. Automation, particularly through the use of CI/CD pipelines, plays a crucial role in this process by making the transition between experimentation and production phases more continuous. Monitoring is also an essential part of this process: A model deployed in production needs to be continuously assessed so as to detect major drifts that may reduce the predictive performance of the model and thus necessitate further adjustments, such as retraining or fine-tuning the model.

Another major goal of MLOps is to promote reproducibility, ensuring that any ML experiment can be reliably reproduced with the same results. MLOps tools thus facilitate thorough logging of ML experiments, including data preprocessing steps, model hyperparameters, and training algorithms. Data, models, and code are versioned, enabling teams to revert to previous versions if an update does not perform as expected. Finally, these tools help to produce detailed specifications of the computing environment used to produce these experiments—such as versions of libraries—and often rely on containers to help replicate the same conditions in which the original model was developed.

Finally, MLOps aims at fostering collaborative work. ML-based projects generally involve a wider range of profiles: business units and data science teams on the one hand, developers and operations teams on the other. Like DevOps, MLOps thus emphasizes the need for a collaborative culture and to avoid working in silos. MLOps tools generally include collaborative features, such as centralized stores for ML models or ML features which facilitate the sharing of components between team members and limit redundancy.

8.4.2.3 Implementation with MLflow

Numerous tools have been developed to implement the MLOps approach in actual projects. All of these frameworks aim at enforcing, in some form, the core principles described above. In this project, we chose to rely on a popular open-source framework named MLflow.[18] This choice does not indicate any inherent superiority of MLflow over alternative software, but reflects a set of good properties associated with MLflow that made it a very relevant solution for our application. First, it covers the entire life cycle of ML projects, while other tools may be more specialized in some parts of it. Second, it exhibits great interoperability as it is well-interfaced with popular ML libraries—such as PyTorch, Scikit-learn, XGBoost, etc.—and supports multiple programming languages—including Python, R, and Java—thus covering the spectrum of programming languages commonly used at INSEE. Finally, it proved to be very user-friendly and thus encouraged adoption among the project members and facilitated continuous collaboration between them.

MLflow provides a cohesive framework to operationalize MLOps principles effectively within ML projects. Data scientists can encapsulate their work in *MLflow Projects* that package together ML code and its dependencies, ensuring that each project is reproducible and can be consistently re-executed. A project relies on an *MLflow Model*, a standard format that is compatible with most ML libraries and offers a normalized way of serving the model, e.g., via an API. This interoperability and standardization are instrumental in supporting continuous improvement of the project, as models trained with a variety of packages can be readily compared or switched by one another without breaking any code. As experiments with various models progress, the Tracking Server logs detailed information about each run—hyperparameters, metrics, and outputs artifacts and metrics—which there again promotes reproducibility but also facilitates the model selection phase through a user-friendly interface. After this experimentation phase, selected models are integrated into the *Model Registry*, where they are versioned and staged for deployment. This registry serves as a centralized model store that enables the

[18] https://github.com/MLflow/MLflow

different project members or teams to collaboratively manage the life cycle of the project.

8.4.3 *Facilitating Iterative Development with Cloud Technologies*

While continuous improvement is a fundamental principle of MLOps, it is also a very demanding one. In particular, it requires designing and building our project as an integrated pipeline whose various stages are mainly automated, from data ingestion to monitoring the model in production. In this context, iterative development is essential in order to build a minimum viable product that is then refined and improved over time. This section shows how cloud-native technologies, through the Onyxia project, were instrumental in building the project from the start as a collection of modular connected components, thus greatly enhancing the capacity for continuous refinement over time.

8.4.3.1 A Flexible Development Environment

In an ML project, the flexibility of the development environment is essential. First, due to the diversity of tasks to be performed—data collection, preprocessing, modeling, evaluation, inference, monitoring, etc. Second, because ML is a fast-evolving field, it is preferable to build an ML app as a collection of modular components so as to be able to update components without disrupting the entire pipeline. As discussed in Sect. 8.2.2, cloud-native technologies enable the creation of modular and scalable development environments.

However, as also discussed in Sect. 8.3, access to such resources is not enough. An ML project requires a wide variety of tools to comply with MLOps principles—data storage, interactive development environments to experiment freely, automation tools, monitoring tools, etc. While these tools can be installed on a Kubernetes cluster, making them available to data scientists in an integrated and preconfigured manner is essential to facilitate their adoption. Through its catalog of services and the automatic injection of configuration in the services, Onyxia enables building projects that rely on multiple cloud-native components that can communicate readily with each other.

The way model training was carried out for this project emphasizes the flexibility provided by Onyxia in the experimentation phase. All the code performing the training is written in Python in the context of a VSCode service. As personal S3 credentials are injected in each service at startup, the various users of the projects can interact directly with the training data stored on a S3 bucket on MinIO, Onyxia's default object storage solution. All experiments performed for the model selection phase are logged on a shared instance of MLflow, which stores logged data on a PostgreSQL instance automatically launched on Kubernetes and artifacts (trained models and associated metadata) on MinIO. The model was trained using grid search for hyperparameter tuning and evaluated through cross-validation, a combination that is known to provide a better evaluation of the generalization performance of the model but also requires a lot of computing resources, due to the combinatorial nature of testing many hyperparameter combinations (Bischl et al. 2023). In our case, we leveraged Argo Workflows, an open-source workflow engine designed to orchestrate parallel jobs on Kubernetes, each job being specified as an independent container. It then becomes straightforward to compare performances of the different trained models and select the best one using the comparison and visualization tools available in the UI of MLflow.

In summary, the training stage was made efficient and reproducible, thanks to the use of numerous modularly connected components—a distinctive feature of cloud-native technologies—readily made available to data scientists by Onyxia.

8.4.3.2 Deploying a Model

Once candidate models have been optimized, evaluated, and a best-performing model has been selected, the next step is to make it available to the application end users. Simply providing the trained model as an artifact or even just the code to train the model is not a convenient way to serve it, as it assumes that users have the resources, infrastructure, and knowledge required for training it under the same conditions. The goal is therefore to make the model available in a simple and interoperable manner, in the sense that it should be possible to query it with various programming languages. Furthermore, it should be possible for other applications to query the model in a programmatic way.

Against that background, we opted to serve the model through a REST API. This technology has become a standard way to serve ML models as they offer several benefits. First, they fit very well a cloud-oriented environment: Similarly to the other components of our stack, it makes it possible to query the model using standard HTTP requests, which contributes to the modularity of the system. It also means that they are interoperable: As they rely on standards technologies for queries (HTTP requests) and responses (generally, a JSON-formatted string), they are mostly agnostic to the programming language used to request them. Finally, they offer great scalability because of their stateless design.[19] As each request must contain all the information needed to understand and process the request, REST APIs can easily be duplicated on different machines to balance a challenging load—a process known as horizontal scaling.

[19] Stateless design refers to a system architecture where each request from a client to the server must contain all the information needed to understand and process the request. This means that the server does not store any information about the client's state between requests, allowing each request to be handled independently. This design simplifies scaling and enhances the robustness of the system, as any server can handle any request without relying on prior interactions.

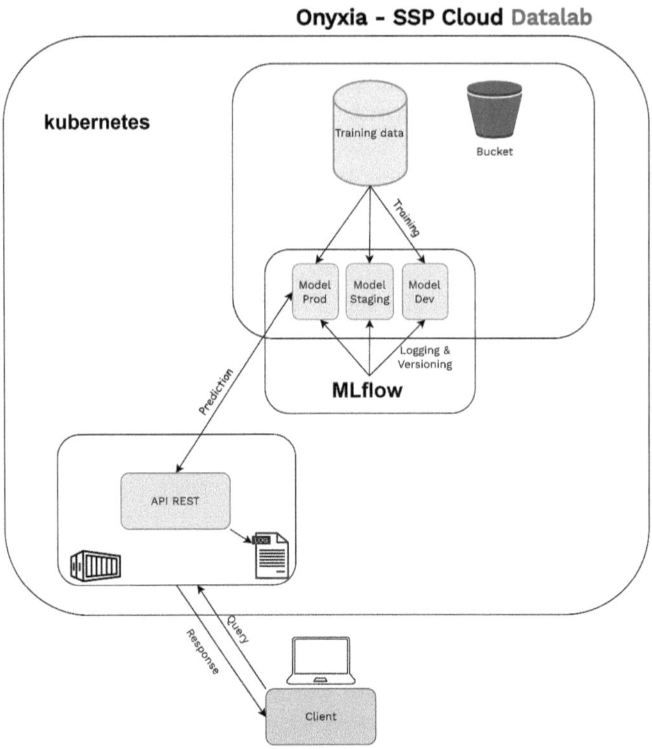

Fig. 8.9 A cloud-native approach to serve an ML model using a REST API

We developed the API serving the model with FastAPI,[20] a fast and well-documented web framework for building APIs with Python. The API code and required software dependencies are encapsulated into a Docker image so that it can be deployed as a container on the Kubernetes cluster. An important benefit of using Kubernetes is the ability to scale the API—through the number of API pods effectively deployed—to the demand and provide automatic load balancing. Upon startup, the API automatically retrieves the correct model from the MLflow model registry, which acts as a proxy from the actual artifact of the production model, stored on MinIO. Finally, as the application code is packaged using MLFlow's standardized API—enabling for instance to integrate the preprocessing step directly to each API call—the inference code can remain mostly uniform regardless of the underlying ML framework used. This deployment process is summarized in Fig. 8.9.

[20] https://fastapi.tiangolo.com

8.4.3.3 Building an Integrated Pipeline

The architecture we built at this stage already reflects some important principles of MLOps. The use of containerization to deploy the API as well as the use of MLflow to track the experiments while developing the model ensures reproducibility of the predictions. Using the central model registry provided by MLflow facilitates the management of the life cycle of the models in a collaborative way. Furthermore, the modularity of our architecture leaves room for further improvement as modular components can be easily added or modified without breaking the structure of the application as a whole. As we shall see in subsequent sections, this property was essential in building the application iteratively, enabling to add a monitoring layer (Sect. 8.4.3.4) and an annotation component (Sect. 8.4.3.5) to promote continuous improvement of the model.

However, the ability to refine the base architecture iteratively also requires more continuity in the process. At this stage, the deployment process involves several manual operations. For instance, adding a new feature to the API would require to build a new image, tag it, update the Kubernetes manifests used to deploy the API and enforce them on the cluster to replace the existing one with minimum downtime. Similarly, a change of the model served through the API would require a very simple modification of the code but several manual steps to update the version on the cluster. As a result, data scientists are not fully autonomous when it comes to prototyping and testing updated versions of the model or the API, which limit the potential for continuous improvement.

In order to automate this process, we built a CI/CD pipeline—a concept already presented in Sect. 8.4.2.1—integrating these various steps. Figure 8.10 illustrates our specific implementation of a CI/CD pipeline. Any change in the code of the API repository triggers a CI build process (implemented with GitHub Actions) of the associated docker image, which is then released on a public container registry (DockerHub). This image can then be fetched and deployed by the container orchestrator (Kubernetes), by specifying and applying manually new manifests to update the Kubernetes resources of the API. However, the downside of this approach is that it limits reproducibility of the deployment, since each resource is handled

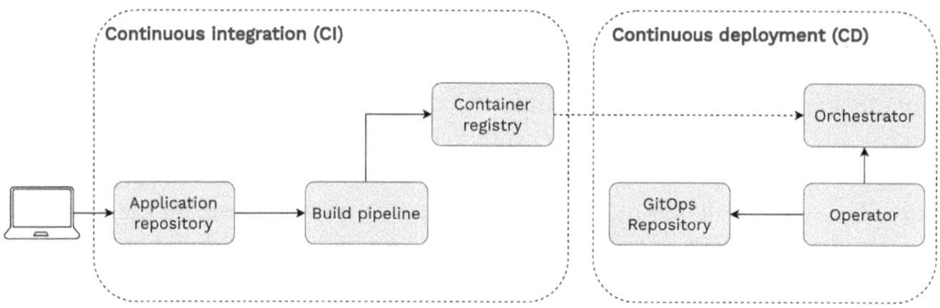

Fig. 8.10 The CI/CD pipeline implemented in the project

independently by the orchestrator, so that the life cycle of the API deployment as a whole is not managed. To overcome this shortcoming, we integrate the deployment part in a CD pipeline based on the GitOps approach: The resources manifest of the API are stored on a Git repository. The state of this "GitOps" repository is monitored by a Kubernetes operator (ArgoCD), so that any change in the application manifests is directly propagated to the deployment on the cluster. In this integrated pipeline, the only action needed for the data scientist to trigger an update of the API is to change the tag of the API image indicating the version to be deployed.

8.4.3.4 Monitoring a Model in Production

Once the initial development phase of the project has been completed—including training, optimization, and deployment of the model to be served to users—it is crucial to understand that the data scientist's responsibilities extend further. Traditionally, the role of the data scientist is often limited to training the model and selecting the model to deploy, with the deployment task being delegated to the IT department. However, a specificity of ML projects is that, once in production, the model has not yet reached the end of its life cycle, and it must be continuously monitored to prevent undesirable performance degradation. Continuous monitoring of the deployed model is extremely important to ensure the conformity of results to expectations, anticipate changes in data, and iteratively improve the model.

The concept of monitoring can take on different meanings depending on the context of the involved team. For IT teams, it primarily involves verifying the technical effectiveness of the application, including aspects such as latency, memory consumption, or disk usage. Conversely, for data scientists or business teams, the focus is more on methodological monitoring of the model. However, real-time tracking of the performance of an ML model is often a complex task, given that ground truth is usually not known at the time of prediction. Therefore, it is common to use proxies to detect any signs of performance degradation. Two main types of degradation of an ML model are generally distinguished. The first one is data drifts, which occur when the data used during inference in production exhibits significant differences compared to the data used during training. The second one is concept drifts, which occur when a change in the statistical relationship between the features and the target variable is observed over time.

In the context of our project, the objective is to achieve the highest rate of correctly classified textual description of economic activities while minimizing the number of textual descriptions requiring manual intervention. Thus, our goal is to distinguish correct predictions from incorrect ones without prior access to ground truth. To accomplish this, we use a confidence index defined as the difference between the two highest confidence scores of the top results returned by the model. For a given textual description, if the confidence index exceeds a determined threshold, the textual description is automatically coded. Otherwise, the textual description is manually coded by an INSEE agent. This manual coding task is still informed by the ML model: Through an application that queries the API, the agent is shown the five most probable codes according to the model.

Defining the threshold for automatic coding of textual descriptions was thus crucial in this process and involved making a trade-off between achieving a high automatic coding rate and maximizing coding performance. To monitor the behavior of our model in production, we developed an interactive dashboard that enables visualization of several metrics of interest for the business teams. Among these metrics are the number of requests per day and the rate of automatic coding per day based on a given confidence index threshold. This visualization allows business teams to understand the rate of automatic coding they would have obtained if they had chosen different thresholds. This dashboard also represents the distribution of obtained confidence indices and compares temporal windows in order to check for changes in the distributions of predictions returned by the model.[21] Finally, confidence indices can be analyzed at finer levels of granularity based on the aggregation level of the statistical classification, to determine which classes are most difficult to predict and which have more or less occurrences.

Figure 8.11 shows the components that were added to the project architecture so as to provide the monitoring dashboard described above. First, we set up a simple extract-transform-load (ETL) process in Python (second box of the bottom row), which fetches the API logs periodically and transforms them into partitioned Parquet files.[22] Then, we use the Quarto framework[23] to build an interactive dashboard (third box of the bottom row). To compute the various metrics presented in the dashboard, the Parquet files are queried using the SQL language through the DuckDB engine. Like the API, the dashboard is built and deployed as a container on the Kubernetes cluster, and this process is automated using a CI/CD pipeline. The annotation component (fourth box of the bottom row) is discussed in the next section.

[21] These metrics are not currently displayed in our monitoring dashboard and are not tracked continuously. They are only computed when we retrain the model to ensure that the distribution predicted by the new model is consistent. However, we plan to add these metrics to the dashboard for the NACE project once we implement automated retraining via triggers. Changes in distributions are typically checked by computing statistical distances—such as the Bhattacharyya distance, the Kullback-Leibler divergence, the Hellinger distance—and/or by performing statistical tests—such as the Kolmogorov-Smirnov or the chi-squared test.

[22] Ideally, existing frameworks should be preferred over custom-made solutions to prioritize standardized routines. At the time of building this component of the pipeline, we found that existing, cloud-native frameworks for log analytics had important limitations. This constitutes an area of improvement for the project.

[23] Successor to R Markdown, Quarto has become an essential tool of our data stack. It unifies the functionality of several very useful packages from the R Markdown ecosystem while providing native support of several programming languages, including Python and Julia in addition to R. It is increasingly used at INSEE as a way to produce reproducible documents and output them in a variety of formats.

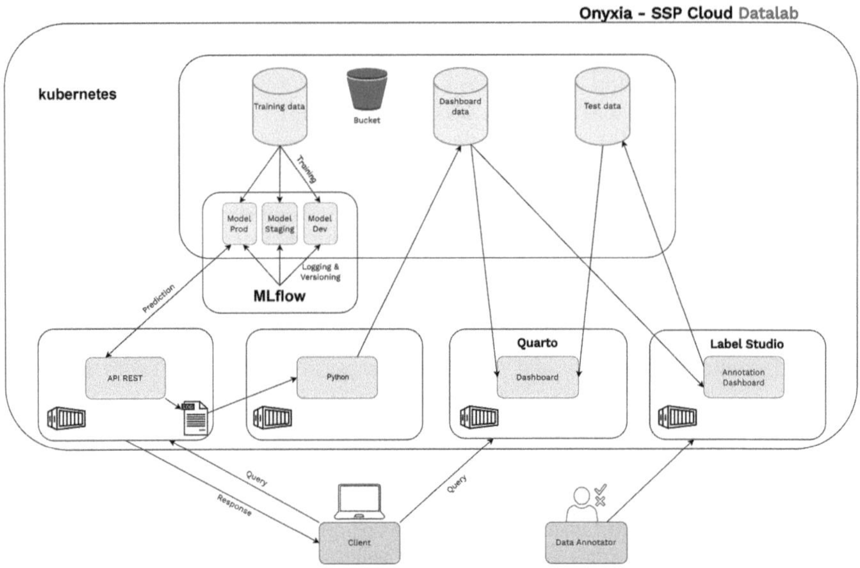

Fig. 8.11 Our implementation of a complete MLOps architecture

8.4.3.5 Promoting Continuous Improvement of the Model

The monitoring layer of our application provides an important and detailed view into the system performance. Due to the dynamic nature of real-life data, ML models' performance often declines over time. To promote continuous improvement of the model, it is thus essential to envision strategies to overcome these performance losses. A frequently used strategy is periodic retraining of the model, requiring the collection of new training data.

Several months after the first version of the model was deployed in production, the need to implement a continuous annotation process became increasingly apparent for two key reasons. First, a gold-standard sample was not accessible at the time of the experimentation phase, so we relied on a subset of the training dataset to perform evaluation, knowing the labeling quality was not optimal. Continuously collecting a gold-standard sample would thus enable us to get an unbiased view of the model's performance in production on real data, particularly on data that has been automatically coded. Another reason is the redesign of the NACE statistical classification in 2025. From 2025 onward, NSOs will be required to use the latest version of NACE, namely NACE Rev. 2.1. This revision brings substantial changes that will require an adaptation of the model, and thus the collection of new training set. Annotation of the old training dataset according to the new statistical classification will be necessary.

Against that background, an annotation campaign has been initiated in early 2024 to continuously build a gold-standard dataset. The annotation campaign is carried out on the SSP Cloud using Label Studio, an open-source annotation tool that provides a user-friendly interface and is available in Onyxia's catalog. Figure 8.11 illustrates how the labeling component (fourth box of the bottom row) could be readily integrated in the project architecture, thanks to its modular nature. In practice, we create a pool of text descriptions randomly sampled from the data passed through the API over the past three months. This sample is then sent to annotation by NACE experts using the UI of Label Studio. The annotation results are automatically saved on MinIO, transformed into Parquet format. Then, these gold-standard data are directly integrated into the monitoring dashboard to compute and observe various model performance metrics. These metrics give us a much clearer picture on the actual performance of the model on production data, and in particular on its shortcomings. In parallel, we will launch an annotation campaign in the upcoming months to construct a new training set tailored to the recently updated NACE statistical classification. Leveraging both the updated training data and performance metrics derived from the gold-standard sample, we aim to iteratively enhance the model's accuracy through periodical and targeted retraining in the near future.

8.5 Discussion

The development of data science methods offers considerable potential for official statistics. However, our ability to create value from these new methods essentially depends on our capacity to produce production-grade systems that serve their purpose in a robust way. This evolution calls for deep reflection on what constitutes a modern, scalable data science infrastructure for official statistics. This chapter presents the Onyxia project, the proposal for such a platform that we are developing at INSEE. By exploiting cloud-native technologies that have become standards in the data ecosystem, it aims to increase statisticians' autonomy in the orchestration of their statistical treatments, while promoting reproducibility of produced statistics. As cloud technologies are notoriously difficult to configure, the core value of Onyxia's lies in making them accessible to statisticians via a user-friendly interface and a catalog of preconfigured services to cover most common uses. Through an internal project aiming at revising the NACE classification process using machine learning methods, we illustrate how Onyxia enables to iteratively build production-grade machine learning projects that promote continuous improvement, a fundamental principle of the MLOps approach.

Initially developed as an internal project, Onyxia has gained recognition beyond the scope of INSEE or the French administration. Convinced of the potential of cloud technologies to foster autonomy and leverage the full potential of data science, several organizations now have a production instance of Onyxia running, and multiple others are in the process of either testing or implementing one. Besides, the choice of Onyxia as the reference data science platform in the context of the AIML4OS project should further facilitate its adoption within the ESS. This trend is naturally very beneficial to the Onyxia project, as it moves from a project developed in open-source—but mainly at INSEE—to a full open-source project with a growing base of contributors. This in turn facilitates its adoption by other organizations, since it gives more guarantees on its sustainability independently of INSEE's strategy. The governance of the project is currently evolving to reflect this trend. For instance, with the organization of monthly community calls and the creation of a public channel and roadmap for the project.[24]

Despite this success, we observe several limitations to the widespread adoption of the project in organizations. First, it is essential to remind that the fundamental choice made by organizations that adopt Onyxia is not the software itself, but the underlying technologies: containerization (through Kubernetes) and object storage. These technologies can represent substantial entry costs for organizations, as they demand a significant commitment to developing and maintaining skills which are not readily found in NSOs. Yet, the general trend toward cloud-native solutions among data-centric organizations suggests a favorable shift that could mitigate these challenges over time.

Similarly, the transition toward cloud-native technologies induces entry costs for statisticians. First, they often deal with a loss of references regarding where computations actually happen: While they may be accustomed to performing computation on centralized servers rather than a personal computer, the container adds a layer of abstraction that makes the location hard to grasp at first. But the major perceived change in this paradigm is the loss of data persistence. In traditional setups—either a personal computer or a server accessed through a virtual desktop—the code, the data, and the computing environment are kind of mixed in a black-box fashion. On the contrary, containers have no persistence by design. While object storage provides this persistence, a proper use of these infrastructures for statistical projects requires a variety of tools and corresponding skills: using a version control system for the code, interacting with the object storage API to store the data, providing configuration files or secrets as inputs, etc. In a way, these entry costs can be seen as the "price" of autonomy: Thanks to cloud-native technologies, statisticians now have access to scalable and flexible environments that enable them to experiment more freely, but this autonomy requires a significant skills upgrade which may be overwhelming at first and limit adoption. However, our experience at INSEE suggests that this effect can largely be mitigated through a combination of training statisticians to develop best practices and accompanying statistical projects when transitioning to cloud infrastructures.

While Onyxia has significantly democratized access to cloud-native technologies for statisticians, the actual integration of data science methods in the statistical production of NSOs encompasses broader challenges, organizational in nature. A major hindsight from the deployment of our first ML model in production is the necessity to overcome skill compartmentalization across IT, business, and

[24] All information is available on the GitHub depository of the project: https://github.com/InseeFrLab/onyxia.

innovation teams. By nature, production-grade ML projects involve a wide range of skills—knowledge of the business domain, model training and fine-tuning, deployment and monitoring—and thus effective collaboration between professionals with different work cultures, programming languages, etc. Our experience shows that cloud technologies, by fostering autonomy of data scientists, give more continuity to ML projects and facilitate this much needed collaboration between various profiles. However, fully addressing these challenges involves measures that go beyond the technical domain. For instance, embedding some data science capabilities directly within business teams, in complement of centralized innovation teams, could foster better alignment with project objectives. Also, recruiting profiles that are not typically present in NSOs, such as data engineers or ML engineers, could bring new essential skills that lie at the intersection of statistical methodology and computer techniques. Ultimately, the transition toward a data science-driven approach in statistical production should rely on a balanced strategy that couples technical solutions such as Onyxia with comprehensive organizational adjustments, fostering a culture of collaboration, continuous learning, and innovation.

References

D. Abadi, P. Boncz, S. Harizopoulos, S. Idreos, S. Madden, et al., The design and implementation of modern column-oriented database systems. Found. Trends Databases **5**(3), 197–280 (2013)

A.I. Abdelaziz, K.A. Hanson, C.E. Gaber, T.A. Lee, Optimizing large real-world data analysis with parquet files in R: a step-by-step tutorial. Pharmacoepidemiol. Drug Safety33:e5728 (2023)

A. Ashofteh, J.M. Bravo, Data science training for official statistics: a new scientific paradigm of information and knowledge development in national statistical systems. Stat. J. IAOS **37**(3), 771–789 (2021)

D. Aymon, D.-T. Lam, L. Marti, P. Maury-Laribière, C. Choirat, R. Fondeville, Lomas: A platform for confidential analysis of private data (2024). https://doi.org/10.48550/arXiv.2406.17087

O. Bentaleb, A.S. Belloum, A. Sebaa, A. El-Maouhab, Containerization technologies: taxonomies, applications and challenges. J. Supercomput. **78**(1), 1144–1181 (2022)

F. Beuter, J. Gussenbauer, E. Minther, V. Szabo, S. Wegner, Approaches to automated NACE coding of German business activity descriptions, in *Foundations and Advances of Machine Learning in Official Statistics*, ed. by F. Dumpert, Chap. 10 (Springer, Berlin, 2025)

B. Bischl, M. Binder, M. Lang, T. Pielok, J. Richter, S. Coors, J. Thomas, T. Ullmann, M. Becker, A.-L. Boulesteix, D. Deng, M. Lindauer, Hyperparameter optimization: foundations, algorithms, best practices, and open challenges. Wiley Interdiscip. Rev. Data Mining Knowl. Discovery **13**(2), e1484 (2023)

H.W. Chesbrough, *Open Innovation: The New Imperative for Creating and Profiting from Technology* (Harvard Business Press, Brighton, 2003)

F. Comte, A. Degorre, R. Lesur, SSPCloud: a creative factory to support experimentations in the field of official statistics. Courrier des statistiques **7**, 68–85 (2022)

T.H. Davenport, D. Patil, Data scientist. Harv. Bus. Rev. **90**(5), 70–76 (2012)

J. Dean, S. Ghemawat, MapReduce: simplified data processing on large clusters. Commun. ACM **51**(1), 107–113 (2008)

P. Descy, V. Kvetan, A. Wirthmann, F. Reis, Towards a shared infrastructure for online job advertisement data. Stat. J. IAOS **35**(4), 669–675 (2019)

DGINS, Bucharest memorandum on official statistics in a datafied society (2018). https://ec.europa.eu/eurostat/documents/13019146/13239158/The+Bucharest+Memorandum+

on+Trusted+Smart+Statistics+FINAL.pdf/59a1a348-a97c-4803-be45-6140af08e4d7?t=1539760880000

B. Dhyani, A. Barthwal, Big data analytics using Hadoop. Int. J. Comput. Appl. **108**(12), 1–5 (2014)

European Commission, European statistics code of practice – revised edition 2017 (2018). https://ec.europa.eu/eurostat/web/products-catalogues/-/ks-02-18-142

EUROSTAT, ESSnet Big Data 2 – final technical report (2021). https://wayback.archive-it.org/12090/20221110013641/https://ec.europa.eu/eurostat/cros/system/files/wpa_deliverable_a5_final_technical_report_2021_06_29.pdf

A.S. Foundation, Apache parquet (2013). https://parquet.apache.org/

A.S. Foundation, Apache arrow (2016). https://arrow.apache.org/

S. Ghemawat, H. Gobioff, S.-T. Leung, The google file system, in *Proceedings of the Nineteenth ACM Symposium on Operating Systems Principles* (2003), pp. 29–43

T. Gjaltema, High-level group for the modernisation of official statistics (HLG-MOS) of the United Nations economic commission for Europe. Stat. J. IAOS **38**(3), 917–922 (2022)

S. Gokhale, R. Poosarla, S. Tikar, S. Gunjawate, A. Hajare, S. Deshpande, S. Gupta, K. Karve, Creating Helm charts to ease deployment of enterprise application and its related services in Kubernetes, in *2021 International Conference on Computing, Communication and Green Engineering (CCGE)*, IEEE (2021), pp. 1–5

A. Joulin, E. Grave, P. Bojanowski, T. Mikolov, Bag of tricks for efficient text classification (2016). https://arxiv.org/abs/1607.01759

A. Kowarik, M. Six, Quality guidelines for the acquisition and usage of big data with additional insights on web data, in *4th International Conference on Advanced Research Methods and Analytics (CARMA 2022)*. Editorial Universitat Politècnica de València (2022), pp. 269–269

M. Leclair et al., Using scanner data to calculate the consumer price index. Courrier des statistiques **3**, 61–75 (2019)

L. Leite, C. Rocha, F. Kon, D. Milojicic, P. Meirelles, A survey of DevOps concepts and challenges. ACM Comput. Surv. **52**(6), 1–35 (2019)

Y. Li, M. Yu, M. Xu, J. Yang, D. Sha, Q. Liu, C. Yang, Big data and cloud computing, *Manual of Digital Earth* (Springer, Singapore, 2020), pp. 325–355

Q. Li, H. Peng, J. Li, C. Xia, R. Yang, L. Sun, P.S. Yu, L. He, A survey on text classification: From traditional to deep learning. ACM Trans. Intell. Syst. Technol. **13**(2), 1–41 (2022)

L. Liu, Computing infrastructure for big data processing. Front. Comput. Sci. **7**, 165–170 (2013)

S. Luhmann, J. Grazzini, F. Ricciato, M. Mészáros, J.-M. Museux, M. Hahn, Promoting reproducibility-by-design in statistical offices, in *2019 New Techniques and Technologies for Statistics (NTTS) conference*, Brussels (2019). https://doi.org/10.5281/zenodo.3240198

M. McNutt, Reproducibility. Science **343**(6168), 229–229 (2014)

M. Mesnier, G.R. Ganger, E. Riedel, Object-based storage. IEEE Commun. Mag. **41**(8), 84–90 (2003)

E. Meyer, P. Rivière, SICORE, un outil et une méthode pour le chiffrement automatique à I'INSEE, in *Actes de la 4ème Conférence Internationale des Utilisateurs de Blaise*, Paris (1997), pp. 280–293. http://www.blaiseusers.org/1997/papers/meyer97.pdf

D. Moreau, K. Wiebels, C. Boettiger, Containers for computational reproducibility. Nat. Rev. Methods Primers **3**(1), 50 (2023)

J. Opara-Martins, R. Sahandi, F. Tian, Critical analysis of vendor lock-in and its impact on cloud computing migration: a business perspective. J. Cloud Comput. **5**, 1–18 (2016)

J. Opara-Martins, M. Sahandi, F. Tian, A holistic decision framework to avoid vendor lock-in for cloud SaaS migration. Comput. Inf. Sci. **10**(3), 29–29 (2017)

M. Raasveldt, H. Mühleisen, DuckDB: An embeddable analytical database, in *Proceedings of the 2019 International Conference on Management of Data* (2019), pp. 1981–1984

F. Ricciato, F. De Meersman, A. Wirthmann, G. Seynaeve, M. Skaliotis, Processing of mobile network operator data for official statistics: The case for public-private partnerships, in *104th DGINS Conference* (2018)

F. Ricciato, A. Wirthmann, K. Giannakouris, M. Skaliotis, et al., Trusted smart statistics: Motivations and principles. Stat. J. IAOS **35**(4), 589–603 (2019)

Y. Saidani, F. Dumpert, Quality dimensions and quality guidelines for machine learning in official statistics, in *Foundations and Advances of Machine Learning in Official Statistics*, ed. by F. Dumpert, Chap. 4 (Springer, Berlin, 2025)

A. Saiyeda, M.A. Mir, Cloud computing for deep learning analytics: a survey of current trends and challenges. Int. J. Adv. Res. Comput. Sci. **8**(2), 68–72 (2017)

D. Salgado, L. Sanguiao-Sande, S. Barragán, B. Oancea, M. Suarez-Castillo, A proposed production framework with mobile network data, in *ESSnet Big Data II - Workpackage I - Mobile Network Data* (2020). https://www.researchgate.net/publication/352150368_A_proposed_production_framework_with_mobile_network_data

S. Samundiswary, N.M. Dongre, Object storage architecture in cloud for unstructured data, in *2017 International Conference on Inventive Systems and Control (ICISC)*, IEEE (2017), pp. 1–6

C.M. Schweik, Free/open-source software as a framework for establishing commons in science, in *Understanding Knowledge as a Commons: From Theory to Practice* (The MIT Press, Cambridge, 2006)

S. Vale, International collaboration to understand the relevance of big data for official statistics. Stat. J. IAOS **31**(2), 159–163 (2015)

R. Vaño, I. Lacalle, P. Sowiński, R. S-Julián, C.E. Palau, Cloud-native workload orchestration at the edge: a deployment review and future directions. Sensors **23**(4), 2215 (2023)

W. Yung, S.-M. Tam, B. Buelens, H. Chipman, F. Dumpert, G. Ascari, F. Rocci, J. Burger, I. Choi, A quality framework for statistical algorithms. Stat. J. IAOS **38**(1), 291–308 (2022)

Q. Zhang, L. Liu, C. Pu, Q. Dou, L. Wu, W. Zhou, A comparative study of containers and virtual machines in big data environment, in *2018 IEEE 11th International Conference on Cloud Computing (CLOUD)*. IEEE (2018), pp. 178–185

Part IV
Use Cases and Insights

Chapter 9
Domain Adaptation of a BERT Model for Analyzing Job Advertisements at the German Federal Employment Agency

Lars Fiedler, Barbara Hofmann, Koen Loogman, and Tobias Scherl

9.1 Introduction

Every year millions of job vacancies are posted on the webpage of the German Federal Employment Agency (FEA). Employers can submit their job ads to the FEA using different channels. While there are ways to post a job without involvement of the FEAs caseworkers, at the time that the project we are referring to was initiated, a large portion of job ads was sent to the caseworkers via email. The caseworkers then transferred the information from the job ad texts into structured data what was required by the IT system to match job-seeking individuals to jobs.

Handling one job ad took a caseworker several minutes. Facing more than one million job ads per year that had to be handled in such a way, in 2021, the FEA decided to initiate a project to reduce the amount of time caseworkers had to spend on that task. The aim of the project was to build a piece of software including machine learning models to take over the task of extracting information from these job ad texts and to propose the extracted information to the caseworkers who had to correct or verify the proposed information. The expectation was to significantly reduce the amount of time that caseworkers had to spend on transferring job ads into the IT system. While we are not aware of a project in other countries with the same focus, there is a growing number of projects in the public sector using machine learning to improve services (European Commission, Joint Research Centre 2021).

L. Fiedler (✉) · B. Hofmann · T. Scherl
IT-Systemhaus der Bundesagentur für Arbeit, Nuremberg, Germany
e-mail: Lars.Fiedler@arbeitsagentur.de; barbara.hofmann3@arbeitsagentur.de; Tobias.Scherl2@arbeitsagentur.de

K. Loogman
Master student of FOM Hochschule für Oekonomie & Management, Nuremberg, Germany
e-mail: koen.loogman@fom-net.de

F. Dumpert (ed.), *Foundations and Advances of Machine Learning in Official Statistics*, Society, Environment and Statistics,
https://doi.org/10.1007/978-3-032-10004-7_9

We present insights about the use of different machine learning approaches from this project. The task at hand is the prediction of different entities and classes related to job advertisements. There is empirical evidence that domain adaptation of language models yields performance gains (Gururangan et al. 2020), e.g., in the biomedical discipline (Lee et al. 2020). We used various natural language processing approaches and finally adapted a gBERT-base model to our specific text domain of job advertisement texts following Gnehm et al. (2022).

9.2 Tasks

In our project, we had about 40 different attributes that should be predicted by the models. The attributes were predicted by separate models without taking into account potential interdependencies. In this chapter, we focus only on six attributes listed in Table 9.1 that we investigated in more detail. We present comparisons of models on two types of tasks. First, classification of job advertisement texts—either multi-label, multi-class, or binary classification—and second, Named Entity Recognition (NER). The classification tasks are "type of job" (binary classification), "collective wage agreement" (binary classification), "type of contract" (multi-class), and "type of application" (multi-label classification). We chose to present two NER tasks, one of which is a model extracting "contact details" (first name, second name, email address, and telephone number) and the other is a model extracting the "type of collective agreement."

Table 9.1 Task and description

Model task general	Model task specific	Label description
Binary classification	Type of job	Regular employment vs. apprenticeship/dual study program
Binary classification	Bound by collective wage agreement	Employer bound by a collective agreement
Multi-label classification	Application type	Application via mail, in person, in written form, via a portal, etc.
Multi-class classification	Contract type	Permanent, fixed-term, or not specified
NER	Collective wage agreement	Legally binding agreement between employers and trade unions. Example values like TVÖD SuE, IG Metall
NER	Contact details	First and second name of contact, email address, and telephone number

9.3 Models

Training

As base model, we use gBERT-base model from deepset (Chan et al. 2020). It is a BERT-based transformer model specifically developed to process and understand German language. The architecture is similar to the original BERT model, but it is solely trained on German data. We adapted a gBERT-base model to our specific text domain of job advertisement texts following Gnehm et al. (2022). We used the same hyperparameters as Gnehm et al. (2022). The model was adapted to the domain by training it with 580,000 job advertisements. It is important to note that training the domain-adapted model does not require labeled data. Instead, it only uses the texts of the job advertisements as training data. This training took around 1 week until our final model called SteA-BERT was created. Afterward, we used the adapted model to train attribute-specific models for attributes like “application type” or “contract type.” This process is visualized in Fig. 9.1.

To see the gain of the domain adaptation, we also trained the gBERT-base to get the attribute-specific models (Fig. 9.2). Below you will find our examinations for different aspects of the model performance. All of them measure the performance of the attribute-specific models, to indirectly compare gBERT-base and SteA-BERT.

Examination 1: Training Without Fine-Tuning the Pre-trained Encoder Parameters

To get a better idea of the impact of our domain adaptation, we froze the body of the models for some of the evaluation examinations, meaning that during training, the parameters of the pre-trained encoder were not modified, and only the classification layers (the head of the model) were trained (Fig. 9.3). This can be achieved in PyTorch by setting the attribute “requires grad” of the encoder Neural Network (NN) to False.

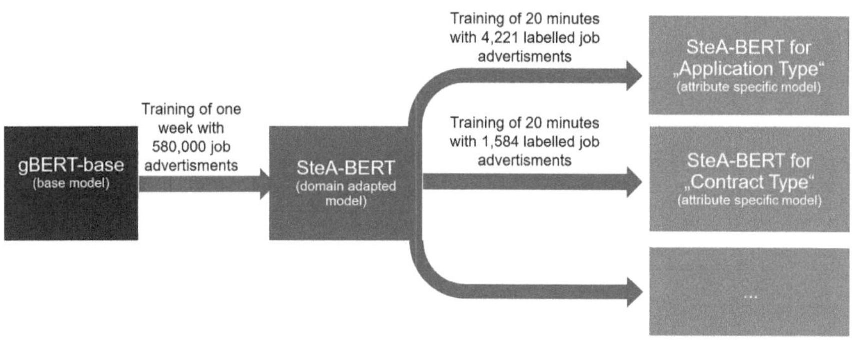

Fig. 9.1 Visualization of the domain adaptation and training process for attribute-specific models

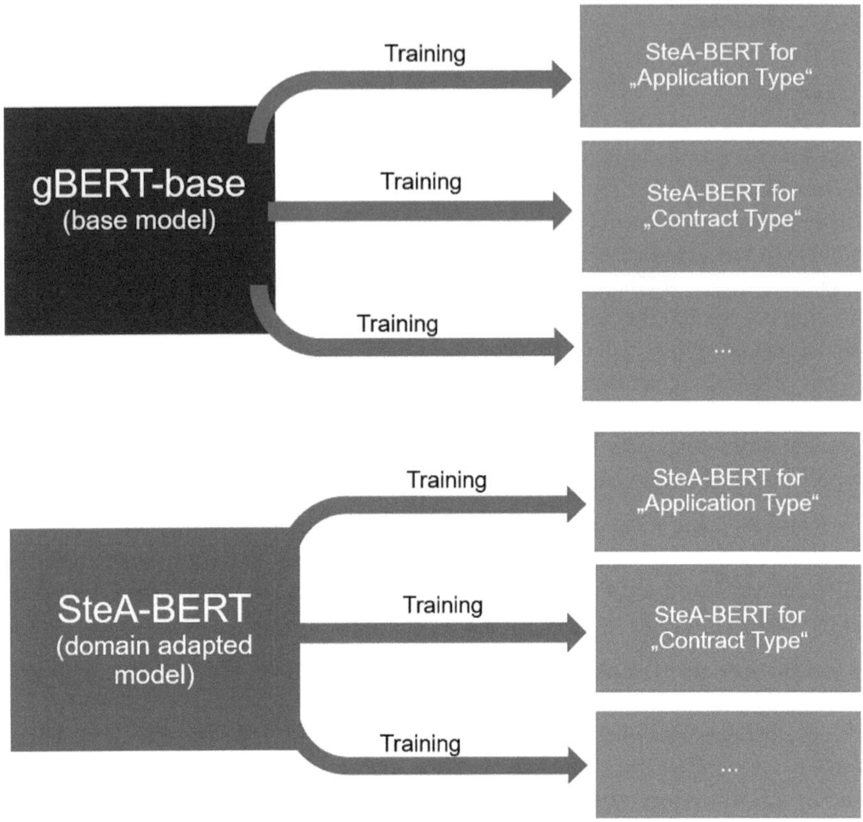

Fig. 9.2 Training the attribute-specific models based on gBERT-base and based on SteA-BERT

Examination 2: Training with Fine-Tuning the Pre-trained Encoder Parameters

Training with fine-tuning the pre-trained encoder parameters is the usual setup to achieve best results. This can be achieved in PyTorch by setting the attribute “requires grad” of the encoder NN to True. Again, we compared the results of gBERT-base and SteA-BERT. In all further examinations, we also used this setup.

Examination 3: Reduced Sample Size

Another gain of domain adaption is that the training of the specific models needs less data. So, we reduced the amount of data for the training of the attribute-specific models to compare the results of gBERT-base and SteA-BERT.

Examination 4: Comparing the Starting Points of Task-Specific Fine-Tuning

During a training run, the model needs a certain number of steps to reach its best results. The number of steps needed affects the duration of a training run. The domain-adapted model should have a better starting point. So, it should converge

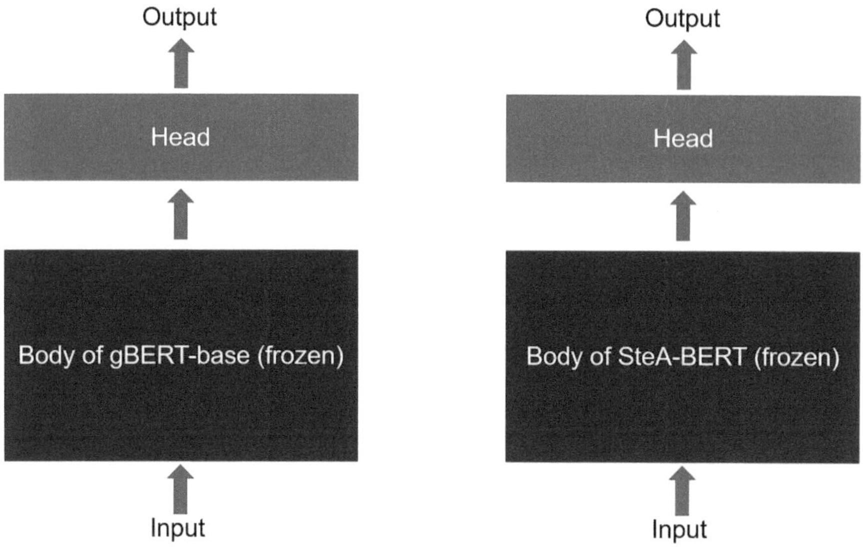

Fig. 9.3 Visualization of the examination where the parameters of the pre-trained encoder (body) were not modified during training and only the classification layers (head) were trained

with less steps. We compared gBERT-base and SteA-BERT to see their starting points and how many steps they needed to converge to their upper limit.

9.4 Data

Data for Base Model (gBERT-Base)
The gBERT-base was trained with some common German data sources (as specified in the paper of deepset (Chan et al. 2020)):

- OSCAR (Open Super-large Crawled ALMAnaCH Corpus) from the Common Crawl project; size: 145 GB
- OPUS: a source for diverse texts in different domains (politics, science, movies, etc.); size: 10 GB
- German Wikipedia: source for general-purpose text; size: 6 GB
- OpenLegalData: a collection of legal texts; size: 2.4 GB

Data for Domain-Adapted Model (SteA-BERT)
The base model mentioned above was then fine-tuned using data from our internal systems. We had a database containing the German job advertisements for several years until February 2022. We removed duplicated and outdated data and only used data that was generated by caseworkers. In the end, we had about 580,000 job advertisements that we used for the domain adaptation of the base model.

Table 9.2 Task and data points

Model task general	Model task specific	N
Binary classification	Type of job	200,000
Binary classification	Bound by collective wage agreement	2,747
Multi-label classification	Application type	4,221
Multi-class classification	Contract type	1,584
NER	Collective wage agreement	2,442
NER	Contact details	2,205

Data for Attribute-Specific Models
The database also contained additional attributes for each job advertisement. These attributes were filled by the caseworkers over the years. So, these data could be used as labels for the training of our models. However, we found out that the quality of the pre-labeled data was not sufficient. One reason was just human mistakes. Another reason was that the caseworkers had different opinions about how to fill the attributes. This led to inconsistencies of the data, which were confusing for the training of our models. So, for most of the attributes, a team of domain experts re-labeled the data manually.

For each attribute, the data was split initially into training (80%), validation (10%), and test (10%) data. Afterward, we ran examinations with different splits for training (60%), validation (20%), and test (20%). The sample sizes varied by attribute and depended on whether pre-labeled data was used or manually re-labeled data. Table 9.2 lists the information about the attributes and the data used.

9.5 Results

The models were trained on the tasks from Table 9.2 using the same hyperparameters, random state, and data distributions with only the base model being different. We use the weighted F1-score to assess model performance.

Examination 1: Training Without Fine-Tuning the Pre-trained Encoder Parameters
Table 9.3 lists the results when the body of the models was frozen. In general, the domain-adapted SteA-BERT delivered better results than its base model gBERT-base (except for the attribute "contract type"). But besides the model for the attribute "type of job," even the domain-adapted results are poor ranging from 0.3% ("collective wage agreement" and "contact details") to 47.1% ("bound by collective wage agreement").

Table 9.3 Weighted F1-scores: results without fine-tuning the pre-trained encoder parameters

Model task	gBERT-base	SteA-BERT
Type of job	95.6%	99.6%
Bound by collective wage agreement	46.2%	47.1%
Application type	17.9%	19.1%
Contract type	36.6%	34.6%
Collective wage agreement	0.0%	0.3%
Contact details	0.1%	0.3%

Table 9.4 Weighted F1-scores: results with fine-tuning the pre-trained encoder parameters

Model task	gBERT-base	SteA-BERT
Type of job	99.9%	99.8%
Bound by collective wage agreement	82.3%	85.8%
Application type	89.7%	89.3%
Contract type	76.7%	84.8%
Collective wage agreement	87.5%	85.5%
Contact details	96.8%	96.7%

Examination 2: Training with Fine-Tuning the Pre-trained Encoder Parameters

Turning to the results of the models with a fine-tuned body in Table 9.4, we find both the performance of the models based on SteA-BERT and the performance of the models based on gBERT-base to improve significantly. In fact, the relative performance gain by domain adaptation shrinks and even turns negative in the cases of the attributes "type of job," "application type," and "collective wage agreement." Since we found that the model for "type of job" performed well with a frozen and a fine-tuned body, we concluded that the sample size per attribute plays a significant role in determining the performance of a model and potentially compensates the impact of domain adaptation. Therefore, we examined the performance of the models when reducing the sample sizes.

Examination 3: Reduced Sample Size

To examine the impact of a reduced sample size, we used (i) 1000 data points as well as (ii) up to 5000 data points that were randomly sampled. To avoid too small validation and test samples, we decided to use different splits for training (60%), validation (20%), and test (20%) data in this examination. We found that smaller sample sizes highlight the benefit of domain adaption. The results of our examinations listed in Tables 9.5 and 9.6 indicate that bigger sample sizes are beneficial for the performance of our models as well.

Table 9.5 Weighted F1-scores with 1000 data points: results with fine-tuning the pre-trained encoder parameters

Model task	gBERT-base	SteA-BERT	Number of data points
Type of job	100.0%	100.0%	1000
Bound by collective wage agreement	85.3%	80.5%	1,000
Application type	84.7%	86.5%	1,000
Contract type	73.8%	79.2%	1,000
Collective wage agreement	78.2%	80.9%	1,000
Contact details	92.3%	95.2%	1,000

Table 9.6 Weighted F1-scores with up to 5000 data points: results with fine-tuning the pre-trained encoder parameters

Model task	gBERT-base	SteA-BERT	Number of data points
Type of job	99.6%	99.8%	5,000
Bound by collective wage agreement	86.6%	88.9%	2,747
Application type	88.7%	89.0%	4,221
Contract type	77.8%	84.0%	1,584
Collective wage agreement	83.8%	85.1%	2,442
Contact details	96.0%	96.3%	2,205

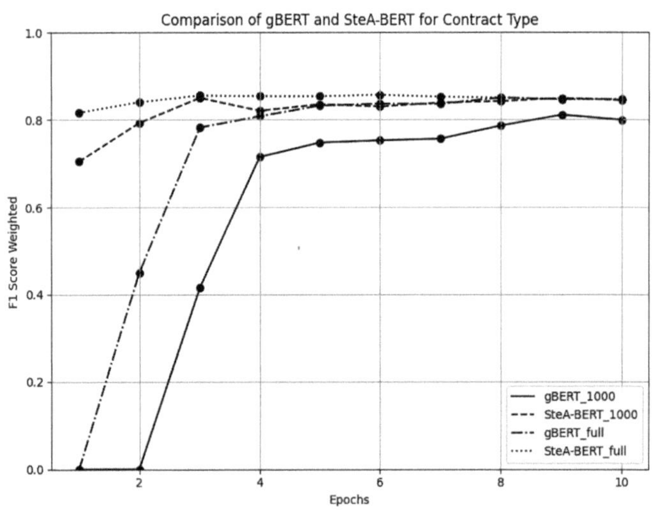

Fig. 9.4 Comparison of training SteA-BERT and gBERT on the task "contract type" with 1000 data points as well as the full sample and fine-tuning of the encoder. Showing the better start of the domain adaptation when fine-tuning a model

Examination 4: Comparing the Starting Points of Task-Specific Fine-Tuning
While we were able to achieve similar results with both models, we can see that the domain-adapted SteA-BERT model gets a better start than gBERT-base as shown in Fig. 9.4. This means we are able to do more training examinations with less training

steps in the same amount of time using our SteA-BERT compared to gBERT-base. This is useful, for example, when you have many training runs to find optimal hyperparameters or to try different modeling approaches. However, the results also suggest that the advantage of better starting points of the SteA-BERT seems to become smaller with bigger sample sizes.

9.6 Conclusion

Our approach showed that the domain adaptation of a pre-trained model can lead to performance boosts when predicting different entities and classes related to job advertisements. We discovered that the impact of adapting to a domain depends on the task and sample size. The results of our examinations indicate that in most of the tasks, bigger sample sizes are beneficial for the performance of our models. Additionally, if there is sufficient high-quality data available, domain adaptation is not necessary, because domain adaptation will occur during the training process. Our results also suggest that even if the task relates to the domain, there is not always a performance gain. In such cases, the model is just introduced to the syntax of a domain but does not benefit from this. However, if a task benefits from adapting to a specific domain, we detected that experimenting becomes faster since the model has a better start, because it is already familiar with the respective domain.

References

B. Chan, S. Schweter, T. Müller, German's next language model (2020). https://arxiv.org/abs/2010.10906

European Commission, Joint Research Centre. Selected AI cases in the public sector (JRC129301) (2021). European Commission, Joint Research Centre (JRC) [Dataset] PID: http://data.europa.eu/89h/7342ea15-fd4f-4184-9603-98bd87d8239a

A.-S. Gnehm, E. Bühlmann, S. Clematide, Evaluation of transfer learning and domain adaptation for analyzing German-speaking job advertisements, in *Proceedings of the Thirteenth Language Resources and Evaluation Conference*, ed. by N. Calzolari, F. Béchet, P. Blache, K. Choukri, C. Cieri, T. Declerck, S. Goggi, H. Isahara, B. Maegaard, J. Mariani, H. Mazo, J. Odijk, S. Piperidis, Marseille (European Language Resources Association, Paris, 2022), pp. 3892–3901. https://aclanthology.org/2022.lrec-1.414

S. Gururangan, A. Marasović, S. Swayamdipta, K. Lo, I. Beltagy, D. Downey, N.A. Smith, Don't stop pretraining: Adapt language models to domains and tasks (2020). https://arxiv.org/abs/2004.10964

J. Lee, W. Yoon, S. Kim, D. Kim, S. Kim, C.H. So, J. Kang, Biobert: a pre-trained biomedical language representation model for biomedical text mining. Bioinformatics **36**(4), 1234–1240 (2020)

Chapter 10
Approaches to Automated NACE Coding of German Business Activity Descriptions

Felix Beuter, Johannes Gussenbauer, Elias Minther, Viktoria Szabo, and Susanne Wegner

10.1 Introduction

Digitalization and automation are crucial for ensuring the future of official statistics. One labor-intensive task in the statistical process involves assigning codes from a standardized classification scheme to unstructured data, often in the form of human language text. This step takes place in Subprocess 5.2 (classify and code) of the Generic Statistical Business Process Model (GSBPM)[1] and is essential for standardizing concepts and compiling statistical data. Notable examples of standard classification schemes include the European Classification of Individual Consumption according to Purpose (ECOICOP) and the International Standard Classification of Occupations (ISCO).[2]

Among these numerous standard classification schemes, the categorization of economic activities is of significant importance in data analysis and interpretation.

[1] See https://unece.org/statistics/modernstats/gsbpm

[2] See https://unstats.un.org/unsd/classifications/ or https://joinup.ec.europa.eu/collection/eurostat/solution/stat-classifications

F. Beuter · E. Minther · S. Wegner (✉)
Federal Statistical Office of Germany, Wiesbaden, Germany
e-mail: Felix.Beuter@destatis.de; Elias.Minther@destatis.de; Susanne.Wegner@destatis.de

J. Gussenbauer
Statistics Austria, Wien, Austria
e-mail: Johannes.Gussenbauer@statistik.gv.at

V. Szabo
Reply, München, Germany
e-mail: v.szabo@reply.de

F. Dumpert (ed.), *Foundations and Advances of Machine Learning in Official Statistics*, Society, Environment and Statistics,
https://doi.org/10.1007/978-3-032-10004-7_10

The NACE (Nomenclature générale des activités économiques dans les communautés européennes) coding system (Eurostat 2008) is widely used for this task; however, manual coding by domain experts is time-consuming and susceptible to inconsistencies. Consequently, there is an increasing demand to automate the NACE coding process to mitigate these issues.

In recent years, many statistical offices have embraced machine learning and natural language processing (NLP) techniques for this task, such as L'Institut national de la statistique et des études économiques (INSEE) (Faria and Seimandi 2023),[3] Statistics Finland (Kärkimaa and Larja 2018), the National Institute of Statistics and Economic Studies of the Grand Duchy of Luxembourg (STATEC) (Huang et al. 2024), the Federal Statistical Office of Germany (Kühnemann et al. 2020), United Kingdom's Office for National Statistics (ONS) (Brown et al. 2021), and the Swiss Federal Statistical Office (Bundesamt für Statistik 2021), among others.[4]

While the NLP community has accumulated ample experience in text classification, the task continues to pose challenges, as detailed in Sect. 10.3. A significant challenge in official statistics pertains to the stringent quality requirements. To tackle this hurdle, automatic coding is usually not applied to all data directly; rather, it forms a segment of the statistical process where an accurate automatic assignment can be achieved. For the remaining data, additional quality assessments and manual coding may be necessary. The techniques presented by Statistics Austria and the Federal Statistical Office of Germany concentrate solely on the automated classification component.[5]

10.2 The NACE Classification

The NACE is a standardized classification system for economic activities within the European Union. NACE coded data is being used in various statistics like censuses or business statistics. It organizes economic data into a hierarchical four-level structure for statistical purposes, facilitating the comparison and analysis of economic activities across member states. The hierarchy is structured as follows:

1. **Section:** The broadest category, represented by alphabetic letters (A-U), describes general sectors of economic activities.
2. **Division:** It is defined by two-digit numeric codes and offers a more detailed categorization within each section.
3. **Group:** It further refines divisions, represented by three-digit numeric codes, indicating specific fields within a division.

[3] See also Sect. 8.4 of Avouac et al. (2025) in this book.

[4] Another example can be found in Sect. 13.2.3 of Barragán et al. (2025) in this book.

[5] The authors of this chapter are listed in alphabetical order.

4. **Class:** The most detailed level, with four-digit numeric codes, delineates precise activities within a group.

This classification system is essential for the harmonization of economic data across the EU, serving a pivotal role in economic research, policymaking, and administrative processes by ensuring consistent data collection and analysis methodologies.

In Germany, a more detailed fifth level is used on top of NACE, known as "Wirtschaftszweigklassifikation." At the time of writing, version 2008 is the current one used, but it will be updated in the near future with the upcoming NACE revision. When classifying, the aim is to predict the code of the deepest level possible, but the practicality of the chosen level depends on the specific use case.

10.3 Specific Challenges

The handling of NACE coding in official statistics presents several challenges. While NLP tasks like sentiment or topic classification typically involve a small number of classes, NACE encompasses hundreds of classes at its most granular level. Standard text classification algorithms and some labeling tools are not tailored to handle such a vast array of classes.

The large number of classes arises from the intricate information requirements, resulting in highly nuanced classes and classes that are inherently difficult to differentiate. Texts may not consistently provide information at this detailed level. A practical example is class 46 ("wholesale trade, except of motor vehicles and motorcycles"), which is notoriously hard to distinguish from class 47 ("Retail trade, except of motor vehicles and motorcycles"), when textual information does not give a clue about the kind of the company's customers. An example would be the textual description "Herstellung und Verkauf von Modellbauteilen" (manufacture and sale of model components). Other texts contain the necessary details but may be ambiguous in itself, for instance, a company describing multiple economic activities that fall into different NACE classes. A common example for this would be a generic description like "Einzel- und Großhandel mit Lebensmitteln aller Art" (retail and wholesale of all types of food).

Using a hierarchical class structure raises questions about how best to approach this issue. Additionally, we note a prevalent issue of highly imbalanced training data across all known use cases.

In summary, the specific challenges for text-to-code tasks in official statistics are:

- A high number of classes
- Specificity of the classes vs. vague or ambiguous descriptions
- Class imbalance
- Hierarchically structured classes

These challenges complicate not only automated but also manual coding. A human coder faces the task of comprehending all classes, determining the appropriate choice in cases of ambiguity or vagueness, and assigning the correct hierarchical level for labeling a text. Automated coding hinges on top-tier training data quality, and any labeling inconsistencies will be mirrored by the machine learning algorithm.

Furthermore, the absence of high-quality NLP tools for the German language poses a challenge that primarily impacts automatic coding. An illustration of this is the lack of a cutting-edge stemmer or lemmatizer for German that matches the quality of tools available for the English language. Additionally, Large Language Models (LLMs) exhibit superior performance in English compared to German.

10.4 Use Case at Statistics Austria

Statistics Austria investigated the NACE classification of companies in the Statistical Business Register (SBR) during the ESSnet Trusted Smart Statistics—Web Intelligence Network, Grant Agreement Number 101035829. One of the analysis tasks involved deriving the NACE code of an enterprise by analyzing text gathered from its website. The analysis focused on classifying the two-digit NACE code. This section delineates the methodological approach employed and presents the outcomes of this task.

10.4.1 Data Acquisition

Text from websites was mostly collected for enterprises which are part of the sampling population of the annual survey on the usage of information and communication technologies (ICT) in Austrian enterprises for the years 2019 until 2021. Additionally, during the 2021 ICT survey, websites of enterprises with a number of employed people ranging from 5 to 9 were also incorporated for the purpose of acquiring website text data. During each survey year Statistics Austria used their internal URL finding procedure to acquire websites for the enterprises in the ICT sample and sampling population.

The URL finding procedure uses the Google Search API, as well as the R programming language (R Core Team 2023) and Selenium.[6] First, a set of candidate URLs is retrieved searching for each enterprise with the Google Search API. The search string contains the name and address of the enterprise taken from the SBR. In the case of one-man businesses the search string does not contain the name of the business, but more details on the business address are used instead. The API configuration includes an extensive list of URLs that are blocked from the search

[6] Selenium is a framework for web browser automation, see also https://www.selenium.dev/

results. The URLs received from Google Search are processed further to remove duplicates and to keep country-coded folders for the regionalized website. After preprocessing each enterprise e_i, $i = 1, \ldots, N$, one receives a list of possible websites $u_{1(i)}, \ldots, u_{n(i)}$ to match with.

The results for ICT 2021 population, which contained roughly 41,000 legal units, yield, after preprocessing, about 90,000 URLs to be scraped. The URLs are scraped using the R package RSelenium; see Harrison (2020).

During the scraping process the crawler scrapes the main page and up to 25 subpages on the same domain while respecting the robots.txt exclusion protocol. The crawler is instructed to look for certain subpages, like "imprint," "contact," "impressum," which can contain the name, contact information, as well as VAT or commercial register numbers (CRNs) of the enterprise who owns the website. In Austria, businesses are legally obliged to identify themselves and in many cases also list their VAT or CRN on their webpages. These register numbers are available in the SBR and represent a reliable source for linking an enterprise to a website.

The text extracted from the crawled webpages of a URL undergoes a filtering process where all HTML tags and embedded code are stripped away, retaining only the textual content. Subsequently, these text snippets are subjected to further processing to extract any VAT or CRN details using regular expressions.

The presence of a VAT or CRN on a website allows for deterministic linking to corresponding enterprises. Approximately 64% of enterprises can be deterministically linked to a website through the identification of their VAT or CRN numbers.

Overall, the data collected contains over 128,000 website-enterprise pairs (e_i, u_i) with about 88,600 websites linked to roughly 62,500 enterprises. As the ICT population does not change drastically throughout the years, the enterprise website pairs were scraped every year.

10.4.2 Data Preprocessing

Before applying a classification model, the text gathered from the website is preprocessed:

- Transform each word with the German morphological lexicon available on https://www.openthesaurus.de/about/download.[7]
- Remove all digits and punctuations.
- Remove characters not part of the German dictionary.
- Remove German stop words.

For this purpose a "word" is defined as a consecutive sequence of characters without any spaces positioned between two spaces or located at the beginning or end of a paragraph or a sentence.

[7] Accessed 17 May 2024.

10.4.3 Feature Selection

After the preprocessing steps have been applied, the scraped text contains over two million different words; thus a feature selection strategy is needed before applying a classification model. Initially the NACE code description including examples, which is used and maintained by the classification unit, was considered as a set of features. In addition, the descriptions for the economic classification by the Austrian chamber of commerce were considered. This set contains roughly 21,000 words, but only about 55% of these words appear in the processed text data. Thus the feature selection strategy proposed by Uysal (2016) was tested. In Uysal (2016), both a global and a local feature selection score are combined to select a balanced set of features for use in a classification model. The global feature selection score determines the overall importance of a feature, and the local feature selection score addresses the issue of choosing a balanced feature set. For the global feature selection score, we tested the *Gini Index* (GI), *Distinguishing Feature Selector* (DFS), and *Information Gain* (IG). For the local feature selection score, we used the *Odds Ratio* (OR). The feature selection strategy was applied to the processed text grouped by the two-digit NACE codes of the respective enterprises.

Initially, the selection strategy was applied to all available data to extract words useful for predicting the two-digit NACE codes. It is important to note that selecting features after splitting the data into training and test sets represents a more realistic scenario, because in that case the test part of the data cannot have an influence on selection the feature subset. However, due to the computational intensity of this feature selection method, performing this step once was a convenient choice for testing the approach.

We applied the selection strategy to select up to 200 and 500 words for each two-digit NACE code. In the following discussion, we will denote these sets of words as W_{200} and W_{500}.

10.4.4 Classifier

For predicting the NACE code, two different models were tested: first, a Neural Network model with the R package Keras and the TensorFlow software, see Allaire and Chollet (2019) and Abadi et al. (2015), and second, the XGBoost algorithm, see Chen and Guestrin (2016), available through the R package xgboost, see Chen et al. (2023). The motivation for testing a Neural Network model was based on the efficient implementation of modern Neural Network software, which is designed to handle thousands of features. Additionally, the use of pretrained word embeddings could potentially improve prediction quality without requiring large amounts of data.

The Neural Network model was applied using two different architectures. The first architecture consisted only of feedforward layers and used one-hot encoded

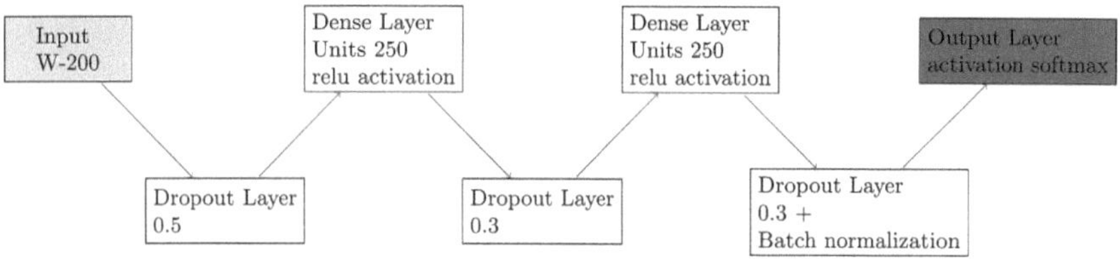

Fig. 10.1 Neural Network architecture: wide

words from the webpages as input, weighted by the term frequency-inverse document frequency (TF-IDF) transformation. The set of words used as features was W_{200}. Figure 10.1 shows the architecture of the Neural Network.

The second model specification builds on the first by incorporating additional input features found with W_{500}. These features are transformed using pretrained word embeddings, and multiple convolutional filters are applied to the word embeddings. The outputs from the feedforward and convolutional layers are concatenated in a penultimate layer and then fed into a final softmax layer. For the word embedding the pretrained embeddings with 300 dimensions from fastText were used; see Joulin et al. (2016). These embeddings are trained on Wikipedia and Common Crawl,[8] an open repository of web-crawled data. Prior to applying these word embeddings the dimensionality reduction algorithm proposed by Raunak (2017) was applied to reduce the dimensionality from 300 to 50. This algorithm applies a combination of Principle Component Analysis algorithms and a so-called postprocessing algorithm, see Mu et al. (2018), to generate embeddings of lower dimensions. The postprocessing algorithm aims to eliminating common dominating directions in the embeddings. The dimensionality reduction algorithm was utilized primarily to reduce training time, and initial tests indicated that the reduced dimensionality resulted in minimal performance loss regarding prediction quality.

Figure 10.2 shows the architecture of the second model specification.

The hyperparameters for both Neural Network models are displayed in Figs. 10.1 and 10.2 and were determined with an additional hyperparameter tuning step with an 80-20 training and test split.

Compared to specifying Neural Network models, the XGBoost algorithm is more straightforward to apply. In addition to being easier to use, XGBoost can handle sparse matrices, which significantly reduces memory consumption during model training. Table 10.1 shows the hyperparameters chosen for the XGBoost algorithm. These parameters were not derived using hyperparameter tuning but were instead taken from another classification task where they proved to be useful.

[8] https://commoncrawl.org/

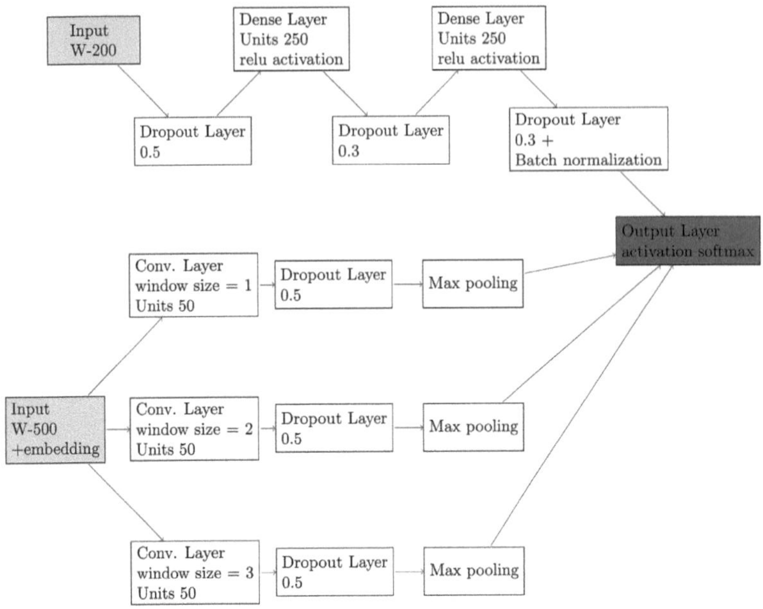

Fig. 10.2 Neural Network architecture: wide and deep

Table 10.1 Hyperparameters choosen for the XGBoost algorithm

Parameter	Value
nrounds	1,000
max_depth	7
eta	0.01
subsample	0.5
colsample_bytree	1
eval_metric	mlogloss
objective	multi:softprob

10.4.5 Results

This subsection presents the performance measures for applying the classifiers in different settings. The models were evaluated using a fivefold cross-validation four times with an 80-20 split, totaling a number of 20 prediction runs. Apart from directly using the classifiers, two additional settings were applied, one that uses the prediction on the first NACE level as predictor and another one that aims to limit the number of noise by selecting text only from subpages on a website which potentially hold information on the enterprise in question.

10.4.5.1 Using Predictions from a Higher Level Hierarchy

The hierarchical structure of the NACE code may potentially improve the prediction quality of the model as it seems intuitively reasonable that modeling higher level NACE codes would be a slightly more straightforward task. To implement this idea a model is trained to predict for each enterprise and URL pair (e_i, u_i) the NACE level 1 code. This was done four times using fivefold cross-validation, and, afterward, the average of the resulting four probability vectors was used as additional input. For the NACE level 1 prediction the feature selection method discussed in Sect. 10.5.2.2 was applied with respect to the different NACE level 1 codes. A total of 500 words per NACE level 1 code for the one-hot encoding scheme and 1,500 words per NACE level 1 code for the word embeddings were used. The model used for the prediction was the Neural Network model shown in Fig. 10.2. For the following results this scenario will be denoted as "Hierarchy."

10.4.5.2 Selecting Certain Subpages of a Website

Because the data available potentially holds a lot of noise, it could be beneficial for the prediction task to select certain parts of the available data prior to the feature selection. Thus only text from the landing page and certain subpages was selected as training input. The subpages were defined by having one of the following words either in its URL or in the text between the hyperreference tags on the landing page:

enterprise, company, unternehmen, home, welcome, ueber, über uns, über, geschichte, about us, uber uns, about, unsere, willkommen, produkt, product, artikel, article, organisation, dienstleistung, angebot, leistung, offer

For the following results this scenario will be denoted as "Selected Links."

Figure 10.3 illustrates the distribution of accuracy, F1 score, and top-5 accuracy[9] for each method across all cross-validation runs. The methods tested include the Neural Network using both One-Hot Encoding and Word Embedding inputs (NNet), the XGBoost model (XGBoost), the use of predicted NACE level 1 probability scores (Hierarchy), and the selection of only a subset of available text data (Selected Links). As our results show hardly any differences between the choice of selection score (DFS, Gini, or Information Gain) the XGBoost model and the Neural Network using predicted NACE level 1 probability scores (Hierarchy) was only used in combination with the DFS. Furthermore the usage of word embeddings and convolutional filters showed hardly any improvements despite the model's input containing much more information in terms of different words compared to using a one-hot encoding scheme. Incorporating predicted NACE level 1 codes as model input (Hierarchy) improves the F1, accuracy, and top-5 accuracy measures. The use of XGBoost yields results similar to those achieved with the Neural Network,

[9] See Powers (2020) for definitions of various evaluation metrics.

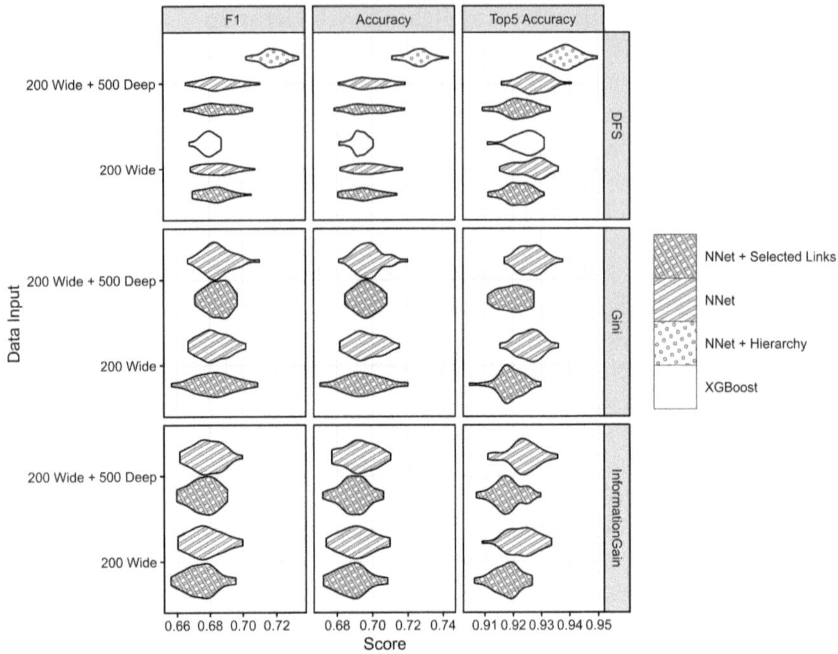

Fig. 10.3 Overall results using the Neural Network models (NNet), the XGBoost model (XGBoost), using predicted NACE level 1 codes (Hierarchy) and only a subset of available text data (Selected Links). The vertical panels indicate different performance measures and the horizontal panels show different feature selection scores

and selecting only a limited amount of text data does not enhance prediction performance. For top-5 accuracy, there even appears to be a negative effect.

Figure 10.4 shows the average scores based on the number of employees, grouped into the classes [0,10], (10,25], (25,50], (50,100], and 100+. The figure clearly demonstrates that predicting the two-digit NACE code becomes increasingly difficult as the size of the company, in terms of employees, grows. Using predicted NACE level 1 code probability scores (Hierarchy) outperforms the other strategies and models for almost all size classes. Notably, the XGBoost model performed significantly worse than the other models for the smallest size class, which could be attributed to the hyperparameters not being fine-tuned for this specific problem.

Figure 10.5 shows the average accuracy achieved (y-Axis) when automatically classifying cases with the highest prediction probability greater than or equal to x (x-axis). For the use of predicted NACE level 1 code probability scores (teal line) classifying 25% of the cases with the highest prediction probability results in a prediction accuracy slightly below 95%. The figure suggests that none of the models can achieve a high accuracy when automatically predicting a sizeable portion of the data. Thus using the predictions to support the manual editing of NACE codes seems to be more effective at this stage.

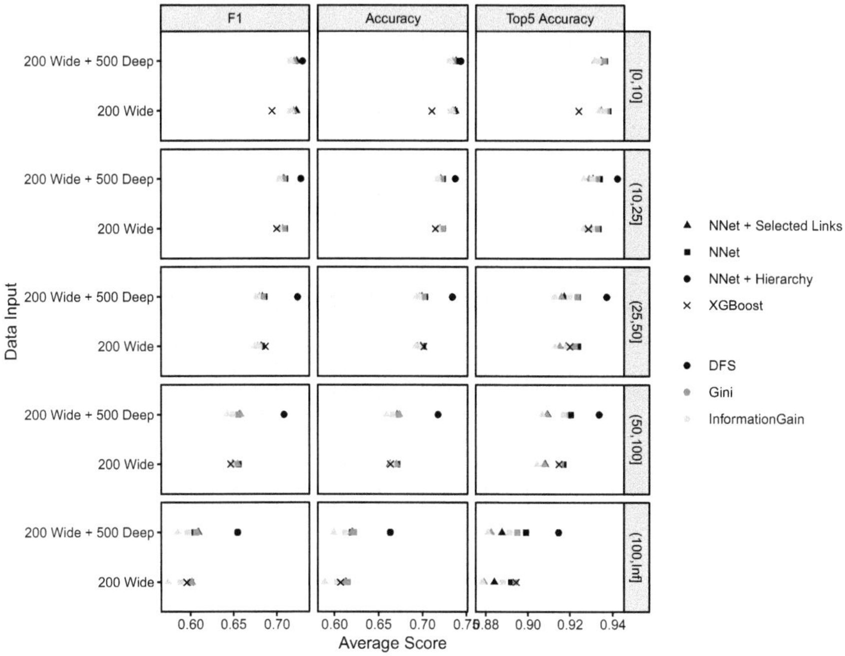

Fig. 10.4 Average scores per model (point shape) and feature selection score used (gray scale). The horizontal panels distinguish the results by the number of employed people for each enterprise

10.4.6 Hierarchical Performance Measures

As the hierarchy of the NACE seems to play an important role, specific evaluation metrics that respect the hierarchical nature of the classification were additionally used. For measuring the prediction quality a class distance (CD) measure following Sun and Lim (2001) was applied. Given an enterprise e_i, let $c_i = (c_{1;i}, c_{2;i}, c_{3;i}, c_{4;i})$ be the true NACE codes on levels 1 to 4 and $c_i^* = (c_{1;i}^*, c_{2;i}^*, c_{3;i}^*, c_{4;i}^*)$ the predicted NACE codes from a classification model. For ease of annotation let $c_{[j];i}$ and $c_{[j];i}^*$ contain the full NACE code up to level j. The category distance between true and predicted NACE codes for level $L = 1, \ldots, 4$ can be defined as

$$Dis(c_i, c_i^*, L) = 2 \cdot \left(\sum_{l=1}^{L} \mathbb{1}_{c_{[l];i}^* \neq c_{[l];i}} \right)$$

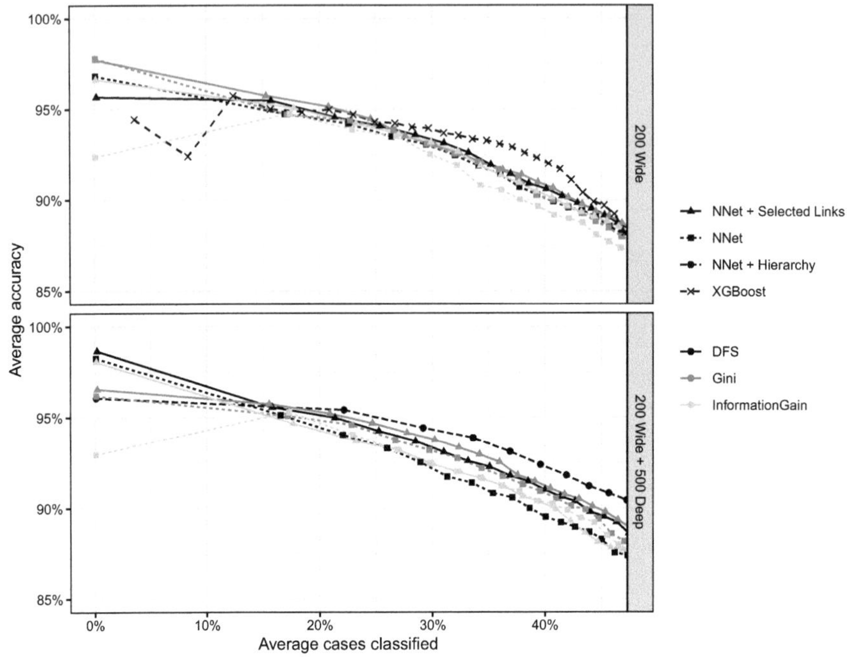

Fig. 10.5 Average accuracy given average classified cases for the different model specifications (linetype and point shape) and feature selection score (gray scale). The horizontal panels display the inputs used

Thinking of the hierarchical classification as a graph of depth L, the $Dis(c_i, c_i^*, L)$ is the shortest path between $c_{[L];i}^*$ and $c_{[L];i}$.

For a given class $\tilde{c}$, the contribution of enterprise e_i and predicted NACE c_i^* being either a false positive (FP) or a false negative (FN) predicted is defined as follows:

$$Conb(e_i, \tilde{c}) = \begin{cases} \min(1, \max(-1, 1 - \frac{Dis(c_i,\tilde{c},L)}{L})) & \text{, if } e_i \text{ is } FP \\ \min(1, \max(-1, 1 - \frac{Dis(c_i^*,\tilde{c},L)}{L})) & \text{, if } e_i \text{ is } FN. \end{cases}$$

Precision ($PR^{CD}(\tilde{c})$) and recall ($RE^{CD}(\tilde{c})$) of a given class $\tilde{c}$ incorporating class distance are then given by

$$PR^{CD}(\tilde{c}) = \frac{\max(0, TP(\tilde{c}) + FpConb(\tilde{c}) + FnConb(\tilde{c}))}{TP(\tilde{c}) + FP(\tilde{c}) + FnConb(\tilde{c})},$$

$$RE^{CD}(\tilde{c}) = \frac{\max(0, TP(\tilde{c}) + FpConb(\tilde{c}) + FnConb(\tilde{c}))}{TP(\tilde{c}) + FN(\tilde{c}) + FpConb(\tilde{c})},$$

$$AC^{CD}(\tilde{c}) = \frac{TP(\tilde{c}) + TN(\tilde{c}) + FpConb(\tilde{c}) + FnConb(\tilde{c})}{TP(\tilde{c}) + FP(\tilde{c}) + TN(\tilde{c}) + FN(\tilde{c})},$$

where $FpConb(\tilde{c})$ and $FnConb(\tilde{c})$ are defined as

$$FpConb(\tilde{c}) := \sum_{e_i \in FP} Conb(e_i, \tilde{c}),$$

$$FnConb(\tilde{c}) := \sum_{e_i \in FN} Conb(e_i, \tilde{c}).$$

The class distance-based F1 score can then be computed as

$$F1^{CD}(\tilde{c}) = 2 \cdot \frac{PR^{CD}(\tilde{c}) \cdot RE^{CD}(\tilde{c})}{PR^{CD}(\tilde{c}) + RE^{CD}(\tilde{c})}.$$

Including the notion of class distance the overall accuracy of a classifier AC^{CD} is proposed as

$$AC^{CD} = \frac{\sum_{\tilde{c}} TP(\tilde{c}) + FpConb(\tilde{c})}{\sum_{\tilde{c}} TP(\tilde{c}) + FN(\tilde{c})}.$$

The above described hierarchical performance measures are applied on the previous cross-validation runs using the data collected during the ICT Survey 2019 until 2021, and results are displayed in Fig. 10.6. Figure 10.6 shows, in the same fashion as Fig. 10.3, the accuracy and micro F1 score based on the class distance. Because the hierarchical performance measures directly punish a method the larger the distance from the predicted value $c^*_{[L];i}$ to the $c_{[L];i}$, Fig. 10.6 shows overall lower values for these performance measures. Nevertheless the ranking between the methods reveals a similar picture as is displayed in Fig. 10.3. The similar result might be caused by predicting only up the NACE level 2 codes and not more detailed.

To gain deeper insights into the behavior of this performance measure, an additional analysis was conducted to predict NACE codes up to level 4. In this scenario, only a single Neural Network and an XGBoost model were tested. Feature selection for both models was performed using the method described in Sect. 10.5.2.2. The Neural Network model was trained separately for each NACE level, with the predicted NACE code from the next higher level used as an additional feature. The XGBoost model was trained multiple times for each NACE level. For each NACE level 1 code, an additional model was trained to predict NACE level 2 codes, and the same approach was applied to predict NACE level 3 and level 4 codes. Predicted values were generated using models for NACE levels 1 through 4 which were applied sequentially, with the predicted NACE code from each previous

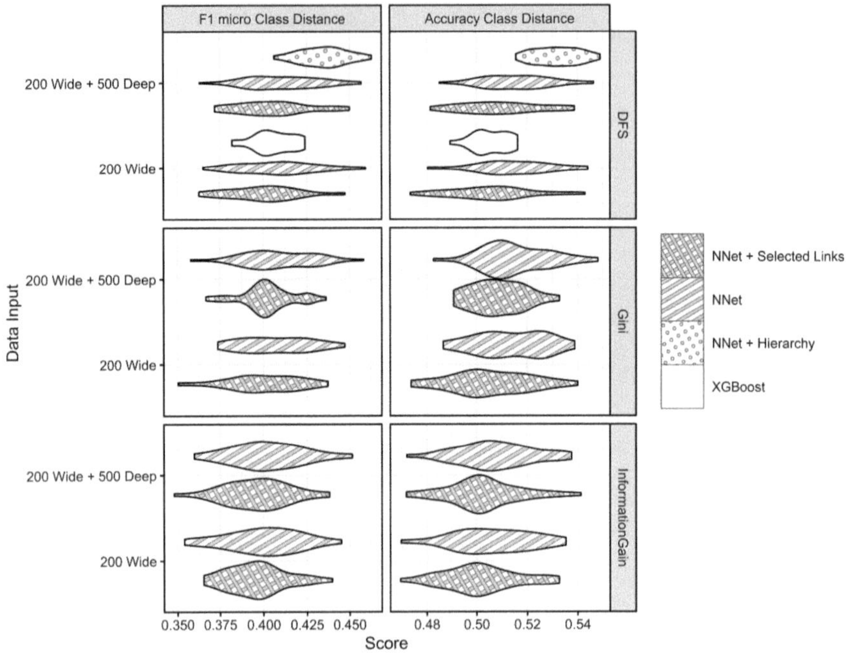

Fig. 10.6 Overall results when using class distance-based evaluation metrics. The hatchings indicate the use of Neural Network models (NNet), the XGBoost model (XGBoost), using predicted NACE level 1 codes (Hierarchy), and only a subset of available text data (Selected Links). The vertical panels indicate different hierarchical performance measures, and the horizontal panels show different feature selection scores

Table 10.2 Comparison of classical evaluation measures and class distance-based evaluation measures for predicting from NACE level 2 to NACE level 4

Measure	Method	NACE2	NACE3	NACE4
Accuracy	NNet + Hierarchy	0.738	0.663	0.615
	XGBoost	0.664	0.547	0.497
Accuracy CD	NNet + Hierarchy	0.567	0.511	0.467
	XGBoost	0.464	0.375	0.340
F1	NNet + Hierarchy	0.735	0.644	0.588
	XGBoost	0.664	0.548	0.501
F1 CD	NNet + Hierarchy	0.482	0.424	0.386
	XGBoost	0.503	0.365	0.325

hierarchy level defining the model for the current hierarchy level. Table 10.2 shows the results of this comparison. Similar to Fig. 10.6 the ranking between the methods does not change when using hierarchical evaluation metrics, implying that both methods make similarly severe errors with respect to the hierarchy of the classification.

10.4.7 Use Case Results

The use case of Statistics Austria involves utilizing text from enterprise websites to predict NACE level 2 codes. The classification methods employed include Neural Networks and the XGBoost algorithm in combination with various feature selection strategies. The results show only limited potential for directly applying the classification methods for predicting NACE codes. It is however more promising to use these methods in combination with a recommender system when manually editing NACE codes for enterprises. With respect to the different methods used no drastic differences were observed regarding various performance measures. The use of specific performance measures that take into account the hierarchical nature of NACE did not reveal additional insights into the performance of the different methods. This only implies that the methods made similar errors with respect to the hierarchy of the classification. Using hierarchy based instead of classical evaluation metrics should nevertheless be the preferred choice as it respects the hierarchical nature of the classification and is thus more useful for the task at hand.

10.5 Use Case at the Federal Statistical Office of Germany

The NACE classification is relevant to several statistics within the Federal Statistical Office of Germany (Destatis), including applications in business registration notices, the business register, and the microcensus. The focus here will be on the classification of business registration notices.

Business registration notices consist of self-descriptions by the registering companies, provided in a free text field within their registration submissions. These self-descriptions are traditionally classified into NACE codes by local trade offices and statistical offices of the federal states. The use case data includes 1,596,352 observations from January 2018 to April 2019, covering all federal states except Berlin. The observations are very unevenly distributed between the respective NACE classes, as can be seen in Fig. 10.7. Not all classes are represented in the observations, and not all observations are classified down to the most detailed NACE level, as can be seen in Table 10.3, where the average number of observations at NACE level 5 is not even a 16th of the number at NACE level 2. Table 10.4 shows some sample observations from the data set with their NACE classification codes.

For example, while class 47 (retail) is most strongly represented with a total of 266,704 observations, class 05 (coal mining) has only 1 observation. The length of the texts also varies significantly, ranging from single-word entries, such as "Friseur" (hairdresser) to one instance with 468 words, with an average of 8 words per description.

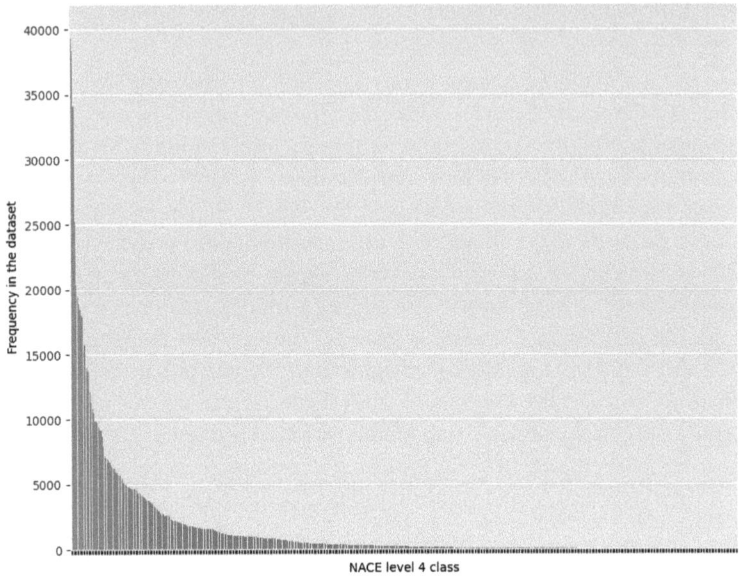

Fig. 10.7 Frequency of NACE level 4 classes in the data set

Table 10.3 Number of observations and classes per NACE level in the data set

Level	Number of observations	Number of classes	Avg. number of obs. per class
2	1,593,810	85	18,750.7
3	1,337,231	245	5,411.9
4	906,002	547	1,648.3
5	857,267	763	1,118.6

Table 10.4 Short excerpt from the training data

Description	NACE code
Fahrzeugvermietung	77.1
Friseursalon	96.02.1
Verhaltensberatung für Hunde	96.09.0
Solarium, Verkauf von Umzugsartikeln	96
IT-Beratung	62.02.0
Promotion	73.11.0

To enrich the data set with high-quality data, a list of keywords provided by the Federal Statistical Office of Germany[10] was used, which includes standard keywords for all classes. The keyword list currently contains a total of 34,528 entries

[10] https://www.klassifikationsserver.de/klassService/thyme/variant/wz2008, accessed 12 June 2024.

and is continuously being expanded. However, it was only used for a subset of our experiments.

10.5.1 Model Specifications

This section presents the methods used to select and evaluate the different models considered for the NACE classification task. Section 10.5.1.1 explains the specific steps applied to the data set to improve its quality. Section 10.5.1.2 describes the classifiers tested and the methods to tune and evaluate them. Finally, the results are presented in Sect. 10.5.1.3.

10.5.1.1 Preprocessing

Data preprocessing is an integral aspect of any classification problem. It involves improving data quality through various techniques and transforming it into a machine interpretable format. The best method was identified by analyzing different techniques described in Sect. 10.5.2.2. The results indicated that methods such as cleaning or character n-grams offered minimal to no improvement. Consequently, with a focus on performance, word tokenization combined with TF-IDF vectorization was chosen. Additionally, the words were lowercased. In summary, in contrast to the Austrian use case, only very few preprocessing steps were applied.

10.5.1.2 Classifiers

Different classifiers were tested during the experiments:[11]

- Multinomial Naïve Bayes using MultinomialNB from the Python library naive_bayes[12]
- Support Vector Classifier using linearSVC from the Python library svm[13]
- Logistic Regression using LogisticRegression from the Python library linear_model[14]
- Random Forest using RandomForestClassifier from Python sklearns ensemble library[15]

---

[11] For further information about the different machine learning classifiers, see Hastie et al. (2009).

[12] See https://scikit-learn.org/1.5/modules/generated/sklearn.naive_bayes.MultinomialNB.html

[13] See https://scikit-learn.org/dev/modules/generated/sklearn.svm.LinearSVC.html

[14] See https://scikit-learn.org/1.5/modules/generated/sklearn.linear_model.LogisticRegression.html

[15] See https://scikit-learn.org/dev/modules/generated/sklearn.ensemble.RandomForestClassifier.html

- Neural Net: A simple Feedforward Neural Network (FNN) was implemented with the number of features as the input layer, the number of classes as the output layer, and a ReLu activation function. Batch normalization was applied. Enlarging the network by adding hidden layers quickly led to overfitting and did not improve the results. The keyword list has not been included in this experiment. The FNN was implemented using the Python PyTorch library.[16]

Evaluation Strategy

Various evaluation strategies from different settings were used:

1. **Setting 1:** First, an 80-20 train-test split was applied to the data set. Therefore, classes with only a single observation had to be removed from the data to avoid issues. Since regular fivefold cross-validation would require each class to appear five times, the results were averaged from five different train-test splits with varying random seeds. The keyword list was not used in this setting. It was used mainly in previous experiments (Sects. 10.5.2.1–10.5.2.3) conducted by Herold et al. (2022).
2. **Setting 2:** A stratified 85-15 train-test split was used, with classes containing less than three observations removed.[17] First, model classes, hyperparameters, and preprocessing steps were evaluated using a fivefold cross-validation, where classes occurring fewer than five times were removed. The model with the best settings was then trained in the whole training data set and validated on the test data set. This setting was used for the main investigations, and the results reported in Sect. 10.5.1.3 were achieved using this approach.

Furthermore, different metrics were used to ensure optimal classification quality. In addition to accuracy, a strong focus was placed on the F1 score, evaluated using the macro average. The F1 metric is particularly suited for highly imbalanced data.

Table 10.5 displays the different classification metrics based on the baseline model. The F1 score is considerably lower compared to the accuracy, indicating a discrepancy. This is mainly because classes with more training data are more accurately classified, and, in the case of accuracy, they receive a higher weight due to their frequency compared to the F1 macro score, where all classes are weighted

Table 10.5 Metrics for NACE level 2 classification using Multinomial Naïve Bayes

Precision	Recall	Accuracy	F1	ROC_AUC	MCC	NIR
0.53	0.25	0.71	0.28	0.62	0.68	0.17

[16] See https://github.com/pytorch

[17] The various settings reflect the individual choices of the different researchers at different points in time. While Setting 1 was chosen by Herold et al. (2022), Setting 2 was chosen for Destatis inhouse experiments.

equally. The *no information rate (NIR)* denotes the largest class percentage. Despite the imperfect results of the baseline model, it is thus far better than best guessing the majority class. As there are 85 classes, the NIR is low, but it would even be much lower when considering deeper levels.

Hyperparameters

Hyperparameters are a crucial aspect of classifier design as they define the way the classifier is trained. Thus, they have a significant influence on the classifier's classification quality. According to the evaluation strategies mentioned above (Sect. 10.5.1.2), the hyperparameter tuning was divided into two different settings.

Setting 1
Initially, classes containing only one observation were removed from the data set, followed by the utilization of a CountVectorizer for feature extraction. The following classifiers were considered: Support Vector Classifier and Logistic Regression. In this process, the optimal hyperparameter settings were identified by searching for the highest F1 macro score values across a range of hyperparameter configurations (Tables 10.6 and 10.7) using the Python library TPOT (Tree-based Pipeline Optimization Tool),[18] an automated machine learning tool that optimizes machine learning pipelines using genetic programming.

Table 10.8 presents the best hyperparameter configuration for each model. It can be seen that the linear Support Vector Classifier achieves the best F1 value. However,

Table 10.6 Hyperparameter configurations tested for Logistic Regression

Parameter	Configurations
'penalty'	['l2']
'tol'	[1e-5]
'C'	[2**(-8), 2**(-6), 2**(-4), 2**(-2), 2**0, 2**2, 2**4, 2**6, 2**8]
'max_iter'	[500]
'dual'	[False]

Table 10.7 Hyperparameter configurations tested for the Support Vector Classifier

Parameter	Configurations
'penalty'	['l1', 'l2']
'tol'	[1e-5]
'C'	[2**(-8), 2**(-6), 2**(-4), 2**(-2), 2**0, 2**2, 2**4, 2**6, 2**8]
'max_iter'	[1000]

[18] See https://github.com/EpistasisLab/tpot

Table 10.8 Best model configurations

Model	Parameter	F1
linearsvc	C=1, dual=False, tol=1e-05	0.519475
logisticregression	C=16, max_iter=500, tol=1e-05	0.513003

Table 10.9 Accuracy and F1 score for the best preprocessing and hyperparameter configuration of each model

Classifier	Vector	Preproc.	Token	Acc.	F1	Hyperparameter configuration
Naïve Bayes	TF-IDF	Lower	char(7,7)	0.71	0.295	alpha=1.0, fit_prior=True
Support Vector Classifier	TF-IDF	Lower	char(7,7)	0.80	0.545	loss='hinge', penalty='l2', dual=True, C=1.0, tol=1e-5, max_iter=1000
Random Forest	TF-IDF	Lower	word	0.58	0.364	n_estimators=100, max_depth=100, class_weight='balanced'
Logistic Regression	TF-IDF	Lower	word	0.78	0.549	solver='lbfgs', penalty='l2', max_iter=5000, tol=1e-5, C=16, dual=False
Feedforward Neural Network	TF-IDF	Lower	word	0.78	0.542	Over-/Undersampling=False, num_epochs=5, batch_size=2048

the difference from the best result of Logistic Regression is marginal. Furthermore, no general trend for the selection of individual hyperparameters could be identified.

Setting 2

Various preprocessing and model configurations were tested and compared using fivefold cross-validation for NACE level 2 with Naïve Bayes as the baseline model. The best configurations for each model class are shown in Table 10.9. Logistic Regression and Support Vector Machine classifiers performed the best with F1 scores of around 0.55. Despite the slightly superior accuracy of SVC at 0.8 and a similar F1 score, Logistic Regression with the hyperparameter and preprocessing configurations mentioned below was chosen for further modeling due to the inability of SVC to provide probability outputs, which are critical for nuanced decision-making processes. Furthermore, Logistic Regression offers simplicity and interpretability, which are advantageous for practical implementations. Although Random Forest achieved lower scores, it shows potential for improvement through the testing of additional hyperparameter configurations. TF-IDF outperformed Bag of Words (BoW) in almost all configurations. However, the best tokenization method depended on the specific model class.

Table 10.10 Logistic Regression performance over all NACE levels

Level	F1	Accuracy	Balanced accuracy	Precision	Recall
2	0.56	0.78	0.53	0.63	0.53
3	0.44	0.70	0.40	0.52	0.40
4	0.41	0.68	0.37	0.50	0.37
5	0.41	0.64	0.37	0.50	0.37

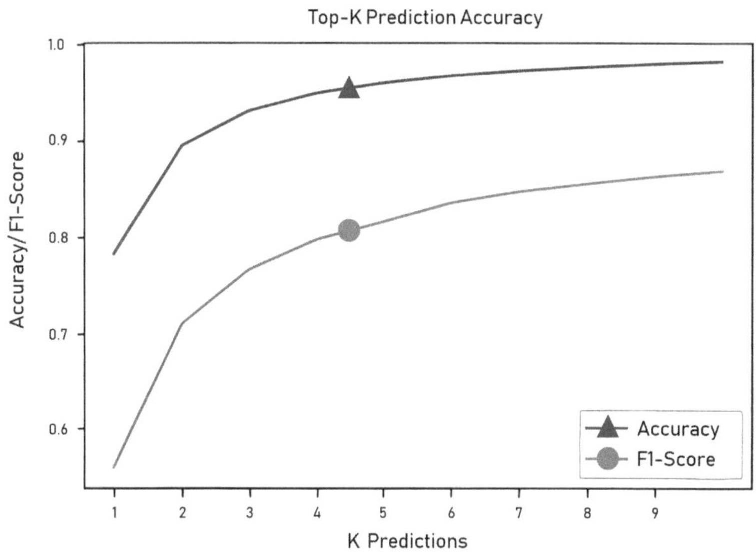

Fig. 10.8 Accuracy and F1 score for top ten class predictions

10.5.1.3 Results

Table 10.10 presents performance metrics for the previously selected Logistic Regression model across different depth levels of the NACE code, from level 2, which was also shown above, to level 5. A significant performance drop can be observed from level 2 to level 3, with the F1 score decreasing to 0.44 and accuracy falling to 0.70. This trend of decreasing performance continues through levels 4 and 5, but not as strongly. These results suggest that the model struggles with finer granularity in the NACE classification, probably due to increasing complexity and data sparsity at more detailed class levels.

One of the most significant challenges, which intensifies at lower NACE levels, is differentiating between individual classes. As a result, considering the top two predicted classes instead of just the best one can significantly improve the outcomes. Figure 10.8 illustrates how accuracy and the F1 score are improved when not only the best prediction but also the top ten predictions are taken into account. In general, considering the ten best predictions, the accuracy can reach 0.98 and the F1 score up to 0.87. The discrepancy between accuracy and the F1 score is substantially reduced when multiple predictions are considered, indicating that smaller classes particularly benefit from this approach. Even when considering the top two predictions, both metrics increase by more than ten percentage points: accuracy from 0.78 to 0.89 and F1 score from 0.56 to 0.71.

This issue is well illustrated by examining individual classes. For example, class 56 (gastronomy), which includes observations such as "Verkauf von Wurstwaren, Gastronomie (Verkauf von Speisen und Getränken, auch alkohol. Getränke) Imbiss" (sale of sausage products, gastronomy [sale of food and beverages, including alcoholic beverage] snack bar), achieves a fairly high F1 score of 0.94. This particular observation was accurately identified by the model as class 56 (gastronomy) with a probability of 0.97. However, the observation "Backwarenverkauf und Kaffeeausschank" (sale of baked goods and serving coffee) was categorized by the model as class 47 (retail) instead of class 56 (gastronomy), which would be the actual ground truth class in the data. The phrase "selling baked goods and serving coffee" actually spans two classes: (1) "retail for baked goods" and (2) "gastronomy with serving coffee." Without additional information, the model cannot determine which class predominates; therefore, its classification into class 47 (retail) would logically not be completely incorrect, yet it is treated as incorrect in all summary metrics for the model. There are numerous similar borderline cases that do not clearly fit into a single class. Therefore, achieving an accuracy or even more so an F1 score close to one will not be possible.

10.5.2 Further Experiments

Addressing the challenges of hierarchy and data imbalance, this section explores various methods aimed at enhancing classification performance in automated NACE coding. The techniques discussed include data augmentation and different feature extraction methods. Furthermore, the idea of using the hierarchy of the NACE code to optimize classification quality and enhance interpretability is proposed, but in a slightly different way than in the Austrian use case. Additionally, it delves into the approach of using Large Language Models (LLMs) to optimize NACE coding.

In Sect. 10.5.2.1, a variety of different augmentation and sampling methods are introduced to address data imbalance and clarify the decision boundaries. Section 10.5.2.2 describes the tokenization process using character n-grams and vectorization methods, such as bag of words (BoW) or TF-IDF. In Sect. 10.5.2.3 a number of different approaches to use the NACE hierarchy are presented. Finally, Sect. 10.5.2.4 presents a hybrid approach combining Generative LLMs with a top-k selection.

Each method addresses specific challenges encountered in NACE coding, offering insights into enhancing the overall classification process.

10.5.2.1 Data Augmentation

The effect of data augmentation was tested using three different classifiers: Naïve Bayes, Support Vector Classifier, and Logistic Regression. Different augmentation and sampling methods were conducted so that the number of each class was proportional to the number of the most represented class. The F1 score was calculated for different augmentation fractions (percentage proportion to the most represented class).

The following methods were tested (Chawla et al. 2002; Wei and Zou 2019):

- **Random oversampling:** Underrepresented classes are sampled more frequently.
- **SMOTE:** New samples are linear combinations of the count vectorized data. To ensure that enough neighbors are available, the data is oversampled to 10% of the most frequent class.
- **Random insertion:** Words are randomly duplicated.
- **Random deletion:** Words are randomly deleted.
- **Synonym replacement:** Individual words are replaced by their synonym looked up on the website Openthesaurus for German synonyms.

The augmentation fraction was analyzed in two distinct ranges: [0, 0.2] with a step size of 0.025 and [0.2, 1] with a step size of 0.1.

The following results were obtained:

- **Naïve Bayes:** The various methods of data augmentation yield similar results. In particular, a significant improvement can be observed in the range of augmentation fractions between 0.075 and 0.1. This is especially noteworthy in the context of the low computational power required by Naïve Bayes, as it allows the classifier to achieve the performance of plain Logistic Regression and SVC through the use of additional augmentation (Fig. 10.9).

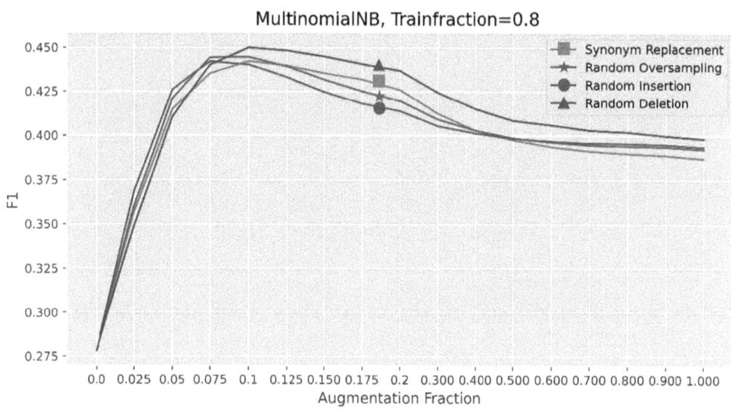

Fig. 10.9 F1 score for Naïve Bayes with data augmentation

- **Linear Support Vector Classifier:** The highest F1 score was achieved with an augmentation fraction at 0.025. In particular, with an augmentation fraction that is higher than this, the score decreased significantly, especially with methods such as synonym replacement and random deletion (Fig. 10.10).
- **Logistic Regression:** The highest F1 score was reached with an augmentation fraction between 0.025 and 0.05, with random insertion performing the best (Fig. 10.11).

In general, random insertion and random oversampling perform better than synonym replacement and random deletion. This is likely because random insertion and random oversampling do not discard any data that could be relevant. Furthermore, data augmentation had a considerable effect on training time, with the synonym

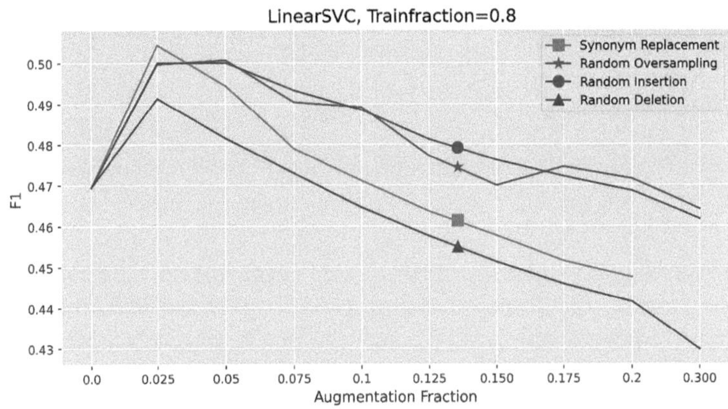

Fig. 10.10 F1 score for Linear Support Vector Classifier with data augmentation

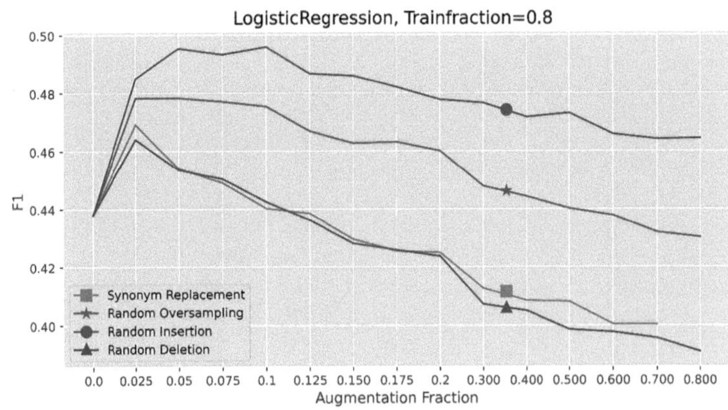

Fig. 10.11 F1 score for Logistic Regression with data augmentation

replacement process in particular being more complex to implement and computationally intense.

10.5.2.2 Feature Extraction

Feature extraction involves selecting and transforming textual data into a structured format that algorithms can process. The idea is to transfer as much information as possible from the text to the representation in order to achieve optimal classification results (Jurafsky and Martin 2008). The feature extraction process typically occurs in two steps:

1. **Tokenization:** Tokenization describes the creation of the vocabulary by splitting the text corpora. Words or character n-grams were tested for the following evaluation:
   - One-word tokenization
   - Two-word tokenization
   - Character n-grams with n $\in \{2, 3, 4, 5, 6, 7, 8, 9, 10\}$, across words
   - Character n-grams with n $\in \{2, 3, 4, 5, 6, 7, 8, 9, 10\}$, not across words
2. **Vectorization:** The vectorization step includes two possible procedures for making text computationally readable. The first is the conventional BoW approach, which counts the frequency of each word's occurrence. The second is the aforementioned term frequency-inverse document frequency (TF-IDF) vectorizer, which adjusts these BoW counts relative to the frequency of the word in the entire document.

Due to the size of the data set, training interruptions occurred frequently, resulting in the inclusion of only partial data points during evaluation. Nevertheless, general trends are still identifiable. In order to examine the influence of individual feature extraction methods independently, the default values of the Python library scikit-learn (sklearn)[19] were used primarily for the selection of hyperparameters.

Results

The F1 scores for word tokenization can be found in Table 10.11, which serves as a baseline for evaluating results with enhanced feature extraction methods.

Table 10.12 lists the top five models that use character n-grams for tokenization. The results were slightly better when using n-grams compared to word tokens. The data set's text is not structured around keywords. Instead, it is written in a more natural language, which boosts TF-IDF performance due to its ability to downweight common words. TF-IDF appears to be the most effective approach in

[19] See https://github.com/scikit-learn

Table 10.11 Word tokenization comparison

Model	Method	Tokenizer	F1
SVC	CountVectorizer	word	0.527982
SVC	TF-IDF	word	0.543232
LogisticRegression	CountVectorizer	word	0.524389
MultinomialNB	CountVectorizer	word	0.278255
MultinomialNB	TF-IDF	word	0.230299

Table 10.12 Character n-gram tokenization comparison

Model	Method	Tokenizer	n-gram	F1
SVC	TF-IDF	char	7	0.581615
LogisticRegression	CountVectorizer	char	5	0.574495
SVC	TF-IDF	char	6	0.572351
SVC	TF-IDF	char	8	0.570064
SVC	TF-IDF	char	9	0.568654

Table 10.12, but for a more comprehensive comparison, additional data points from some more experiments would be required.

10.5.2.3 Hierarchical Classification

A hierarchical approach to classification was tested due to the taxonomy of the NACE codes. According to Silla and Freitas (2011), various approaches can be used:

- **Flat approach:** The hierarchical structure is ignored, and only the leaf nodes are predicted.
- **Local approach:** a top-down approach to explore the hierarchy
- **Global (or Big-Bang) approach:** training a single classifier in a way that considers the hierarchy

The tests performed focused on the local approach, differentiating between:

- **Number hierarchy:** The hierarchy is defined by the number sequence. The first level consists of single-digit numbers and the second of two-digit numbers. The initial prediction is made for the first digit, and a subsequent prediction is made for the second digit.
- **Count hierarchy:** A hierarchy was constructed based on the frequency of classes. Classes that occurred with less than 10% of the most frequent class were categorized as "low," while the remaining classes were categorized as "high."

The following base estimators were used:

- **Multinomial Naïve Bayes:** [MultinomialNB(alpha=1, fit_prior=True, class_prior=None)]

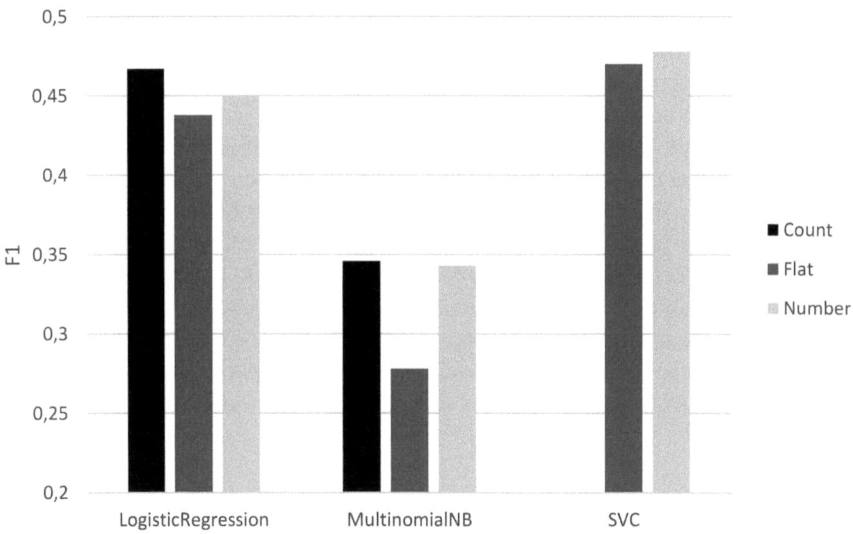

Fig. 10.12 Comparison of different hierarchical models for each classifier. The SVC run for the count hierarchy was not successful and is therefore not included in the results

- **Support Vector Classifier:** [SVC(C=0.1, kernel="linear", probability=True, tol=0.0001)]
- **Logistic Regression:** [LogisticRegression(penalty="l2", dual= False, tol= 0.0001, C=1, fit_intercept=True, intercept_scaling=1, solver="lbfgs", max_iter =500)]

Results

In general, hierarchical classification offers a slight improvement over flat classification (Fig. 10.12). However, the resource requirements are considerably higher because multiple classifiers need to be trained. In addition, the implementation is more complex.

10.5.2.4 Large Language Models

In recent years, the field of natural language processing has witnessed remarkable advances, largely attributed to the emergence of Large Language Models (LLMs). These models, powered by deep learning techniques and trained on vast amounts of text data, have demonstrated unprecedented capabilities in understanding, generating, and processing natural language.

This section delves into the utilization of Generative Large Language Models for automated NACE coding in official statistics. Generative LLMs, such as GPT models, have demonstrated remarkable capabilities in understanding and generating human-like text across various domains. We explore how these models can be harnessed to optimize the NACE coding process, improve efficiency, and enhance the quality of the resulting classifications.

Advantages of such pretrained models lie in the absence of the need for custom training or fine-tuning procedures. Due to their extensive pretraining, LLMs already possess comprehensive world knowledge that can be applied to a specific task without the need for extensive model adaptation beforehand. The inherent capability of Generative LLMs to understand and generate human-like text stems from their pretraining on vast corpora of text data, encompassing diverse linguistic patterns and semantic relationships. This preexisting knowledge enables Generative LLMs to grasp the nuances of economic activity descriptions encoded in NACE classifications and generate accurate coding outputs without the need for task-specific training.

Hybrid Model

To put LLMs to use, they are integrated into the context of a hybrid model, where a Logistic Regression model is used as a preselector. This section describes a hybrid system comprising this baseline model combined with an LLM. Figure 10.13 illustrates the general structure and workflow of the classification performed by the hybrid system. The system consists of the original baseline model, which selects a subset of classes from the activity description (a top-k selection from the classes

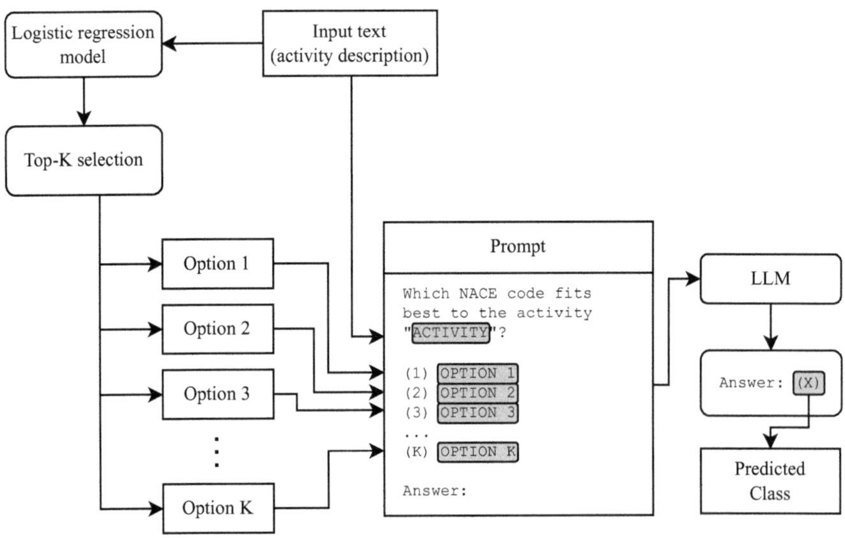

Fig. 10.13 Overview of the hybrid system architecture consisting of a Logistic Regression and a Large Language Model

with the highest probabilities), and a generative language model that determines the final class from this preselection. In the formulation of the prompt, the activity description is initially provided, followed by the titles of the k classes listed for selection one by one. Finally, at the end of the prompt, the LLM receives instructions to select one of the listed classes.

The Logistic Regression model serves as a preselector by narrowing down the candidate classes based on their probabilities, thus reducing the search space for the LLM and improving its efficiency in determining the final class. On the other hand, the LLM enhances the classification process by generating text sequences that capture the semantic nuances of the activity description and its corresponding class, thus refining the classification decision.

Choosing the Prompt

To gain insight into the importance of prompts for the performance of the hybrid system, several experiments are conducted in which the LLM is isolated and tested. These experiments are structured as follows: Unlike what is shown in Fig. 10.13, the preselected classes from 1 to K are not determined by the baseline model. Instead, $K - 1$ random incorrect classes and one correct known class are presented to the LLM for selection. This adjustment significantly reduces the difficulty level of classification, as the LLM less frequently receives very similar classes. Instead, the correct class typically stands out strongly from the others and is easier to identify. By simplifying the problem in this manner, the influence of the respective parameters of the LLM can be more easily examined and interpreted. Using this setup we were able to expose important parameters by measuring performance across 50 examples:

- **Chain-of-Thought prompting (Wei et al. 2023):** By simply adding a phrase such as *"First indicate for each option on a scale from 1 to 9 how well it fits the job description"* before prompting for the answer, we were able to detect an increase in accuracy.
- **Translation:** Utilizing an English prompt instead of a German one resulted in improved performance, even when the class labels and activity descriptions remained in German.
- **Model size:** We conducted experiments with various model sizes of multiple Llama derivatives, such as Vicuna, and observed an increase in performance correlated with the augmentation of model parameters.

Using these three adaptations in conjunction with the LLM model Solar-70B,[20] the system achieved a classification accuracy of approximately 80% in identifying the correct class out of nine random classes.

[20] https://huggingface.co/upstage/SOLAR-0-70b-8bit

Results

To evaluate the overall system, we conducted an experiment that included the classification of 150 additional observations. In this experiment, we used the global classifier approach as discussed in Sect. 10.5.2.3. The hybrid system correctly classified only 22% of the examples, which is a lower performance compared to using Logistic Regression alone. Manual investigation of misclassifications revealed that the LLM faced difficulties when the classes were very similar, a situation that frequently occurred. However, the reasoning behind the LLM's classifications was mostly valid, as the errors primarily arose due to the high similarity between classes rather than shortcomings in the model's logic.

10.5.3 Use Case Results

In our use case, it became evident that even a basic machine learning algorithm such as Logistic Regression achieves satisfactory top-k results. Subsequent experiments yielded only marginal score enhancements at most. Given that more intricate methods introduce additional complexity and computational overhead, a straightforward approach appears to be adequate. This method can effectively aid in the labeling process by offering recommendations for potential NACE classes.

Although the top-k results were promising, pinpointing the accurate top one result remains challenging, particularly at lower hierarchical levels with a large number of classes, where the top one scores are still unconvincing. Establishing high-confidence predictions is essential to improve automation rates within the statistical process.

The low scores reflect the challenges of categorizing vague or ambiguous texts into highly specific classes, a task that proves daunting even for human annotators. Enhancing the quality of training data and subsequently improving automated coding results require a uniform labeling process, clear guidelines, and effective communication among diverse labeling experts. However, achieving this is complex given that the data was independently labeled by various decentralized institutions. One potential solution involves collecting additional data and strategically utilizing it, such as by reevaluating its overall quality.

10.6 Conclusion

It is evident that, despite the use of two completely different data sets from Statistics Austria and Destatis, similar conclusions can be drawn from the attempt to classify NACE codes. For the use case of Statistics Austria the prediction of NACE level 2 codes was of interest, whereas the use case of Destatis aimed to predict up to NACE level 5 codes. In both scenarios a fully automatic classification with acceptable error margins currently seems unfeasible. Moreover, the results of both use cases suggest that the choice of the classification algorithm does not heavily influence the prediction quality for this classification task.

This shifts more significance to training data collection, preprocessing, and ultimately pressing demand for cleaner data sets to improve classification accuracy. However, it is inherent to the NACE classification that, particularly at the deeper levels, where the differences between categories become finer and more granular, accurate classification becomes increasingly difficult. This difficulty also extends to human classification, given that businesses are complex and not always easy to categorize.

These inconsistencies in the training data significantly degrade the quality of the classification. Consequently, it would be beneficial to focus future research on improving the preprocessing of training data to enhance overall data quality and, consequently, the quality of the classification.

References

M. Abadi, A. Agarwal, P. Barham, E. Brevdo, Z. Chen, C. Citro, G.S. Corrado, A. Davis, J. Dean, M. Devin, S. Ghemawat, I. Goodfellow, A. Harp, G. Irving, M. Isard, Y. Jia, R. Jozefowicz, L. Kaiser, M. Kudlur, J. Levenberg, D. Mané, R. Monga, S. Moore, D. Murray, C. Olah, M. Schuster, J. Shlens, B. Steiner, I. Sutskever, K. Talwar, P. Tucker, V. Vanhoucke, V. Vasudevan, F. Viégas, O. Vinyals, P. Warden, M. Wattenberg, M. Wicke, Y. Yu, X. Zheng, TensorFlow: Large-scale machine learning on heterogeneous systems (2015). https://www.tensorflow.org/. Software available from tensorflow.org

J. Allaire, F. Chollet, *keras: R Interface to 'Keras'* (2019). https://CRAN.R-project.org/package=keras. R package version 2.2.4.1

R. Avouac, T. Faria, F. Comte, A cloud-native data science platform for official statistics, in *Foundations and Advances of Machine Learning in Official Statistics*, ed. by F. Dumpert, Chap. 8 (Springer, Berlin, 2025)

S. Barragán, A. Pérez-Bote, C. Sáez, D. Salgado, L. Sanguiao-Sande, Streamlining business functions in official statistical production with machine learning, in *Foundations and Advances of Machine Learning in Official Statistics*, ed. by F. Dumpert, Chap. 13 (Springer, Berlin, 2025)

E. Brown, H. Sands, L. Ferguson, Automated coding of Standard Industrial and Occupational Classifications (SIC/SOC). Technical report, ONS Data Science Campus, 2021. https://datasciencecampus.ons.gov.uk/projects/automated-coding-of-standard-industrial-and-occupational-classifications-sic-soc/, cited 16 May 2024

Bundesamt für Statistik, Das Dateninnovationsprojekt 'NOGAuto'. Technical report, Bundesamt für Statistik, 2021. https://dam-api.bfs.admin.ch/hub/api/dam/assets/18465454/master, cited 16 May 2024

N.V. Chawla, K.W. Bowyer, L.O. Hall, W.P. Kegelmeyer, SMOTE: synthetic minority over-sampling technique. J. Artif. Intell. Res. **16**, 321–357 (2002)

T. Chen, C. Guestrin, XGBoost: A Scalable Tree Boosting System, in *Proceedings of the 22nd ACM SIGKDD International Conference on Knowledge Discovery and Data Mining* (Association for Computing Machinery, New York, 2016), pp. 785–794

T. Chen, T. He, M. Benesty, V. Khotilovich, Y. Tang, H. Cho, K. Chen, R. Mitchell, I. Cano, T. Zhou, M. Li, J. Xie, M. Lin, Y. Geng, Y. Li, J. Yuan, *xgboost: Extreme Gradient Boosting* (2023). https://CRAN.R-project.org/package=xgboost. R package version 1.7.3.1

Eurostat, *NACE Rev. 2 - Statistische Systematik der Wirtschaftszweige in der Europäischen Gemeinschaft*. Amt für amtliche Veröffentlichungen der Europäischen Gemeinschaften (2008)

T. Faria, T. Seimandi, Classifying companies in France using machine learning, in *UNECE Machine Learning for Official Statistics Workshop, 5–7 June 2023, Geneva* (2023). https://unece.org/sites/default/files/2023-04/ML2023_S1_France_Faria_A.pdf, cited 16 May 2024

J. Harrison, *R Bindings for 'Selenium WebDriver'* (2020). https://CRAN.R-project.org/package=RSelenium. R package version 1.7.7

T. Hastie, R. Tibshirani, J. Friedman, *The Elements of Statistical Learning*, 2nd edn. (Springer, Berlin, 2009)

F. Herold, D. Wintergerst, J. Kaltenbacher, P. Meißner, KI-unterstützte Klassifikation deutschsprachiger Texte – Abschlussbericht. Technical report, virtual7 GmbH (2022)

Y.-L. Huang, A. Cannet-Delbosq, M. Walzer, L. Maretti, C. Lamboray, M. Cordy, Y.L. Traons, An overview of STATEC's projects on automatic coding, in *Conference on Foundations and Advances of Machine Learning in Official Statistics, 3–5 April 2024, Wiesbaden* (2024). https://www.destatis.de/EN/About-Us/Events/Machine-Learning/Slides/s3_huang_lamboray.pdf?__blob=publicationFile, cited 16 May 2024

A. Joulin, E. Grave, P. Bojanowski, T. Mikolov, Bag of tricks for efficient text classification (2016). https://arxiv.org/abs/1607.01759

D. Jurafsky, J. Martin, *Speech and Language Processing: An Introduction to Natural Language Processing, Computational Linguistics, and Speech Recognition* (Pearson Prentice Hall, London, 2008)

J. Kärkimaa, L. Larja, How to Make AI do Your Job for Statistical Classification of Industry and Occupation, in *BigSurv18, 26 October 2018, Barcelona* (2018). https://www.bigsurv.org/bigsurv18/uploads/216/385/AI_classification_Karkimaa_Larja.pdf, cited 16 May 2024

H. Kühnemann, A. van Delden, D. Windmeijer, Exploring a knowledge-based approach to predicting NACE codes of enterprises based on web page texts. Stat. J. IAOS **36**(3), 807–821 (2020)

J. Mu, S. Bhat, P. Viswanath, All-but-the-top: simple and effective postprocessing for word representations, (2018). https://arxiv.org/abs/1702.01417

D.M.W. Powers, Evaluation: from precision, recall and F-measure to ROC, informedness, markedness and correlation (2020). https://arxiv.org/abs/2010.16061

R Core Team, *R: A Language and Environment for Statistical Computing*. R Foundation for Statistical Computing (2023). https://www.R-project.org/

V. Raunak, Simple and effective dimensionality reduction for word embeddings (2017). https://arxiv.org/abs/1708.03629

C.N. Silla, A.A. Freitas, A survey of hierarchical classification across different application domains. Data Min. Knowl. Disc. **22**, 31–72 (2011)

A. Sun, E.-P. Lim, Hierarchical text classification and evaluation, in *Proceedings of the 2001 IEEE International Conference on Data Mining* (IEEE Computer Society, Washington, 2001), pp. 521–528

A.K. Uysal, An improved global feature selection scheme for text classification. Expert Syst. Appl. **43**, 82–92 (2016)

J. Wei, K. Zou, EDA: easy data augmentation techniques for boosting performance on text classification tasks, in *Proceedings of the 2019 Conference on Empirical Methods in Natural Language Processing and the 9th International Joint Conference on Natural Language Processing (EMNLP-IJCNLP)*, ed. by K. Inui, J. Jiang, V. Ng, X. Wan (Association for Computational Linguistics, Stroudsburg, 2019), pp. 6382–6388

J. Wei, X. Wang, D. Schuurmans, M. Bosma, B. Ichter, F. Xia, E. Chi, Q. Le, D. Zhou, Chain-of-thought prompting elicits reasoning in large language models (2023). https://arxiv.org/abs/2201.11903

Chapter 11
An Automated Machine Learning Pipeline for Statistical Matching

Theresa Küntzler

11.1 Introduction

The mandate of official statistics is to provide data and information on a wide range of topics while, among other things, ensuring a high standard of quality (Saidani et al. 2023; European Statistical System Committee 2019; Saidani and Dumpert 2025). At the same time, a certain degree of economy is advisable in order to reduce the financial burden on the public and the demands on the information providers. In addition to high quality standards and the need for prudent budgeting, a third growing demand placed on official statistics is that of timeliness. Particularly in the aftermath of the covid pandemic, the need for fast and robust information and for official statistics to adapt to new data requirements has increased (Commission Future Statistics 2024). The merging of multiple data sources, and thus the multiple use of the same data, provides an opportunity to implement the said parsimony, thus allowing an efficient use of resources to meet budgetary constraints. At the same time, the combination of several already available data sources makes it possible to extend the information services provided by official statistics, if necessary in a comparatively short period of time (e.g., D'Alberto and Raggi 2024; Lewaa et al. 2021; D'Orazio 2019a).

Statistical matching is a data merging technique that is particularly useful when the two data sources of interest do not cover exactly the same entities, when there is no common identifier to link observations, or when other linkage techniques are legally restricted. Broadly speaking, statistical matching aims to transfer information on a variable of interest from one file, called the donor file, to a second file, called the recipient file. This is done by estimating a model that predicts

T. Küntzler (✉)
KfW, Frankfurt am Main, Germany

F. Dumpert (ed.), *Foundations and Advances of Machine Learning in Official Statistics*, Society, Environment and Statistics,
https://doi.org/10.1007/978-3-032-10004-7_11

the variable of interest using the donor data, based on features, that are available in both data sets. Often, these can be a series of sociodemographic variables. In a final step, this model is applied to the recipient data (Rässler 2012). Classic modeling strategies involve the use of nearest neighbor approaches (e.g., D'Orazio 2019a; European Commission et al. 2017; Abello and Phillips 2004) or regression models (e.g., Webber and Tonkin 2013). In this chapter, I propose the use of a number of other nonparametric algorithms, commonly known as classical machine learning strategies. These algorithms have been shown to have great potential for predicting nonlinear relationships. However, in the statistical matching literature, especially for official statistics, this class of models is rarely explored and used.

To leverage machine learning for statistical matching, a statistical matching pipeline is proposed.[1] This pipeline automates the entire process from data ingestion, over a series of data preprocessing steps, training, tuning and comparison of multiple potential modeling strategies, estimation of the final model, and the matching itself, and concludes with a comprehensive report on the donor and recipient data, the matching process, and the matched data. The approach includes a number of considerations to assess the quality of the matching process and to ensure the high methodological standards expected in official statistics. In particular, a nested resampling strategy is used to evaluate which algorithm is likely to be best suited for the data and problem at hand (Bischl et al. 2023; Becker et al. 2024). The code of the statistical matching pipeline is available to the public (Küntzler 2024). Early stages of the pipeline were developed in collaboration with Oliver Hauke. Within the same project, the use of federated learning, another modern model estimation strategy, was investigated for its applicability in official statistics (Stock et al. 2023).

The contribution of this chapter is twofold: First, I present the statistical matching pipeline, which automates the matching process to a very high degree, and second, I suggest the usefulness of machine learning techniques for statistical matching and provide evidence comparing the performance of the statistical matching pipeline to other matching strategies and to a human-driven process.

Section 11.2 briefly introduces the method and reviews recent applications in official statistics. In Sect. 11.3, I introduce the statistical matching pipeline step by step, along with technical details on the implementation. Sections 11.4 and 11.5 provide empirical evidence on the performance of the statistical matching pipeline by offering a simulation study and a replication of a nonautomated project. Section 11.6 concludes the chapter.

[1] This project was fully developed at the Federal Statistical Office of Germany.

11.2 Literature Review

11.2.1 The Method

The proposed pipeline aims to perform statistical matching. Statistical matching is a technique to add variables to a data set that are initially only available in a second data source. Both sources consist of data collected from different samples drawn from the same population, but which are unlikely to cover the same observations due to large population sizes and small selection probabilities (Marella and Pfeffermann 2019).

Formally, imagine two independent samples, a so-called donor sample A of size n_A and a recipient sample B, of size n_B. Consider that both samples are drawn from the same population of independent records (x_i, y_i, z_i) characterized by a joint density function $f(x, y, z; \Theta)$ depending on a parameter vector Θ. The need for merging arises because not all variables (X, Y, Z) are observed simultaneously in both samples. In particular, sample A contains only observations for its units on variables (X, Z), while sample B includes only (Y, Z) observations (Marella and Pfeffermann 2019; Rässler 2012). However, the association of the variable X and Y is of interest, though not jointly observed. The technique of statistical matching, as applied by the proposed statistical matching pipeline, is to estimate a model that predicts the variable of interest X as observed in the donor data, using only features, that are available in both data sets, the set of Z as described above. In the following, the estimated model is applied on the recipient data B, using the shared features Z to estimate $\hat{X}$ for each observation in B. The result is a so-called synthetic data set $\hat{B}$, consisting of $(Y, Z, \hat{X})$.

For this process to produce valid results, it relies on a central and strong assumption (among multiple assumptions): the conditional independence assumption (CIA). This assumption allows to solve the identification problem, which states that initially the conditional association between the two variables X and Y, which are never observed together, cannot be estimated from the given data (Rubin 1974; Rässler 2003). However, assuming that X and Y are conditionally independent given the jointly observed variables Z, the identification problem can be solved. This assumption allows decomposing the identification of the joint distribution $f(x, y, z; \Theta)$ into smaller estimation problems that can be solved with observed data (Rubin 1974; D'Alberto and Raggi 2024; Rässler 2004, 2003).

A common criticism of applications of statistical matching based on the conditional independence assumption is that while the assumption of conditional independence theoretically solves the identification problem, in practice this assumption rarely holds completely. Moreover, it is impossible to test the conditional independence assumption. Practitioners are left with a series of checks that indicate a successful matching and thus the validity of the conditional independence assumption. Most checks support the fourth (and lowest) level of validity of statistical matching proposed by Rässler (2012, 2004): the preservation of marginal distributions. After the matching process, the marginal and joint distributions of X

and Z should be similar in both files, such that $f_{\hat{X}} = f_X$ and $f_{\hat{X}Z} = f_{XZ}$ when comparing the donor sample to the matched recipient sample. The other three levels of statistical validity remain untestable. The third level requires the preservation of correlation structures, the second level requires the preservation of joint distribution of all variables ($f_{\hat{X},Y,Z} = f_{X,Y,Z}$), while the first and highest level is satisfied by the preservation of individual values.

Beyond the four levels of validity, another interesting indication of the prospects for successful matching is the predictive power, or correlation structure, of the variables Z for X and Y (Rässler 2004; Rodgers 1984). First, and rather obviously, only if Z contains good predictors for X, it is possible to build a good predictive model. If Z contains no or very little information about X, the matching is bound to fail. In addition, Z must also contain relevant information about Y for conditional independence to be achieved or at least approximated. For a case where conditional independence cannot be fully established, Rässler (2004) gives a nice example of how it is possible to construct an interval of plausible correlations ρ_{XY} based on the observed correlation structure of ρ_{ZX} and ρ_{ZY}. For this to be informative, the observed correlations must nevertheless be high (Rässler 2004; Rodgers 1984; Kadane 2001). However, estimating such bounds is problematic or infeasible when the data structure is complex and high dimensional (Rässler 2004).

Once statistical matching is understood as described, the distinction from the related method of record linkage becomes clear: While for both statistical matching and record linkage, the two data sources stem from the same target population, record linkage underlies the assumption that both sources cover the same observations. Thus, the goal of record linkage is to find matches of identical observations in both data sets and generate a complete file with the shared and directly observed information (Rässler 2012; Rodgers 1984). In contrast, the aim of statistical matching is to create a synthetic data set that contains plausible predictions for the observations in the receiver data set based on a model built using the donor data (Eurostat 2023).

Methodologically, statistical matching can also be applied in a mixed scenario, where some observations are covered in both data sets, but both sources also contain different observations. However, linking data from different sources on the same unit of observation may be subject to stronger legal restrictions, which requires careful consideration before applying statistical matching to such a mixed scenario.

The definition of the terms used in this chapter is particularly important because the terminology is not used consistently in the literature. While statistical matching is also referred to as data fusion (Cielebak and Rässler 2014; Lewaa et al. 2021; Rässler 2004) or synthetic matching (D'Orazio 2019b), exact matching is another term used for what is described above as record linkage. Data integration is most often used as an umbrella term for both statistical matching and record linkage (Lewaa et al. 2021) but also appears as a synonym for record linkage only (Cielebak and Rässler 2014). Adding to the complexity is the fact that even literal translations into other languages can lead to a shift in the definition of terms (Cielebak and Rässler 2014). To avoid confusion, I will strictly adhere to the term statistical matching and not use potential synonyms throughout this chapter. The terminology described above and used in this chapter is the most widely accepted in research on the subject (Marella and Pfeffermann 2019; Rässler 2012; Lewaa et al. 2021; Eurostat 2023).

11.2.2 The Evolution of Statistical Matching

The evolution of statistical matching from its beginnings to the present is described in detail by D'Alberto and Raggi (2024), including a comprehensive list of relevant papers and applications. In the following, I will give a short overview: A very early application of merging two data sets in official statistics is the "1966 merge file." Okner (1972) observes a new "effective demand for large amounts of disaggregated economic and demographic information" (Okner 1972, p. 325). To meet this demand, demographic data from the 1967 Survey of Economic Opportunity are merged with data from tax returns, allowing the distribution of income to be observed across a range of demographic groups.

In the 1980s, Rodgers (1984) raises important concerns about the validity of the matched files if certain assumptions are not met, which in turn endangers all subsequent analyses based on the matched files (D'Alberto and Raggi 2024). Despite the methodological criticism, official statistics in the USA and Canada used statistical matching in a number of applications (D'Alberto and Raggi 2024). Around the same time in Europe, until the late 1990s, statistical matching was used mainly in the field of marketing and market research in France and Germany, typically to integrate television viewing with other data (D'Alberto and Raggi 2024; Rässler 2012).

A systematic theoretical framework for statistical matching evolved during the 1990s and early 2000s (D'Alberto and Raggi 2024), strongly advanced by Rässler (2012, 2004, 2003) and D'Orazio et al. (2006a). From then on, statistical matching is commonly used in both research studies and official statistics to integrate data from multiple sources (for overviews, see Beck et al. (2018), D'Alberto and Raggi (2024), and the references therein).

Applications in the last two decades in official statistics cover a variety of topics. Income, household expenditure, and living conditions are of great interest, as the European Union Statistics on Income and Living Conditions (EU-SILC) is a frequent target in statistical matching (European Commission et al. 2017; Beck et al. 2018; Schaller 2021; Webber and Tonkin 2013; Leulescu and Agafitei 2013), as well as other surveys on household expenditure (Sutherland et al. 2002; Abello and Phillips 2004). The use of time and the connection to labor force has been studied, too (Leulescu and Agafitei 2013; Gazzelloni et al. 2008). Additional topics of interest are health expenditures (Gutman et al. 2013; Abello and Phillips 2004) and farming (D'Alberto et al. 2018; D'Orazio and Catanese 2016).

The methods used in the vast majority of these applications are classical matching techniques. The use of modern nonparametric machine learning algorithms to estimate predictions for the matching variable, as proposed in the pipeline presented in this chapter, is hardly tried or applied in the recent literature on applications in official statistics. The only exception is the use of some form of nearest neighbor approaches, also called random hot deck or distance-based hot deck methods, which is a very common nonparametric matching strategy (D'Orazio 2019a; European Commission et al. 2017; Schaller 2021; Webber and Tonkin 2013; Gazzelloni et al. 2008; Abello and Phillips 2004; Leulescu and Agafitei 2013; D'Orazio and Catanese 2016; D'Alberto et al. 2018). Other strategies that can be observed are regression-based models (European Commission et al. 2017; Webber and Tonkin 2013; Leulescu and Agafitei 2013), predictive mean matching (Schaller 2021; Leulescu and Agafitei 2013), Bayesian approaches (Gutman et al. 2013; Rässler 2012), or a mixture of the above (European Commission et al. 2017; Webber and Tonkin 2013). An exception is Leulescu and Agafitei (2013), who test, among other models, a decision tree that performs better than a nearest neighbor approach but is outperformed by multiple imputation strategies. The second exception is D'Orazio (2019a), who evaluates the performance of different machine learning algorithms in a simulation setting.

Overall, the use of modern machine learning algorithms, such as tree-based models and others, is rarely explored in the statistical matching literature, especially for official statistics. However, these models have shown a great potential when it comes to prediction tasks in cases of underlying nonlinear relationships (Hastie et al. 2009). The proposed statistical matching pipeline provides an innovative application of such models, along with a rigorous strategy for model comparison and performance evaluation. Generally, the performance assessment of the matching model needs to be taken into account and reported. Furthermore, the comparison between different models needs to be handled carefully (Bischl et al. 2023). Both are ensured within the proposed pipeline.

11.3 The Statistical Matching Pipeline

The statistical matching pipeline proposed in this chapter automates the matching process to a very high degree. Figure 11.1 sketches the complete process. Given a series of inputs (1), the pipeline performs the following steps: (2) data preprocessing, including simple imputations and more. (3) A variety of possible models is trained and tuned using default hyperparameter search spaces, wrapped in a

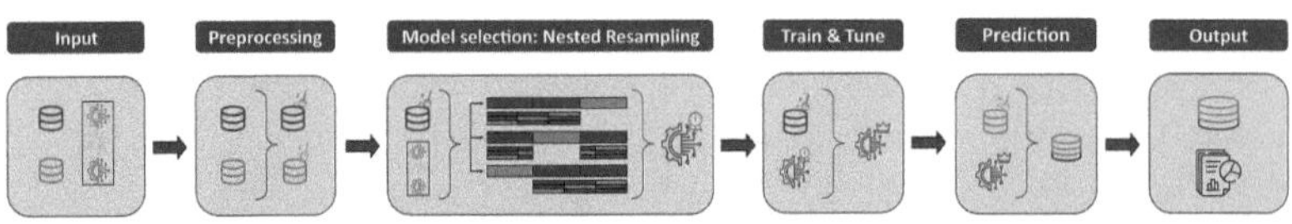

Fig. 11.1 Statistical matching pipeline (Note: most icons designed by Freepik (2023), modified by the author)

nested resampling to then (4) evaluate each model on its performance. (5) The best performing model is retrained and tuned on the full data set and finally (6) used to estimate the variable of interest in the receiving data set. (7) The output of the pipeline is the matched data itself, along with an automatically produced report containing information about the data, different distributions of interest, performance of the pipeline, and the relationship between the variables of interest. In the following, I describe details on the technical implementation, followed by a description of each pipeline step. The complete code of the pipeline is publicly available (Küntzler 2024).

11.3.1 Technical Implementation: mlr3, targets, and markdown

The statistical matching pipeline is implemented in the R language (R Core Team 2021).[2] The pipeline design makes use of three main ecosystems or packages, which are developed to be used within or together with R: mlr3 (Lang et al. 2019; Bischl et al. 2024), targets (Landau 2021b), and R Markdown (Allaire et al. 2024).

The core of the pipeline is built in the mlr3 universe, which provides a "generic, object-oriented, and extensible framework for regression [...], classification [. . .], and other machine learning tasks [...] for the R language" (Kotthoff et al. 2024). mlr3 provides a large ecosystem for building machine learning pipelines that can include preprocessing steps, feature selection, benchmarking to compare multiple algorithms, and a variety of training and tuning techniques (Bischl et al. 2024). The mlr3 universe is characterized by a strictly object-oriented programming paradigm. This comes with the slight disadvantage that it can be difficult to adapt to mlr3 if previous programming experience is mainly based on the R language. Another advantage of mlr3 is that it relies on the R package data.table, which allows fast data operations and computations, but otherwise mlr3 remains very light on dependencies. In addition, it has built-in parallelization opportunities (Lang et al. 2024). Specifically when making use of resampling strategies during tuning, but also in other instances, many processes are independent from each other and computationally demanding at the same time. By parallelization instead of sequential execution, and given the respective hardware, one can reduce the runtime of the matching process remarkably.

[2] All R packages used: DataExplorer (Cui 2020), data.table (Dowle and Srinivasan 2023), dplyr (Wickham et al. 2023), DT (Xie et al. 2023), ggplot2 (Wickham 2016), here (Müller 2020), inspectdf (Rushworth 2022), magittr (Bache and Wickham 2022), Metrics (Hamner and Frasco 2018), mlr3 (Lang et al. 2019), mlr3learners (Lang et al. 2023a), mlr3measures (Lang 2022), mlr3misc (Lang and Schratz 2023), mlr3pipelines (Binder et al. 2021), mlr3tuningspaces (Becker 2023), mlr3viz (Lang et al. 2023c), paradox (Lang et al. 2023b), progressr (Bengtsson 2023), skimr (Waring et al. 2022), stringr (Wickham 2022), tarchetypes (Landau 2021a), targets (Landau 2021b), tibble (Müller and Wickham 2023), twosamples (Dowd 2023), and VIM (Kowarik and Templ 2016).

Since the estimation of multiple machine learning algorithms using resampling techniques can be computationally demanding, the statistical matching pipeline built with mlr3 is additionally wrapped in a so-called targets pipeline. Targets is an R package that allows to write code as a series of steps, with each step defining a single object, called target. A series of such objects or targets forms a pipeline and thus can be used to build a machine learning pipeline. Dependencies between targets are automatically detected and can be visualized as a network (Landau 2021b). For example, a dependency between two targets is between the donor data and a graph showing the correlation matrix of said donor data. Targets monitors each element for changes in either the element itself or its dependencies. When a targets pipeline runs, only those targets that have either changed or depend on changed targets are executed. All other targets are not executed but loaded from a cache. In the example above, if the donor data changed, the graph would be re-rendered based on the updated data. Another graph, representing unchanged receiver data, would not be executed again. This setup of an mlr3 pipeline wrapped around a targets pipeline makes the computationally intensive exercise much more efficient, since unchanged elements are skipped during a new estimation run. How to combine an mlr3 machine learning pipeline with a targets pipeline is shown by Schratz (2022).

The third major technical design element is the use of R Markdown to build the report. R Markdown provides dynamic documents that can be structured along sections, combining code snippets, code output, and rendered elements such as graphics. A parameterized R Markdown file (Xie et al. 2018) builds the skeleton of the report. A number of parameters control the appearance and behavior of the report. These parameters can be as simple as the headline of the report, but parameters are also used to pass information about the optimization criteria from the mlr3 pipeline to the report. To take full advantage of the technical setup, most report elements, such as graphs and tables, are separate targets and are therefore built during pipeline execution. As long as their dependencies do not change, the elements remain in the target cache and can be called from the R Markdown document that generates the report. In addition, the R Markdown document of the report itself is also a target, allowing changes to the report structure to be implemented without having to rerun the entire matching process.

11.3.2 Input

The pipeline requires a number of inputs that cover the researcher's matching strategy decisions and control subsequent behavior. The most obvious inputs are the donor A and recipient B data files. It is required that both data sources be harmonized beforehand. Harmonization ensures that all Z variables contained in both the donor and receiver data inherit the same meaning. This includes, for example, that they are based on the same operationalization, that factors cover the same classes, and that metric variables are measured with the same unit. This process will most likely be done manually beforehand. Since the files themselves can amount to large files, the pipeline input is reduced to a file location. The pipeline entails a function to read multiple file formats, which are RDS, csv, DTA, sas7bdat, and xls(x).

Additionally, the pipeline requires the specification of the target variable X in the donor data A, the second variable of interest Y, contained in B, and a list of variable names that can be used for modeling Z, which are contained in both data sets.

Beyond the data, a series of methodological information needs to be present. A specification of the task type, i.e., classification versus regression, along with the value of the positive class for binary classification is necessary.

At the heart of the pipeline is the evaluation of multiple algorithms for the given problem. Obviously, a specification of algorithms to be tested is required. In the mlr3 universe, the algorithms are also called learners. Along with the learners, respective hyperparameter search spaces are necessary for tuning. Since the pipeline builds on the default hyperparameter search spaces supported in the mlr3 universe (Becker 2023; Binder et al. 2020; Bischl et al. 2023; Kühn et al. 2018), the selection of models to choose from is restricted by the models covered with default hyperparameter spaces. This restriction is acceptable, though, since the most widely chosen learners are included. At the time of writing, one can choose among implementations of k-nearest neighbor (Schliep and Hechenbichler 2016), a decision tree (Therneau and Atkinson 2022), a random forest (Wright and Ziegler 2017), a support vector machine (Meyer et al. 2023), generalized linear models with elastic net regularization (Friedman et al. 2010), and extreme gradient boosting (Chen et al. 2023). Additionally, a featureless algorithm can also be added for a baseline.

The resampling and tuning process also needs specification that requires user input. Nested resampling is divided into inner and outer resampling. In short, the inner resampling includes the more well-known resampling procedure, while the outer resampling adds another layer that allows comparison across multiple algorithms. See Sect. 11.3.4 for more details. To specify the inner resampling, it is necessary to first decide on the strategy itself. All possibilities offered by mlr3 (Lang et al. 2019) can be used here, including cross-validation or bootstrapping. Furthermore, the tuning strategy has to be chosen. Again, all options offered in the mlr3 universe can be used (Becker et al. 2023), among them, for example, a random search. The last input for the inner tuning is the definition of a terminator. Possible criteria are a certain clock time, a maximum number of evaluations, or reaching a certain performance level (Becker et al. 2023). Finally, the resampling strategy for the outer resampling must be defined.

Two other optional features may require input if chosen. First, the pipeline can be enriched with individual preprocessing scripts for each of the input data sets. This can be useful, if some parts of the harmonization or other data cleaning steps should only apply to the data used as pipeline input, but not to the original data stored in the given file location. This preprocessing can also be used to select a sample of the data for test runs or when technical limitations require that the data be reduced.

The second optional feature that can be used is the simulation option. If only a simulated matching is desired, the pipeline can be run on a single input data set. In this case, the data are split into simulated donor and recipient sets. This split requires additional input on the proportion of data to be assigned to each simulated data set.

11.3.3 Data Preprocessing

The data preprocessing encompasses multiple steps. First and optional is the run of individual preprocessing scripts for donor and recipient data as described above in Sect. 11.3.2. This allows to include, for example, certain aspects of the harmonization or to draw a sample from the data.

Next follow two more technical implementations: The hyperparameters as set in the input are assigned to the respective learners. Additionally, a so-called fallback learner is attached to each learner. A fallback learner comes into action when the original learner returns an error. A classic example for when certain algorithms might fail is during resampling, when the test set contains factor levels that were not seen during training, and therefore a prediction is not possible. Having set a fallback learner prevents the failing and allows to make a prediction based on said fallback. mlr3 allows featureless fallback learners, which predict the majority class for classifications and the average value for regression problems. Alternative procedures to deal with failing algorithms are either to ignore those runs, which leads to overestimation of the performance since only runs without problems are considered for the estimation, or to penalize failing using the worst possible prediction, which likely leads to an underestimation of the performance (Lang et al. 2024).

The last step in preprocessing is applying mlr3's robustify-pipeline (Thomas 2024). This option equips each learner with a series of preprocessing steps. Those steps include, among others, the removing of constant features, imputation of missings features, and one-hot encoding of factors. The robustifying steps are adjusted to the respective algorithm and task at hand, such that, for example, imputation of missings is only applied if the learner cannot handle missing values.

11.3.4 Model Selection: Nested Resampling

In order to select the most appropriate algorithm for the problem at hand, it is necessary to compare the performance of different trained and tuned algorithms on the data in a valid way. Nested resampling provides a method to do just that: Estimate the performance, in terms of generalization error, of a trained and tuned algorithm on a specific data set. Becker et al. (2024) and Bischl et al. (2023) explain the method and its necessity in detail.

In short, nested resampling allows an algorithm to be first trained and tuned by cross-validation (inner resampling). To estimate the generalization error or, in other words, the performance, it is still necessary to test the tuned algorithm on new data

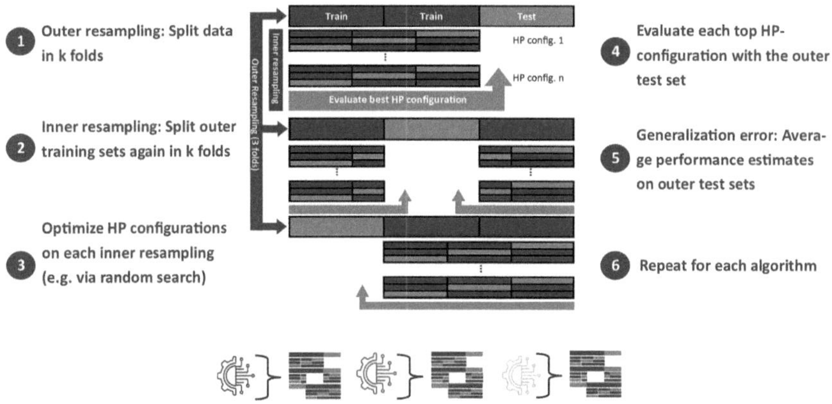

Fig. 11.2 Nested resampling (Note: figure after Becker et al. (2024, Figure 4.5). Icons denoting the algorithms are designed by Freepik (2023), modified by the author)

that were not involved in the tuning process. Thus, it is necessary to set aside a portion of the data before tuning. To avoid the possibility of a particularly lucky or unlucky split, an outer layer of cross-validation (outer resampling) is added to the process.

Figure 11.2 graphically depicts the method in six steps. First, the complete data are divided into k_{outer} outer folds, in Fig. 11.2 $k_{outer} = 3$. Second, each outer training batch is split again into k_{inner} folds, again $k_{inner} = 3$ in Fig. 11.2. Third, based on the inner resampling, multiple hyperparameter configurations can now be tested, e.g., via grid search or random search. Fourth, the best hyperparameter configuration from an inner resampling is tested on the unseen outer test set. Fifth, to obtain an unbiased generalization error, the evaluations on the different outer test sets are combined, usually by taking the mean. Finally, this procedure is performed for each possible algorithm. This allows an unbiased comparison of different algorithms given the specific data and modeling task.

11.3.5 Training and Tuning of the Final Model

Once it is determined which algorithm is expected to perform best on the given data and problem, a final model is built. This starts another round of training and tuning of that algorithm. The cross-validation pattern and the search algorithm for tuning are kept constant from the inner resampling. However, for this final training and tuning, the full donor data A is used, so that the algorithm can learn the most from the given data. Note that the generalization error to be reported for the model remains the result of the nested resampling and should not be based on the one-stage resampling of this step.

11.3.6 Prediction

The final step of the matching procedure is the application of the trained and tuned model from the previous step to the receiver data B. Since the model is built solely on features represented in both data sets Z, the model can be used to predict the variable of interest X in the recipient data. It is important to note that, unlike other matching techniques, this technique creates a prediction of X for each observation in B, rather than finding the closest possible observation in A and recycling the corresponding observed value of X. The result is a synthetic data $\hat{B}$ containing $(Y, Z, \hat{X})$.

11.3.7 Output: Data and Report

The output of the statistical matching pipeline is twofold: Obviously, the matched data $\hat{B}$ containing $(Y, Z, \hat{X})$ is given. The second output is a comprehensive report on the two data sets and the matching procedure. The aim of the report is to equip the researcher, who uses the pipeline, to decide whether the underlying assumptions are likely to hold and whether the result is valid and reliable or not. It is stressed that this final decision needs to be an informed decision made by the applying researcher and cannot be left to the automated process itself.

The report is divided into several sections. First, it contains exploratory data analyses for both the donor and recipient data individually. This information is mainly dedicated to evaluate the data quality itself. Each data set is presented in an overview table (Waring et al. 2022), which shows, e.g., variable types, factor levels, and distribution information of numerical features. Furthermore, the proportion of missing values per characteristic and a correlation matrix are given.

A second section provides similar information, but in a comparable manner. Several figures show side by side, for both donor and recipient data, the types of variables, the proportions of missing values in the variables, the distributions across categories for categorical data and the distributions of binned data for numerical variables, a comparison of correlations between numerical variables, and information about the memory usage of the data (Rushworth 2022).

The third section of the report is dedicated to the comparability of the two data sets. These analyses should help to decide whether the assumption that both data sources cover the same population is valid. The distribution of each feature included in the set of common variables used for model estimation Z is plotted in a common graph. Numerical variables are represented by overlaying density plots. Categorical features are represented by bar plots. To adjust for varying sample sizes, proportions per category are plotted instead of absolute numbers.

Information about the nested resampling process is covered in the fifth section of the report. It begins with a box plot showing the distribution of the performance estimates of the outer resampling runs (Lang et al. 2023c). Similar information is also covered in a table that gives an overview of the average performance estimate for each algorithm, the number of outer resampling iterations, and the number of warnings and errors that occurred during the process. The performance estimate shown in this table is also the basis for the general comparison between the algorithms. A second table contains details about the inner resampling (Becker et al. 2023). It shows for each resampling iteration the tested hyperparameters and the corresponding performance measure.

The penultimate section concerns the final model. Again, a table shows the selected hyperparameters and the performance estimate based on the resampling. Note again that this measure is potentially biased, and the unbiased estimate is the result of the nested resampling from the previous step.

The main part of the report ends with a graphical illustration of the relationship between the two variables of interest X and Y. A technical appendix contains information about the pipeline in the form of a network graph of the various targets (Landau 2021b).

11.3.8 Limitations and Potential Developments

The process as described has a number of limitations. Most importantly, I want to reiterate that the decision of whether the matching is successful and reliable enough to be used for further analysis is in the hands of the researcher using the pipeline. The machine cannot make this final decision. Part of this decision is the assessment of the conditional independence assumption. The method as implemented in the statistical matching pipeline relies on this assumption to hold (Rubin 1974; D'Alberto and Raggi 2024; Rässler 2004, 2003).

Potential future developments include methods to relax the conditional independence assumption and estimate probable intervals for the values of interest (D'Orazio et al. 2006a,b; Rässler 2004). This is closely related to additional uncertainty measures for the generated estimates (Conti et al. 2017, 2013, 2012) and the use of additional information from other samples. Another interesting area of potential further development for the pipeline is the possibility to adapt the matching process to complex and informative samples (Marella and Pfeffermann 2019; Nalenz et al. 2024; D'Alberto and Raggi 2024).

A more technical and less methodological potential development is the possibility of including models with user-specified search spaces. This would allow on the one hand to individualize the search spaces to a problem if there is information to be exploited. On the other hand, this would open up the possibility to try algorithms that do not (yet) have default search spaces in mlr3 (Becker et al. 2024). Additionally, although the current version of the output report is comprehensive, there is always more to be added. Comparisons between the donor data and the matched data of marginal distributions of the target variable X and a feature variable from Z would be an informative addition. It would also be of interest to add information on the preprocessing, which is applied based on mlr3's robustify-pipeline (Thomas 2024), since the steps are chosen dynamically, based on the learner and the data.

11.4 Simulation Study: European Social Survey

11.4.1 Design

To test the performance of the statistical matching pipeline, I perform a simulated statistical matching case. Following D'Orazio (2019a), I use the European Social Survey (ESS) Round 7 (Norwegian Agency for Shared Services in Education and Research, Norway 2014) to simulate the matching by splitting the data set into two and reuniting it using the pipeline. The ESS is a biennial cross-national survey conducted as face-to-face interviews, covering attitudes, beliefs, and behaviors (European Social Survey 2024).

This use case has three advantages: First, the use of a simulated case provides known observations for the data to be generated by the matching procedure. This allows the estimated performance measures and relationships in the matched data to be compared to actual observed cases, which can serve as a form of "ground truth." Second, by using the same case as D'Orazio (2019a), it is possible to compare the results of his study with the results based on the pipeline. Third, the ESS data provide a use case that is of interest to many European countries, rather than providing an example that is specific to a single country.

For the simulation, the relationship of interest is between (a) a binary indicator of whether or not the respondent and his or her household can live comfortably on their household income and (b) the total net income of the household, given as a decile. The impression of living comfortably on the household income functions as the X-variable in a hypothetical donor data set A, while the total net income will be the variable of interest in the simulated receiver data set B as Y. In addition, a set of 11 sociodemographic variables[3] forms the set of Z, which is present in both the donor and the receiver data set.

The data set used in the following includes all of the variables identified above as X, Y, and Z. Additionally, only complete cases in the variables of interest X and Y are included. This results in a data set with 31,827 observations.

As potential algorithms to be used during the matching, the pipeline is given a k-nearest neighbor (Schliep and Hechenbichler 2016), a decision tree (Therneau and Atkinson 2022), and a random forest (Wright and Ziegler 2017), along with

[3] The sociodemographic variables that make up Z are the following: respondent's gender, age, highest education level, main activity during the last 7 days (e.g., work, education, etc.), whether the respondent has a paid employment or not, the employment relation, whether the respondent lives with husband/wife/partner or not, legal marital status, whether a child lives in the respondent's household or not, the highest education level of the respondent's parents, and the area of living.

the default search spaces as suggested by Becker (2023). For the outer resampling strategy, I apply a threefold cross-validation. For the inner resampling, I also apply threefold cross-validation, paired with a tuning using random search of six hyperparameter configurations each time. The random search suggests hyperparameter configurations randomly selected within the boundaries given by the default search spaces (Becker 2023). The measure to be optimized is accuracy.

Based on the prepared data and the pipeline set up as described, I measure the performance of the pipeline by conducting the following steps 100 times:

1. Split the data set randomly, with two-thirds of the observations as a donor data set A and the other third of the observations as the receiver data set B.
2. Remove the household total net income from the donor set A and the impression of living comfortably on household income from the receiver set B, such that the two data sets contain $A(X, Z)$ and $B(Y, Z)$.
3. Run the pipeline, to estimate a model for X given the data A and use the model to predict X in B.
4. Evaluate the matching.

11.4.2 Evaluation

The first step of the evaluation is to asses which algorithms are chosen by the pipeline as the best fit for the problem in each run. In the majority of runs (91), nested resampling resulted in a random forest for the final matching model. In the remaining runs (9), the matching was based on a decision tree. The third option, a k-nearest neighbor model, was never selected. This is an interesting result, since k-nearest neighbor methods are often used to solve matching problems (D'Orazio 2019a; European Commission et al. 2017; Schaller 2021; Webber and Tonkin 2013; Gazzelloni et al. 2008; Abello and Phillips 2004; Leulescu and Agafitei 2013; D'Orazio and Catanese 2016; D'Alberto et al. 2018).

To assess the quality of the actual matching, three questions are of interest and discussed in the following:

1. How well does the matching reproduce the observed values of the impression of living comfortably on household income, X, in the receiver data set?
2. To what extent is the marginal distribution of X preserved in the receiver data set?
3. To what extent is the association of interest between X and Y reproduced in the matched data set?

The first question considers how well the unknown values of X are reproduced by the model. Rässler (2012) suggests a "hit rate" (Rässler 2012, p. 30), given as the share of correctly reproduced values. Thus, I use accuracy as a measure for this level. Two accuracy values are of interest. First, it is possible to calculate the $accuracy_{measure}$ between observed and predicted values of the impression of living comfortably on the household income in each sample. Second, in the resampling process the $accuracy_{estim}$ is estimated based on cross-validation. Therefore, it is also of interest how well the estimated accuracy matches the observed accuracy.

Over the 100 samples, the average accuracy when comparing the matched to the observed data is $\overline{accuracy}_{measure} = 0.694$ ($\bar{\sigma} = 0.006$). Overall, I consider this a moderate accuracy. However, the results fall between the best (Naive Bayes model: 0.791) and the second-best (AdaBoost model: 0.688) accuracy in the study by D'Orazio (2019a).

The average estimated accuracy is $\overline{accuracy}_{estim} = 0.693$ ($\bar{\sigma} = 0.003$). The small difference between the measured and the estimated accuracy of -0.001 can be considered random error and supports confidence in the estimation procedure.

To assess the second question, the preservation of the marginal distribution of X, I follow D'Orazio (2019a) and use the total variance index or dissimilarity index (Agresti 2013, pp. 352 – 353) to estimate the distance between the observed and matched distribution of X. This index is bounded by 0 and 1, where 0 indicates complete similarity and 1 denotes complete dissimilarity. In the binary case at hand, the dissimilarity index is given by

$$\hat{\Delta} = \frac{1}{2} \sum_{i=1}^{2} \mid p_i^{obs} - p_i^{pred} \mid, \tag{11.1}$$

where p_i^{obs} denotes the share of observations in the observed data belonging to the i's category (here either 1 or 0) and p_i^{pred} denotes the same share in the matched data.

The average dissimilarity between the matched and observed X in the 100 samples is $\bar{\hat{\Delta}} = 0.165$ ($\bar{\sigma} = 0.027$). The similarity between the two marginal distributions can be considered moderate. Considering the moderate accuracy, the two results are coherent. Nevertheless, the simulations by D'Orazio (2019a) resulted in dissimilarity scores of 0.078 and below, which shows that better results are possible given the data.

The most important question to answer is how well the matching reproduces the relationship of interest: the relationship between the impression of living comfortably on household income (X) and the total household net income (Y). The purpose of statistical matching is to infer a relationship between variables that are not commonly observed. Thus, the ability to reproduce this relationship is the reason for doing the matching and is therefore of the highest interest.

To measure the relationship of interest, I use Cramer's V. Figure 11.3 shows the results. The average Cramer's V over the 100 sample runs is $\bar{\varphi}_{matched} = 0.33$ ($\bar{\sigma} = 0.024$). The observed Cramer's V over the complete data is $\varphi_{observed} = 0.398$. In Fig. 11.3, this underestimation is easy to see. Looking at the observed values in the 100 samples, they also show a slight but negligible understatement of Cramer's V.

The overall tendency to underestimate the relationship of interest is also apparent, and even more pronounced, in the results by D'Orazio (2019a). The dot-dashed lines

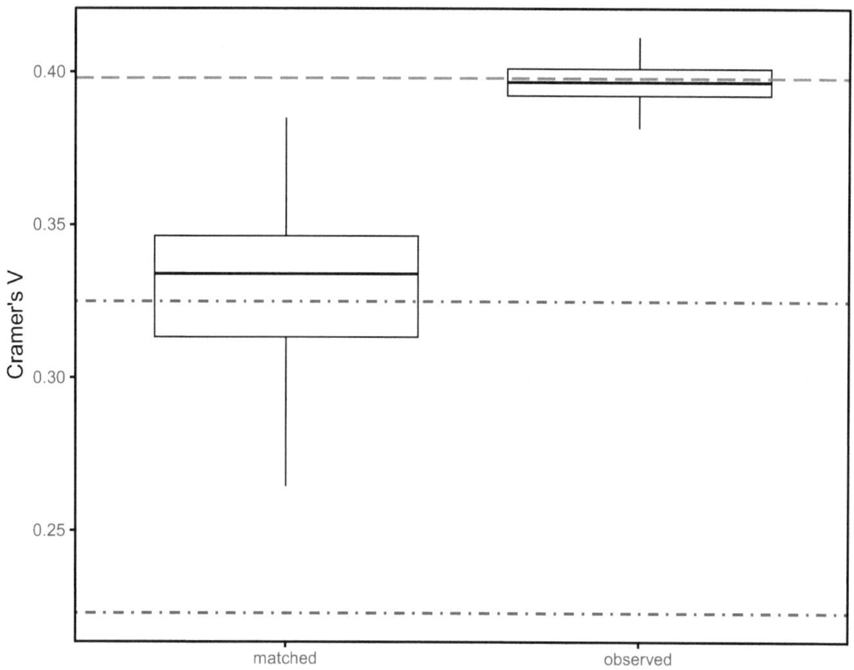

Fig. 11.3 Distribution of Cramer's V on 100 pipeline runs (Note: The box plots depict the distribution of Cramer's V in the 100 data splits as a result of the matching procedure using the pipeline and as observed in the sample data, respectively. In addition, the dashed line shows the observed value of Cramer's V in the full data. The two dot-dashed lines show the best and second-best results of D'Orazio (2019a) for reference)

in Fig. 11.3 show the best and second-best results of the cited study. The arithmetic mean and the median reported in the study at hand are closer to the observed value of Cramer's V than the best model reported by D'Orazio (2019a). All 100 matched pipeline results return better estimates of the relationship of interest than the second-best result shown by D'Orazio (2019a).

To summarize, the simulation study described above is based on data from the ESS Round 7 (Norwegian Agency for Shared Services in Education and Research, Norway 2014). The data are slightly preprocessed and then randomly split 100 times into donor A and receiver B data sets. Both data sets, along with a series of specifications for the estimation process, are used as input to run the statistical matching pipeline. During each run, the respective donor set is used to estimate a model of the matching variable X (impression of living comfortably on household income). This model is used to predict X in the receiver data set B, so to estimate the relationship between X and Y in B.

The results of the pipeline give a generally good estimate of the relationship of interest. However, there is a small but systematic underestimation. Another result apparent in this simulation is generally small variances of all estimates.

This is a result of the nested resampling implemented in the statistical matching pipeline. Such a result increases confidence in individual matching procedures in real use cases. Another notable result of the simulation is the comparatively low implementation time and runtime of the algorithm. Under the technical conditions described above, the total runtime for 100 pipeline runs, including the tuning of multiple algorithms via nested resampling, is about 7 hours. Manual programming and evaluation of the algorithms would require disproportionately more time. In addition, the pipeline's results are comparable to or better than current state-of-the-art matching techniques, such as those implemented by D'Orazio (2019a) for the same problem.

11.5 Replication Study: Microcensus and Central Register of Foreigners (Germany)

The following study evaluates how well the automated pipeline performs in comparison to a more manual and human-led approach in a real use case. The large amount of automation that comes with applying the statistical matching pipeline to the case suggests potential for a severe reduction of manual labor. However, at the same time, the accompanying degree of generalization and standardization bears the risk of losing some performance due to fewer fine-tuning and thus a less specific model. For this comparison, colleagues, who recently worked on a matching project at the Federal Statistical Office (Smilde-Becker and Eberle 2024), kindly allowed me to replicate their work using the statistical matching pipeline.

The original study (Smilde-Becker and Eberle 2024) aims to estimate the labor force participation of refugees in Germany. Especially in the last decade, Germany and many other European countries have been confronted with a large amount of refugees and people in refugee-like situations (Eberle 2019) from, among others, Syria, Iraq, Afghanistan, and, more recently, Ukraine. Their integration into the labor market is of great political interest. In order to monitor and possibly make policy decisions, it is important to have adequate data on this issue.

The German Microcensus provides a wide range of socioeconomic information on the general population, including information on immigrants in general. However, it does not reliably distinguish refugees and persons in a refugee-like situation from other immigrants. This distinction is covered by the national Central Register of Foreigners (Bundesverwaltungsamt 2024). The Central Register of Foreigners covers all people of foreign nationality, who are or have been residing in Germany for 3 months or more, and includes, among other things, information on legal residence status. Yet, the Central Register of Foreigners does not contain the socioeconomic information on employment or housing conditions. Thus, the merging of both data sources, Microcensus and Central Register of Foreigners, is desirable.

As Smilde-Becker and Eberle (2024) note, the data sets "neither share a common personal identifier nor sufficient information for probabilistic linkage" (p. 2). Thus, statistical matching is the only option to generate a (synthetic) common data set.

In this use case, the aim is to analyze the unobserved relationship between refugee status (X) and employment status (Y). The Central Register of Foreigners data serve as the donor data set A, where the refugee status X is directly observed. Consequently, the Microcensus is used as receiver data set B. A set of socioeconomic variables,[4] which are observed in both data sets, serves as predictors Z. Smilde-Becker and Eberle (2024) provided the harmonized data for the years 2019 and 2021. The replication is done for both years independently, but following an identical design. After harmonization, the Central Register of Foreigners includes 11.2 million people in 2019 and 11.6 million people in 2021, including about 1.8 million people seeking humanitarian protection in 2019 and 1.9 million in 2021. The Microcensus data are reduced to foreigners and include about 72,000 observations in 2019 and 85,500 observations in 2021.

Due to a lack of hardware capacity, it was not possible at the time to run the statistical matching pipeline on the full Central Register of Foreigners data. As a resort, for each year, five distinct samples of one million observations are drawn. The pipeline is applied to all five data sets, and average results are reported. The algorithms to be tested are a decision tree (Therneau and Atkinson 2022) and a random forest (Wright and Ziegler 2017). A nearest neighbor approach was eliminated at an earlier stage due to poor performance and a very long run time on a smaller sample of data. As outer resampling strategy, I set a threefold cross-validation. The inner resampling is set to a fourfold cross-validation with four hyperparameter configurations tested in each fold, randomly selected for each fold. Since it is of particular interest to identify refugees among the foreigners, who form a much smaller class, the parameter to be optimized is precision, given as

$$\widehat{prec} = \frac{true\ positives}{true\ positives + false\ positives}. \tag{11.2}$$

All five pipeline runs result in the selection of the random forest model to be the best fit for the given data. In the original study, the authors choose another tree-based modeling approach, namely a C5.0 classification tree (Kuhn and Johnson 2013), including boosting.

The most relevant metric, precision, is nearly the same in both studies. For 2021, Smilde-Becker and Eberle (2024) report a $\widehat{prec}_{original} = 0.77$, while I find a replicated average $\widehat{\bar{prec}}_{replication-2021} = 0.78$. For the 2019 data, the original precision remains at $\widehat{prec}_{original} = 0.77$. In the replication I find $\widehat{\bar{prec}}_{replication-2019} = 0.8$. These results show that the statistical matching pipeline allowed to fully replicate the human-led optimization in this case study.

[4] The sociodemographic variables that make up Z are the following: gender, age, nationality, age at immigration, year of immigration, marital status, and district of residence.

Theoretically, one could imagine information leakage between the original project and the replication study. To prevent such leakage, the authors of the original study took care to share only the harmonized data, which was not processed in any way. Before the pipeline was implemented, it was known that a tree-based model performed best in the manual case. However, no hyperparameter settings were disclosed. In addition, several models were evaluated with the pipeline as described above and without influencing the automated process. All in all, everyone involved was aware of the potential negative impact of information leakage, and all measures were taken to avoid it.

11.6 Conclusion

In this chapter, I first argue that modern machine learning algorithms can be advantageous when performing statistical matching. Second, I present an automated statistical matching pipeline that implements a workflow that allows statistical matching to be performed with automated machine learning while maintaining high methodological standards. Statistical matching applications, especially in official statistics, typically rely on some version of a nearest neighbor approach (D'Orazio 2019a; European Commission et al. 2017; Schaller 2021; Webber and Tonkin 2013; Gazzelloni et al. 2008; Abello and Phillips 2004; Leulescu and Agafitei 2013; D'Orazio and Catanese 2016; D'Alberto et al. 2018) or on parametric models (European Commission et al. 2017; Webber and Tonkin 2013; Leulescu and Agafitei 2013; Gutman et al. 2013; Rässler 2012). The potential of other nonparametric modeling strategies is, to the best of my knowledge, rarely explored or used.

The proposed statistical matching pipeline automates the process from (1) data input, (2) data preprocessing, (3) training, tuning, and comparing a set of models, selecting the most likely best algorithm for the task and data at hand, (4) retraining the most promising model, (5) performing the actual matching, and (6) returning the matched data along with a comprehensive output report on the data, the process, and the matching result. The selection of the algorithm to be used is based on nested resampling, to ensure accurate and comparable estimations of the generalization error (Bischl et al. 2023; Becker et al. 2024). This is combined with the use of a so-called fallback learner, which prevents over- or underestimation of a model's performance, when single runs during resampling fail (Lang et al. 2024).

In an empirical study, I test the statistical matching pipeline in a simulated matching case study, using data from the European Social Survey (ESS) Round 7 (Norwegian Agency for Shared Services in Education and Research, Norway 2014). The simulated relationship of interest is between (a) a binary indicator of whether or not the respondent and his or her household can live comfortably on their household income and (b) the total net income of the household, given as a decile. The data are split into two data sets, and the pipeline is used to merge the data back together. This case is both relevant to many European national statistical institutes and allows to compare results to other state-of-the-art matching techniques applied to the same data (D'Orazio 2019a). Results show a moderate accuracy of single observations, when compared to the original observations. However, the estimated accuracy is highly consistent. When it comes to the estimation of the relationship of interest, I find good results, despite a small but consistent underestimation of Cramer's V. In comparison to other modern matching techniques, the use of machine learning algorithms within the statistical matching pipeline leads to either comparable or improved results.

In a second study, I replicate a matching application, which was first developed with a much more manual approach at the Federal Statistical Office of Germany (Smilde-Becker and Eberle 2024). In this case, Germany's Central Register of Foreigners (Bundesverwaltungsamt 2024) serves as donor data to complement the Microcensus with information on refugee status of immigrants. Based on the harmonized data, which was kindly provided by the authors of the original study, it was possible to replicate their model with nearly the same precision within a few days, instead of several months of development. This shows the efficiency of the statistical matching pipeline, along with the great potential to save resources and time.

All in all, the statistical matching pipeline using automated machine learning presented in this chapter endorses the great potential of matching techniques for official statistics. The method itself makes it possible to take full advantage of the data and information already available in institutions of official statistics. This reduces the burden on those providing information. It also reduces the resources needed to collect more data than is necessary. The proposed statistical matching pipeline elevates matching techniques in several ways. First, the extensive automation substantially speeds up the matching process. This allows for a parsimonious allocation of resources in matching projects. In addition, it opens up the possibility of responding in a timely manner to new information needs that arise in the political, economic, or public spheres. Second, the use of modern machine learning algorithms makes it possible to better estimate nonlinear relationships. Third, the carefully chosen methodological standards for model estimation and model selection strongly support the high quality and trustworthiness requirements of official statistics.

Acknowledgments Some icons are designed by Freepik.

References

R. Abello, B. Phillips, Statistical matching of the HES and NHS: an exploration of issues in the use of unconstrained and constrained approaches in creating a basefile for a microsimulation model of the pharmaceutical benefits scheme. Technical report, Australian Bureau of Statistics (2004)

A. Agresti, *Categorical Data Analysis*, 3rd edn. (John Wiley & Sons, Hoboken, 2013)

J. Allaire, Y. Xie, C. Dervieux, J. McPherson, J. Luraschi, K. Ushey, A. Atkins, H. Wickham, J. Cheng, W. Chang, R. Iannone, *rmarkdown: Dynamic Documents for R* (2024). https://github.com/rstudio/rmarkdown. R package version 2.27

S.M. Bache, H. Wickham, *magrittr: A Forward-Pipe Operator for R* (2022). https://magrittr.tidyverse.org, https://github.com/tidyverse/magrittr

M. Beck, F. Dumpert, J. Feuerhake, Machine learning in official statistics (2018). https://arxiv.org/abs/1812.10422

M. Becker, *mlr3tuningspaces: Search Spaces for 'mlr3'* (2023). https://mlr3tuningspaces.mlr-org.com, https://github.com/mlr-org/mlr3tuningspaces

M. Becker, M. Lang, J. Richter, B. Bischl, D. Schalk, *mlr3tuning: Hyperparameter Optimization for 'mlr3'* (2023). https://mlr3tuning.mlr-org.com, https://github.com/mlr-org/mlr3tuning

M. Becker, L. Schneider, S. Fischer, Hyperparameter optimization, in *Applied Machine Learning Using mlr3 in R*, ed. by B. Bischl, R. Sonabend, L. Kotthoff, M. Lang (CRC Press, Boca Raton, 2024). https://mlr3book.mlr-org.com/hyperparameter_optimization.html

H. Bengtsson, *progressr: An Inclusive, Unifying API for Progress Updates* (2023). https://progressr.futureverse.org, https://github.com/HenrikBengtsson/progressr

M. Binder, F. Pfisterer, B. Bischl, Collecting empirical data about hyperparameters for data driven AutoML, in *Democratizing Machine Learning Contributions in AutoML and Fairness* (2020), p. 93

M. Binder, F. Pfisterer, M. Lang, L. Schneider, L. Kotthoff, B. Bischl, mlr3pipelines – flexible machine learning pipelines in R. J. Machine Learning Res. **22**(184), 1–7 (2021). https://jmlr.org/papers/v22/21-0281.html

B. Bischl, M. Binder, M. Lang, T. Pielok, J. Richter, S. Coors, J. Thomas, T. Ullmann, M. Becker, A.-L. Boulesteix, et al., Hyperparameter optimization: foundations, algorithms, best practices, and open challenges. Wiley Interdiscip. Rev.: Data Min. Knowl. Discovery **13**(2), e1484 (2023)

B. Bischl, R. Sonabend, L. Kotthoff, M. Lang (eds.), *Applied Machine Learning Using mlr3 in R* (CRC Press, Boca Raton, 2024). https://mlr3book.mlr-org.com

Bundesverwaltungsamt, Auslaenderzentralregister (2024). https://www.bva.bund.de/DE/Das-BVA/Aufgaben/A/Auslaenderzentralregister/azr_node.html

T. Chen, T. He, M. Benesty, V. Khotilovich, Y. Tang, H. Cho, K. Chen, R. Mitchell, I. Cano, T. Zhou, M. Li, J. Xie, M. Lin, Y. Geng, Y. Li, J. Yuan, *xgboost: Extreme Gradient Boosting* (2023). https://github.com/dmlc/xgboost. R package version 1.7.5.1

J. Cielebak, S. Rässler, Data fusion, record linkage und data mining, in *Handbuch Methoden der empirischen Sozialforschung*, ed. by N. Baur, J. Blasius (eds.) (Springer, Berlin, 2014), pp. 367–382

Commission Future Statistics, Recommendations of the commission future statistics (2024). https://www.destatis.de/DE/Ueber-uns/Leitung-Organisation/KomZS/abschlussbericht-en.pdf?__blob=publicationFile

P.L. Conti, D. Marella, M. Scanu, Uncertainty analysis in statistical matching. J. Off. Stat. **28**(1), 69–88 (2012)

P.L. Conti, D. Marella, M. Scanu, Uncertainty analysis for statistical matching of ordered categorical variables. Comput. Stat. Data Anal. **68**, 311–325 (2013)

P.L. Conti, D. Marella, M. Scanu, How far from identifiability? A systematic overview of the statistical matching problem in a non parametric framework. Commun. Stat. Theory Methods **46**(2), 967–994 (2017)

B. Cui, *DataExplorer: Automate Data Exploration and Treatment* (2020). http://boxuancui.github.io/DataExplorer/. R package version 0.8.2

R. D'Alberto, M. Raggi, Integrating rather than collecting: statistical matching in the data flood era. Stat. Pap. **65**, 2135–2163 (2024)

R. D'Alberto, M. Zavalloni, M. Raggi, D. Viaggi, AES impact evaluation with integrated farm data: combining statistical matching and propensity score matching. Sustainability **10**(11), 4320 (2018)

M. D'Orazio, E. Catanese, Evaluating revenues and economic growth for farms producing renewable energies: an investigation based on integration of FSS and EOAH 2013 survey data, in *Proceedings of the Seventh International Conference on Agricultural Statistics* (2016), pp. 1–8

M. D'Orazio, Statistical learning in official statistics: the case of statistical matching. Stat. J. IAOS **35**(3), 435–441 (2019a)

M. D'Orazio, Statistical learning in official statistics: the case of statistical matching (2019b). https://www.researchgate.net/publication/331772580_Statistical_Learning_in_Official_Statistics_the_case_of_Statistical_Matching

M. D'Orazio, M. Di Zio, M. Scanu, *Statistical Matching: Theory and Practice* (John Wiley & Sons, Hoboken, 2006a)

M. D'Orazio, M. Di Zio, M. Scanu, Statistical matching for categorical data: displaying uncertainty and using logical constraints. J. Off. Stat. **22**(1), 137–157 (2006b)

C. Dowd, *twosamples: Fast Permutation Based Two Sample Tests* (2023). https://twosampletest.com, https://github.com/cdowd/twosamples

M. Dowle, A. Srinivasan, *data.table: Extension of 'data.frame'* (2023). https://r-datatable.com

J. Eberle, Ein Konzept zur Quantifizierung des Bestands an Ausländerinnen und Ausländern, die sich aus humanitären Gründen in Deutschland aufhalten. WISTA Wirtschaft und Statistik **71**(1), 19–34 (2019)

European Commission, Eurostat, P. Serafino, R. Tonkin, Statistical matching of European Union statistics on income and living conditions (EU-SILC) and the household budget survey. Statistical Working Papers (2017)

European Social Survey, About ESS (2024). https://europeansocialsurvey.org/about-ess

European Statistical System Committee, Quality assurance framework of the European statistical system (2019). https://ec.europa.eu/eurostat/documents/64157/4392716/ESS-QAF-V2.0-final.pdf

Eurostat, Statistical matching methods (method) (2023). https://wayback.archive-it.org/12090/20231230081135/https://cros-legacy.ec.europa.eu/content/statistical-matching-methods-method_en

Freepik, Freep!k (2023). www.freepik.com

J. Friedman, R. Tibshirani, T. Hastie, Regularization paths for generalized linear models via coordinate descent. J. Stat. Software **33**(1), 1–22 (2010)

S. Gazzelloni, M.C. Romano, G. Corsetti, M. Di Zio, M. D'Orazio, F. Pintaldi, M. Scanu, N. Torelli, Time use and labour force: a proposal to integrate the datathrough statistical matching, in *Time Use in Daily Life: A Multidisciplinary Approach to the Time Use's Analysis* (Istat-Produzione libraria e centro stampa, 2008), pp. 297–320

R. Gutman, C.C. Afendulis, A.M. Zaslavsky, A bayesian procedure for file linking to analyze end-of-life medical costs. J. Am. Stat. Assoc. **108**(501), 34–47 (2013)

B. Hamner, M. Frasco, *Metrics: Evaluation Metrics for Machine Learning* (2018). https://github.com/mfrasco/Metrics. R package version 0.1.4

T. Hastie, R. Tibshirani, J.H. Friedman, *The Elements of Statistical Learning: Data Mining, Inference, and Prediction*, 2nd edn. (Springer, Berlin, 2009)

J.B. Kadane, Some statistical problems in merging data files. J. Off. Stat. **17**(3), 423–433 (2001)

L. Kotthoff, R. Sonabend, N. Foss, B. Bischl, Introduction and overview, In *Applied Machine Learning Using mlr3 in R*, ed. by B. Bischl, R. Sonabend, L. Kotthoff, M. Lang (CRC Press, Boca Raton, 2024). https://mlr3book.mlr-org.com/introduction_and_overview.html

A. Kowarik, M. Templ, Imputation with the R package VIM. J. Stat. Software **74**(7), 1–16 (2016)

M. Kuhn, K. Johnson, Classification trees and rule-based models, in *Applied Predictive Modeling* (Springer, Berlin, 2013)

D. Kühn, P. Probst, J. Thomas, B. Bischl, Automatic Exploration of Machine Learning Experiments on OpenML (2018). https://arxiv.org/abs/1806.10961

T. Küntzler, Statistical matching pipeline (2024). https://gitlab.opencode.de/oc00004345291/statistical-matching-pipeline

W.M. Landau, *tarchetypes: Archetypes for Targets* (2021a). https://docs.ropensci.org/tarchetypes/, https://github.com/ropensci/tarchetypes

W.M. Landau, The targets R package: a dynamic make-like function-oriented pipeline toolkit for reproducibility and high-performance computing. J. Open Source Software **6**(57), 2959 (2021b)

M. Lang, *mlr3measures: Performance Measures for 'mlr3'* (2022). https://mlr3measures.mlr-org.com, https://github.com/mlr-org/mlr3measures

M. Lang, P. Schratz, *mlr3misc: Helper Functions for 'mlr3'* (2023). https://mlr3misc.mlr-org.com, https://github.com/mlr-org/mlr3misc

M. Lang, M. Binder, J. Richter, P. Schratz, F. Pfisterer, S. Coors, Q. Au, G. Casalicchio, L. Kotthoff, B. Bischl, mlr3: a modern object-oriented machine learning framework in R. J. Open Source Software **4**(44), 1903 (2019)

M. Lang, Q. Au, S. Coors, P. Schratz, *mlr3learners: Recommended Learners for 'mlr3'* (2023a). https://mlr3learners.mlr-org.com, https://github.com/mlr-org/mlr3learners

M. Lang, B. Bischl, J. Richter, X. Sun, M. Binder, *paradox: Define and Work with Parameter Spaces for Complex Algorithms* (2023b). https://paradox.mlr-org.com, https://github.com/mlr-org/paradox

M. Lang, P. Schratz, R. Sonabend, M. Becker, J. Richter, *mlr3viz: Visualizations for 'mlr3'* (2023c). https://mlr3viz.mlr-org.com, https://github.com/mlr-org/mlr3viz

M. Lang, S. Fischer, R. Sonabend, Advanced technical aspects of mlr3, in *Applied Machine Learning Using mlr3 in R*, ed. by B. Bischl, R. Sonabend, L. Kotthoff, M. Lang (CRC Press, Boca Raton, 2024). https://mlr3book.mlr-org.com/advanced_technical_aspects_of_mlr3.html

A. Leulescu, M. Agafitei, Statistical matching: a model based approach for data integration. Eurostat-Methodologies and Working Papers (2013)

I. Lewaa, M.S. Hafez, M.A. Ismail, Data integration using statistical matching techniques: a review. Stat. J. IAOS **37**, 1391–1410 (2021)

D. Marella, D. Pfeffermann, Matching information from two independent informative samples. J. Stat. Plan. Infer. **203**, 70–81 (2019)

D. Meyer, E. Dimitriadou, K. Hornik, A. Weingessel, F. Leisch, *e1071: Misc Functions of the Department of Statistics, Probability Theory Group (Formerly: E1071), TU Wien* (2023). R package version 1.7-13

K. Müller, *here: A Simpler Way to Find Your Files* (2020). https://here.r-lib.org/, https://github.com/r-lib/here

K. Müller, H. Wickham, *tibble: Simple Data Frames* (2023). https://tibble.tidyverse.org/, https://github.com/tidyverse/tibble

M. Nalenz, J. Rodemann, T. Augustin, Learning de-biased regression trees and forests from complex samples. Mach. Learn. **113**, 3379–3398 (2024)

Norwegian Agency for Shared Services in Education and Research, Norway, European Social Survey Round 7 Data. Data Archive and distributor of ESS data for ESS ERIC (2014). https://doi.org/10.21338/NSD-ESS7-2014

B. Okner, Constructing a new data base from existing microdata sets: the 1966 merge file, in *Annals of Economic and Social Measurement*, vol. 1 (NBER, Cambridge, 1972), pp. 325–362

S. Rässler, A non-iterative bayesian approach to statistical matching. Stat. Neerl. **57**(1), 58–74 (2003)

S. Rässler, Data fusion: identification problems, validity, and multiple imputation. Aust. J. Stat. **33**(1&2), 153–171 (2004)

S. Rässler, *Statistical Matching: A Frequentist Theory, Practical Applications, and Alternative Bayesian Approaches* (Springer, Berlin, 2012)

R Core Team, *R: A Language and Environment for Statistical Computing*. R Foundation for Statistical Computing (2021). https://www.R-project.org/

W.L. Rodgers, An evaluation of statistical matching. J. Bus. Econ. Stat. **2**(1), 91–102 (1984)

D.B. Rubin, Characterizing the estimation of parameters in incomplete-data problems. J. Am. Stat. Assoc. **69**(346), 467–474 (1974)

A. Rushworth, *inspectdf: Inspection, Comparison and Visualisation of Data Frames* (2022). https://alastairrushworth.github.io/inspectdf/. R package version 0.0.12

Y. Saidani, F. Dumpert, Quality dimensions and quality guidelines for machine learning in official statistics, in *Foundations and Advances of Machine Learning in Official Statistics*, ed. by F. Dumpert, Chap. 4 (Springer, Berlin, 2025)

Y. Saidani, F. Dumpert, C. Borgs, A. Brand, A. Nickl, A. Rittmann, J. Rohde, C. Salwiczek, N. Storfinger, S. Straub, Qualitätsdimensionen maschinellen Lernens in der amtlichen Statistik. AStA Wirtschafts- und Sozialstatistisches Archiv **17**(3), 253–303 (2023)

J. Schaller, Datenfusion von EU-SILC und Household Budget Survey – ein Vergleich zweier Fusionsmethoden. WISTA Wirtschaft und Statistik **73**(4), 76–86 (2021)

K. Schliep, K. Hechenbichler, *kknn: Weighted k-Nearest Neighbors* (2016). https://github.com/KlausVigo/kknn. R package version 1.3.1

P. Schratz, *mlr3-targets* (2022). https://github.com/mlr-org/mlr3-targets

M. Smilde-Becker, J. Eberle, Estimating labour force participation of refugees by statistical matching of administrative and survey data. Working Paper (2024). https://unece.org/sites/default/files/2024-04/0.Workshop.Panel1%20GER%20Eberle.pdf

J. Stock, O. Hauke, J. Weißmann, H. Federrath, The applicability of federated learning to official statistics, in *International Conference on Intelligent Data Engineering and Automated Learning* (Springer, Berlin, 2023), pp. 70–81

H. Sutherland, R. Taylor, J. Gomulka, Combining household income and expenditure data in policy simulations. Rev. Income Wealth **48**(4), 517–536 (2002)

T. Therneau, B. Atkinson, *rpart: Recursive Partitioning and Regression Trees* (2022). https://github.com/bethatkinson/rpart, https://cran.r-project.org/package=rpart

J. Thomas, Preprocessing, in *Applied Machine Learning Using mlr3 in R*, ed. by B. Bischl, R. Sonabend, L. Kotthoff, M. Lang (CRC Press, Boca Raton, 2024). https://mlr3book.mlr-org.com/preprocessing.html

E. Waring, M. Quinn, A. McNamara, E. Arino de la Rubia, H. Zhu, S. Ellis, *skimr: Compact and Flexible Summaries of Data* (2022). https://docs.ropensci.org/skimr/, https://github.com/ropensci/skimr/

D. Webber, R. Tonkin, Statistical matching of EU-SILC and the household budget survey to compare poverty estimates using income. *Expenditures and Material Deprivation, Luxembourg: Publications Office of the European Union* (2013)

H. Wickham, *ggplot2: Elegant Graphics for Data Analysis* (Springer, Berlin, 2016). https://ggplot2.tidyverse.org

H. Wickham, *stringr: Simple, Consistent Wrappers for Common String Operations* (2022). https://stringr.tidyverse.org, https://github.com/tidyverse/stringr

H. Wickham, R. François, L. Henry, K. Müller, D. Vaughan, *dplyr: A Grammar of Data Manipulation* (2023). https://dplyr.tidyverse.org, https://github.com/tidyverse/dplyr

M.N. Wright, A. Ziegler, ranger: A fast implementation of random forests for high dimensional data in C++ and R. J. Stat. Software **77**(1), 1–17 (2017)

Y. Xie, J. Allaire, G. Grolemund, *R Markdown: The Definitive Guide* (Chapman and Hall/CRC, Boca Raton, 2018). https://bookdown.org/yihui/rmarkdown

Y. Xie, J. Cheng, X. Tan, *DT: A Wrapper of the JavaScript Library 'DataTables'* (2023). https://github.com/rstudio/DT. R package version 0.29

Chapter 12
Big Data and Machine Learning at Istat

Mauro Bruno, Elena Catanese, Erika Cerasti, Massimo De Cubellis, Fabrizio De Fausti, Marco Di Zio, Gerarda Grippo, Giuseppe Lancioni, Giulio Massacci, Stefano Mugnoli, Francesco Ortame, Angela Pappagallo, Francesco Pugliese, Alessandra Righi, Alberto Sabbi, Francesco Sisti, Donato Summa, and Luca Valentino

12.1 Introduction

The National Italian Institute of Statistics (Istat) has faced significant challenges over the past decade in response to the changes brought by digital transformation, the increasing amount of data available, and evolving European and international landscapes. This chapter outlines Istat's journey, from initial experimentation to establishing a dedicated production process for smart statistics. The resulting products are derived from big data sources and use innovative methodologies. We will describe the organizational solutions put in place, analyze their impact, and outline the additional requirements necessary to support this transformation.

M. Bruno (✉) · E. Catanese · E. Cerasti · M. De Cubellis · F. De Fausti · M. Di Zio · G. Grippo · G. Lancioni · G. Massacci · S. Mugnoli · F. Ortame · A. Pappagallo · F. Pugliese · A. Righi · A. Sabbi · F. Sisti · D. Summa · L. Valentino
Istat, Rome, Italy
e-mail: mbruno@istat.it; catanese@istat.it; erika.cerasti@istat.it; decubell@istat.it; defausti@istat.it; dizio@istat.it; grippo@istat.it; lancioni@istat.it; giulio.massacci@istat.it; mugnoli@istat.it; francesco.ortame@istat.it; angela.pappagallo@istat.it; frpuglie@istat.it; righi@istat.it; sabbi@istat.it; francesco.sisti@istat.it; donato.summa@istat.it; luvalent@istat.it

F. Dumpert (ed.), *Foundations and Advances of Machine Learning in Official Statistics*, Society, Environment and Statistics,
https://doi.org/10.1007/978-3-032-10004-7_12

12.1.1 Background

When the EU Scheveningen Memorandum[1] first formalized the need to look at big data[2] sources for official statistics in 2013, Istat gradually became aware of the new opportunities and challenges offered by big data and began its active participation in the European context with the new data sources (ESSNet Big Data I and II projects). The increased use of big data requires a new approach to statistical production processes that aligns with the agenda of Istat's modernization project. This project has been implemented since 2014 and involves the use of innovative sources, in addition to traditional statistical information, to obtain more significant insights.

Research and innovation are the pillars of Istat's big data and modernization strategy.[3] The research was organized and strategic, with both thematic and methodological components recognized as growth elements for the Institute. Methodological research activities are characterized by a constant drive to innovate methods and tools to improve process efficiency and data production procedures, increase the quality of the statistics produced, reduce the costs of statistical processes, and disseminate new statistical data. Investments in thematic research have represented a tool for enriching knowledge of the phenomena and, through analyses to deepen the understanding of social and economic phenomena, have expanded the Institute's information supply.

Integrating traditional and new sources, focusing on digital technologies and new computational models, requires a paradigm shift in methodology from traditional sample survey-based estimates to a model applicable to a multisource environment. To support this evolution, the Institute relied on experts gathered in two big data committees, the first set in 2013 and the second in 2016, which have played a key role in assisting Istat in developing a roadmap for using big data in official statistics and monitoring investments. Additionally, they have facilitated the sharing of knowledge and experiences, promoted and supported new partnerships, and enhanced the role of research within the Institute.

Another organizational innovation conducive to developing research and innovation in the realm of big data involved establishing a new research infrastructure within the Institute. This infrastructure has been monitored by a governance body (the Research Committee) since 2017 to ensure the quality and coordination of activities. The Research Committee comprises two bodies with predominant roles in guidance and scientific support: the Scientific Committee for Thematic Research—

---

[1] European Statistical System—ESS, Directors General of the National Statistical Institutes—DGINS. 2013. Scheveningen Memorandum. Big data and official statistics. Scheveningen—The Hague, The Netherlands, 25th–27th September 2013: DGINS Conference.

[2] Set of data from different sources that is so large and complex that it requires new technologies, such as artificial intelligence, to be processed.

[3] https://www.istat.it/it/files/2010/12/Programma_modernizzazione_Istat2016.pdf

the Advisory Committee for Statistical Methodologies (Advisory Board)—and Thematic Laboratories and the Innovation Laboratory.[4]

During this period, Istat, a European pioneer in harnessing big data to produce official statistics, initiated a strategic evaluation to implement the recommendations provided. The objective is to move from the exploratory phase to a more systematic and developed use of big data. The initial success of experimental statistics and the outcomes of the ESSnet big data projects spurred Eurostat to take decisive action. The opportunity arose in 2018 with the signing of the "Bucharest Memorandum on Official Statistics in a Datafied Society (Trusted Smart Statistics)" by heads of statistical offices. Trusted Smart Statistics (TSS), mentioned for the first time in this memorandum, represent statistical products generated from smart systems crafted with tools and methodologies adhering to the Code of Practice and Quality Assurance Framework of official statistics, ensuring verifiability and transparency. National statistical institutes uphold their validity and accuracy while fully respecting the privacy of individuals and other stakeholders. The Bucharest Memorandum outlines the strategic direction that national statistical institutes should take regarding the use of big data sources and the production of smart statistics. It emphasizes the need to implement mature use cases and develop experimental statistics[5] on emerging phenomena. These statistics utilize new data sources and methods to respond more effectively and promptly to users' needs. Since these statistics have not yet reached full maturity in terms of harmonization, coverage, or methodology, they are classified as experimental. This highlights the need for increased awareness that leveraging big data sources requires adjustments to statistical business architecture, processes, production models, IT infrastructure, methodological and quality frameworks, and corresponding governance structures.

The Blue Sky Thinking Network of the UNECE High-Level Group for the Modernisation of official statistics (HLG-MOS) introduced a proposition in 2018 to endorse adopting new techniques for TSS production. The chapter suggests that exploring opportunities offered by machine learning is essential for processing secondary data sources (such as administrative sources, big data, and the Internet of Things), which could also provide added value for primary data. Due to the significant responsibility of upholding the credibility of statistics based on the Fundamental Principles of official statistics, using machine learning (ML) in the national statistical offices requires a more careful approach than in other business environment.

[4] https://www.istat.it/en/about-istat/activities/research/organisation/

[5] https://www.istat.it/en/announcement-and-analisys/experimental-statistics/

12.1.2 Toward a Production System for Trusted Smart Statistics

Since 2018, Istat has embarked on a journey toward TSS, marking a significant milestone as related projects were incorporated into Istat's strategic planning for the first time. As part of this initiative, investments in human capital and recruiting new data scientists were strategically programmed. Istat has fostered an extensive collaboration network with various statistical institutes, universities, and private partners. During this phase, a preliminary set of projects has reached completion, spanning exploration, prototype development, and dissemination activities. The outcomes of these experiments have been successfully integrated into the production processes, marking a pivotal step forward in advancing TSS at Istat.

Istat has established an integrated production system with its conventional data acquisition and production infrastructure to facilitate smart statistics production. This system incorporates the Integrated System of Registers, current surveys, and externally manageable processing procedures using agreed methodologies, algorithms, and reliable software. Implementing such a system involves a multi-tiered data processing workflow organization and adopting standard manufacturing process management models, such as the Generic Statistical Business Process Model (GSBPM). This represents an ongoing, step-by-step process, currently under development, which requires continuous efforts to refine and expand the system. Istat's effort to innovate methods has prompted the adoption of innovative organizational solutions for developing TSS despite the lack of specific legal regulations at European and national levels. To facilitate the integration of the TSS production system without compromising existing production quality, an internal Director Steering Committee supervises strategic analysis of the TSS system. Supported by a Scientific Technical Secretariat, each organizational unit (communication, legal, data collection, IT, human resources) within the Institute actively contributes its expertise to develop the new production system collectively. The Steering Committee identifies the demand for statistical information, prioritizes implementing TSS projects, and supervises experiments and investments.

The Center for the Production of TSS has been established to assist the Steering Committee in the TSS Strategic Analysis process, strategic planning, and activity monitoring. The Center supports TSS experimentation, industrialization, and production and is a cornerstone for the new TSS production system, promoting sustainability and certified quality. The Center's role is to provide an organizational structure for operational activities conducted across various organizational structures. The Center is instrumental in refining experimental products, facilitating communication, and ensuring compliance with regulations. Additionally, it plays a crucial role in fostering collaborative partnerships with national and international research institutions, conducting internal training initiatives, and spearheading endeavors.

The Center strengthened the Institute's investments to integrate big data into production processes. In fact, since February 2018, Istat started systematically integrating scanner data into Consumer Price Indexes, incorporating these data into the provisional estimate of the Consumer Price Index. Furthermore, using OpenStreetMap information has enhanced the investigation of road accidents. Additionally, progress has been made in developing a Social Mood on Economy Index (SMEI) using data from social media platforms. Special attention has also been given to the dissemination of the results. Istat has developed dedicated sections on the institutional website focusing on smart and experimental statistics related to nontraditional sources and methodologies.

Adopting new data processing methods led to a renewal of the skills of the staff producing statistics within the Institute. To extract value from big data, it was necessary to strengthen the interdisciplinary nature of the techniques, in a way that was greater than what had been necessary to apply the techniques used so far. Therefore, ML and data science skills were recognized as a strategic priority. ML presents significant potential for statistical organizations, improving efficiency by automating processes or assisting human analysts. The Institute has systematically integrated machine learning techniques into Istat's production.

The Institute has made a notable organizational effort as exploiting big data sources necessitates investments in finances, time, and, most importantly, competencies. The ongoing transformation highlights the importance of acquiring new skills and resources to enable staff to handle the novel statistical treatment of extensive databases. Drawing from the experience of the Center for the Production of TSS and from achieved results, Istat has recognized the need to allocate adequate resources to these projects and acknowledges the current deficiencies within the Institute.

The agency is focusing on two areas of improvement in human resources: firstly, developing the skills of current employees, and secondly, hiring new talents with expertise in data science. The TSS Center has been working closely with the Human Resources Department to gather information about the company's current capabilities in data science. This initiative aims to identify existing skills that can be utilized in new projects and initiatives. The agency has also actively promoted training programs to enhance employee skills.

Training initiatives have been actively promoted to enhance and redirect existing resources toward innovative projects. Several internal courses focusing on Python software usage and web scraping techniques have been conducted, alongside encouraging researchers' participation in the European Statistical Training Programme (ESTP) courses covering data science topics. The training activity plan includes sessions addressing innovation and research topics, explicitly focusing on TSS to deepen understanding of new data sources' characteristics, potential, limitations, and the requisite processes for TSS production. Training sessions on machine learning and the Python language are also scheduled.

Efforts have been made to broaden researchers' participation in ESSnet projects, recognizing the significant benefits of exchanging views and experiences in statistical production, especially concerning big data processes, analysis, and visualization methods. Pilot projects developed during ESSnet Big Data I and II, ESSnet Smart

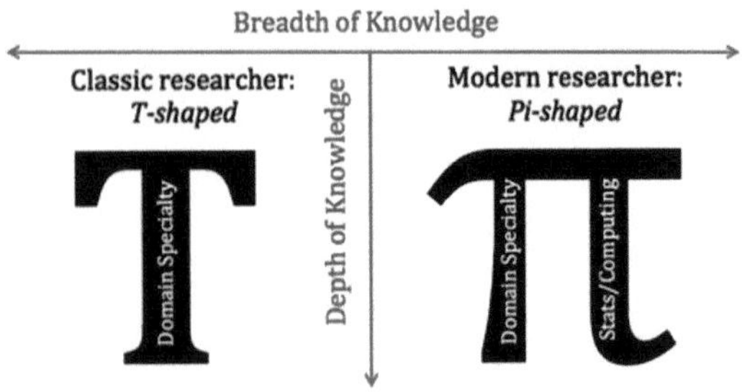

Fig. 12.1 Classic and modern knowledge models

Survey initiatives, or in collaboration with universities in the Innovation Laboratory[6] have facilitated the testing of various approaches, thanks to the concerted efforts of our researchers and fruitful collaborations with colleagues. The "learning by doing" approach has proven particularly effective in building capacity for utilizing big data in official statistics. However, data scientists in official statistics need to blend thematic domain expertise with profound knowledge of statistical/IT methodologies and tools. This evolution from a T-shaped model to a Pi-shaped model, shown in Fig. 12.1, with two distinct areas of expertise, is crucial. The Pi-shaped collaboration model described in the literature (Ceri 2018), which inspired Istat to invest in new resources and improve internal skills, explicitly relies on collaboration between individuals with different specialized expertise.

In recent years, Istat has begun recruiting data scientists[7] and involved them in TSS projects. Additional data scientists have recently been hired to meet the growing demand for these resources when developing new projects.

12.2 Istat's Experience on Methodological Issues of ML Applied to Official Statistics

Istat has started studying the introduction of machine learning techniques in the statistical production processes with the idea of reducing the need for human involvement by automatically learning from and adapting to data, and because machine learning may represent intricate, nonlinear relationships between variables that could be challenging to express and to discover with more conventional parametric models. In addition, the context is favorable. Studies on machine learning are also increasing with interest in nontraditional data sources, for instance, the use of remote sensing data (Mugnoli et al. 2024), signals data such as the ones sent by automatic identification systems of the vessels, or natural language processing of X (formerly Twitter) data (Catanese et al. 2022).

[6] https://www.istat.it/listituto/attivita/ricerca/organizzazione/laboratorio-innovazione/

[7] People with a strong statistical profile who also have IT skills that enable them to operate on infrastructures and with specific software for processing complex datasets.

A vast amount of literature on machine learning methods is available; nevertheless, their application in the NSI statistical production processes deserves further studies and reflections because of the peculiarities characterizing official statistics. An essential characteristic of machine learning (ML) methods is that they are mainly developed for prediction purposes. In contrast, most of the NSIs' production is concerned with statistics that estimate finite population parameters such as, for instance, the inflation rate, the total number of people in the country, and poverty indicators. These are totals, means and percentiles, and more in general quantities related to a set of statistical units; thus, the interest is not on the single unit. The prediction of values and estimation of aggregated parameters are undoubtedly related. However, they still present some differences that should be considered when using machine learning methods and in the uncertainty quantification of the parameter estimates. An interesting discussion on prediction and estimation can be found in Efron (2020). A more focused example of such a problem is in Larbi et al. (2023), which studies ML for the treatment of unit nonresponse in surveys. They observe that "the most predictive nonresponse model may not necessarily yield to the best estimator in terms of mean square error." NSIs generally refer to an estimator's accuracy instead of a prediction's accuracy to evaluate the quality of official statistics.

Given the inherent prediction nature of ML, a natural field of application in NSIs is the treatment of partial nonresponse through imputation, which is characterized by a prediction of the missing values. We observe that, in this case, prediction is not generally the goal of imputation. Indeed, imputation is generally used to estimate some quantity of interest, such as population totals, obtained using observed and predicted data. This is an important characteristic that should be considered when choosing the ML method and how to use it for prediction. In this case, the procedure's evaluation should be driven by focusing, for instance, on the mean squared error of estimates instead of the mean squared error of predictions. There are several studies in this field; some early papers in the official statistics area can be found at the end of the 1990s (Nordbotten 1996; De Waal 2001; Di Zio et al. 2004), and some recent papers are Dagdoug et al. (2023), De Fausti et al. (2022), and Di Zio et al. (2022). Some machine learning methods are tested in those papers with simulated and real data. However, imputing missing values is essentially a tool for obtaining a complete dataset that can be used to make inferences on population parameters more easily. For this reason, making predictions through random imputation can help preserve variability and avoid biased estimates of the probability distribution and parameters such as quantiles (Chen et al. 2019). Methods for random imputation are provided for both quantitative and categorical variables. In the first case, they are obtained by adding a residual term estimated from data (Dagdoug et al. 2023); in the second setting, random imputation can be obtained by sampling at random from the weighted prediction classes (De Fausti et al. 2022).

Another important element is that NSI data is generally gathered through sample surveys. A problem that must be dealt with is the use of sampling weights and, more in general, how to take into account the sampling design. For the sake of truth, this issue must be addressed when models are used, whether they are ML models or not. When the sampling design is not ignorable, the probability of observing a statistical unit is related to the target variable we are interested in; it is essential to include information on the sampling design in the model to avoid biased inferences. This can be done by introducing the variables that determine the sampling design in the model; for example, if stratified random sampling is performed, the stratification variables should be included as covariates in the ML model. Another critical case frequently adopted in the NSI's survey is cluster sampling. This implies dependence on observations within one cluster, leading to violations of the identically distributed and independent assumptions. If we do not consider this element in the ML applications, biased or inefficient estimates can be derived (Kilian et al. 2023; Schulz-Kümpel et al. 2025).

When it is impossible to take the sampling design into account directly, sampling weights should be used. In statistical models, they may be introduced as auxiliary variables (covariates) or by building weighted estimators. Di Zio et al. (2022) discuss using sampling weights when applying multilayer perceptron models (MLP) to real survey data. They carried out a study that compared the results of the MLP procedure with those of the official one. Sampling weights are used in two ways: (1) weights are used to expand the sample that is then treated as the entire population and (2) the loss function used in the training phase of MLP is weighted with sampling weights. In this way, the MLP incurs different misclassification costs for different training examples.

All these questions are nicely discussed in Little (2012, 2022) concerning Bayesian modeling. However, many ideas can be directly interpreted in the ML framework.

A fundamental step of the statistical production process in an NSI is the quality evaluation of the statistical outputs. Statistics are inevitably affected by uncertainty due to sampling and non-sampling errors.[8] Institutes should provide a measure to quantify the uncertainty of numbers disseminated for the correct interpretation and use by the stakeholders and to maintain confidence in the Institute. In the traditional production of statistics, the quality framework is well established. It refers explicitly to measuring the accuracy of estimates such as, for instance, means or totals of the population, measures as mean squared error taking into account (usually) sampling and non-sampling errors.

As noted, machine learning methods are more developed with the aim of prediction, and thus, available accuracy measures reflect this scope; in fact, they are focused on prediction accuracy. Further studies are needed to move quality evaluations toward inference (Larbi et al. 2023). Some answers might directly come from resampling techniques, e.g., bootstrapping, but further investigation needs to be done to see their applicability in official statistics contexts. They may be computationally prohibitive or need special accommodations in the case of the presence of finite populations; see Chen and Haziza (2019) for a thorough discussion

[8] For a discussion of this topic, see also Puts et al. (2025) in this book.

about the latter problem. Also, cross-validation, which is aimed to assess how well a trained model will generalize to an independent dataset, should be revised to take into account the non-exchangeability of the units, which is generally introduced when a complex sampling design is used (Wieczorek et al. 2022). Another technique used in uncertainty quantification in machine learning is conformal prediction (Vovk et al. 2005). It is used to provide a way to quantify the uncertainty of individual predictions made by a model. Unlike traditional machine learning models that typically output a single prediction, conformal prediction gives a prediction set that guarantees a certain level of confidence or significance. Again, further studies should be conducted on conformal prediction to deal with the uncertainty evaluation of a population parameter and with a complex sampling design (Wieczorek 2023).

12.3 Research Projects

In recent years, Istat has invested significant efforts in research projects to integrate new data sources and advanced methodologies, such as machine learning and big data analytics, into producing official statistics. These projects align with a broader strategic vision to modernize statistical processes, improve data quality, and enhance the efficiency and timeliness of statistical outputs. Below, we present some of the key research projects carried out by Istat in this domain.

- **Automatic identification system (AIS) for maritime transport**
  This project integrates AIS data with existing administrative sources to improve maritime traffic statistics. The goal is to enhance the accuracy and timeliness of key indicators, such as vessel arrivals in Italian ports, by reconstructing ship voyages using geospatial and machine learning techniques.
- **Satellite images for urban green area measurement**
  This project uses high-resolution satellite imagery to quantify urban green spaces through vegetation indices like Normalized Difference Vegetation Index (NDVI). The methodology combines remote sensing data with clustering and classification techniques to support environmental monitoring and urban planning.
- **Web intelligence for enterprise data analysis**
  Istat has developed automated methods to extract information from enterprise websites using web scraping and machine learning. Applications include identifying official URLs, detecting e-commerce activities, and analyzing online job postings to enrich business statistics.
- **Analysis of trade data with network analysis techniques**
  The TERRA (imporT ExpoRt netwoRk Analysis) project uses network analysis techniques to study external trade flows and detect key economic patterns. This project identifies central actors in trade networks by analyzing external trade data (Eurostat COMEXT open data), assesses market dependencies, and provides insights into macroeconomic dynamics. The approach enables the development of experimental indicators that enhance understanding global supply chains and trade shocks.

- **Sentiment analysis for economic and social insights**
  Istat applies sentiment analysis techniques to various textual data sources, such as social media, to extract public opinion trends on economic and social issues. Natural language processing and machine learning models classify sentiment polarity and detect emerging themes, providing valuable insights for policymakers and researchers.

12.3.1 Automatic Identification System (AIS)

Istat's TRAMAR (MARitime TRAnsport) survey[9] provides statistics on the ship transport of goods and passengers carried out for commercial purposes in Italian ports. The survey has a census character and refers to ships with a gross tonnage of at least 100 tons moving for commercial reasons. The statistical production of TRAMAR comes through integrating survey data with an administrative source, namely, PMIS (Port Management Information System). PMIS is the digital registry of the Italian General Command of Harbormaster's offices.

Integrating data from various sources is necessary to ensure high-quality statistical outputs. However, relying on a single source does not guarantee complete coverage of the analyzed phenomenon. For example, the PMIS system only includes the most important ports for commercial and passenger traffic (i.e., a list of 38 ports), while the TRAMAR survey covers all ports of interest. Furthermore, due to the extreme complexity and volubility of ship movements, there is no a priori list of the events (the landings and embarkations) or the units (the ships) being surveyed.

The project aims to provide an alternative source for improving the quality of maritime traffic statistics. The automatic identification system (AIS)[10] is a tracking system used on ships for safety and management purposes. It provides the GPS location of all vessels involved in commercial activities and transport of goods and passengers sailing the world's seas regularly and frequently. Storing this data in a big data archive can theoretically allow the reconstruction of a ship's voyage history.

Including the AIS source in the TRAMAR production process can improve the quality and the timeliness of important statistical outputs, for example, the table of vessels arriving in the main ports by type and size of vessels for Eurostat.[11] This table contains the number of arrivals by ports and is often referred to as F2 Table. Still, it does not include data on goods quantity or passenger numbers, which are only collected through the TRAMAR questionnaire. The PMIS source, released monthly through machine-to-machine communication, is available earlier than the TRAMAR survey. However, the F2 Table cannot be generated solely from the PMIS data as it does not cover all ports. The AIS source, which is potentially available in

[9] https://www.istat.it/informazioni-sulla-rilevazione/trasporto-marittimo/

[10] https://www.imo.org/en/OurWork/Safety/Pages/AIS.aspx

[11] https://ec.europa.eu/eurostat/web/transport/information-data/maritime-transport

near real time, could be integrated with the PMIS data to produce an F2 Table that is timely and comprehensive.

The use of AIS data to support official statistics has already been investigated in several works and projects. The *Port Visits Geo-Solution* prototype monitors the ships' movement inside the Piraeus Central Port, defined by a polygon, to compute the number of arrivals and departures. In the *United Nations Global Platform* (UNGP) *Handbook*,[12] several case studies on the use of AIS are described, such as the experimental statistics of the daily number of vessels visiting Danish ports using AIS data published from Statistics Denmark.[13] In the *Port Visits Using Real-Time Shipping Data* statistic[14] from the Irish Central Statistics Office (CSO), two methodologies for generating port visits are compared. The first uses polygons to identify ships inside a port area. This study builds upon previous research, even though none of these studies share our specific goal: the classification of arrival ports given AIS trajectories. Most literature deals with trajectory forecasting and ship behavior prediction, without a specific focus on port classification.

12.3.1.1 Methodology

This section describes the methodology adopted to process AIS data. This process aims to obtain the entire list of voyages of vessels arriving in or departing from Italian ports. The vessels of interest are passenger or commercial vessels, which we will refer to as vessels with a gross tonnage of more than 100 tons.

Data Source

The AIS data used in this study were supplied by the *Task Team on AIS Data of the UN Committee of Experts on Big Data and Data Science for Official Statistics* (UN-CEBD).[15] This data is accessible through the *UN Global Platform* (UNGP),[16] a global repository of live and historical AIS data. The UN-AIS dataset contains regular observations of all kinds of ships, with a gross tonnage greater than 300 tons, from December 1, 2018. On average, the interval between two observations of the same ship is 10 minutes. The available attributes in each AIS observation (Fig. 12.2) include three categories of information: static (MMSI code, IMO code, ship name, and type), dynamic (ship's position coordinates, navigational status, speed, and course), and voyage-related (destination and draft).

[12] https://unstats.un.org/wiki/display/AIS/Case+studies

[13] https://www.statistikbanken.dk/aisdag

[14] https://www.cso.ie/en/statistics/transport/portvisitsusingreal-timeshippingdata

[15] https://unstats.un.org/bigdata/task-teams/ais/index.cshtml

[16] https://unstats.un.org/bigdata/un-global-platform.cshtml

Static data					Dinamic data					Travel related data	
IMO	MMSI	CALLSIGN	VESSEL NAME	VESSEL TYPE	TIME	COORD.	NAVIG. STATUS	SPEED	COURSE	DEST.	DRAFT
8401561	2011011	ZAD4L	FROJDI	Cargo	04/06/2023 19:45	41.1323 16.8530	MOORED	0	258	Ravenna	null

Fig. 12.2 AIS observation attributes

IMO	VESSEL TYPE	DEPARTURE PORT	ARRIVAL PORT	DEPARTURE DATE	ARRIVAL DATE
8401561	Cargo	ITBRI (Bari)	ITRAN (Ravenna)	04/09/2023	05/09/2023
9483712	Passenger	ITGOA (Genova)	ILOLB (Olbia)	05/09/2023	06/09/2023

Fig. 12.3 Dataset produced: the list of vessel trips

The UN-AIS dataset also provides vessel position through the H3 (Hexagonal hierarchical geospatial indexing system)[17] index at multiple resolutions. H3 is a geospatial indexing system developed by Uber Technologies that approximates the GPS coordinates using a hexagonal tessellation of the earth's surface. The H3 index identifies the hexagon containing ship coordinates, with hexagon size depending on the adopted resolution. In addition to the AIS dataset, we used a dataset of world ports taken from Merrien (2021), containing each port's geographical coordinates (latitude, longitude).

Data Processing Pipeline

To achieve the objectives of this work, we need to identify and quantify ships visiting Italian ports. The desired output is a dataset of ships *voyages*, as shown in Fig. 12.3. The three entities defining a *voyage* are the ship, the departure (port and date), and the arrival (port and date). In the following, we will refer to all generic vessel visits in a port as *port calls*.

Figure 12.4 shows the pipeline used to process AIS data. The workflow comprises three primary logical steps, as illustrated in Fig. 12.4, along with their corresponding inputs and outputs. The first step is the vessel selection, which requires access to the AIS database, the list of Italian ports (including port area), and Lloyd's ship register,[18] which provides information such as each vessel's type and gross tonnage. The list of vessels returned in the output is used to select AIS data for processing in the subsequent steps. As mentioned, we are only interested in voyages where at least one of the ports of departure and arrival must be in Italy. To ensure a comprehensive analysis, we must consider the port of departure if the ship arrives in Italy, even if it is foreign. Similarly, we must consider the port of arrival if the ship departs from Italy, even if it is foreign. Therefore, we cannot limit our analysis to AIS records located only in Italian waters. However, if the ship does

[17] https://h3geo.org/

[18] https://www.lr.org/

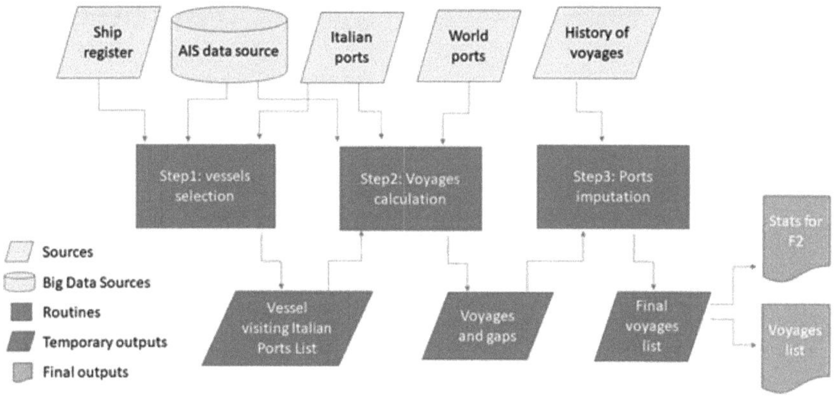

Fig. 12.4 Pipeline for processing AIS data

not visit an Italian port during the specified period, it can be excluded from our analysis. Thus, according to AIS data, the output list of vessels visiting Italian ports will include all ships, except fishing and yacht types, that have visited an Italian port and are present in the ship register with a gross tonnage of more than 100 tons.

The second step is the voyages calculation. An algorithm, described in the following paragraph, computes all trips made by vessels during the specified period. This process requires the AIS database and lists of Italian and non-Italian ports.

In the third and final step, the ports imputation and missing data in the voyages list (the unknown ports) are imputed using the "Port imputation algorithm." The nature of how this data is generated will be clarified in the paragraph "Voyages calculation algorithm."

The processing pipeline produces two final outputs: the voyages list, which follows the format presented in Fig. 12.3, and the statistics for the F2 Table, which displays the number of voyages grouped by arrival port.

Voyages Calculation Algorithm

A departure and arrival are two consecutive port calls of the same vessel. For example, the voyage shown in Fig. 12.5 is defined by the departure port (ITBRI) at time T1 and the arrival port (ITRAN) at time T5. Note that the other AIS observations between T2, T3, and T4 are not required to define the trip. A port call in AIS data is a record where the speed is 0, and the position falls inside a port area. As previously stated, we are only interested in ships that have visited an Italian port at least once and have a gross tonnage exceeding 100 tons. Thus, we only need to select these records to process a ship. Finally, we obtain all the ships' voyages by ordering the port calls in time.

The first problem to be addressed is the identification of port areas. For this purpose, we built each port's area as a set of hexagons of resolution eight, identified by H3 indexes. First, for each port, we selected the hexagon containing

Fig. 12.5 Example of the voyage between the ports *ITBRI* (Bari) and *ITRAN* (Ravenna) in the AIS dataset

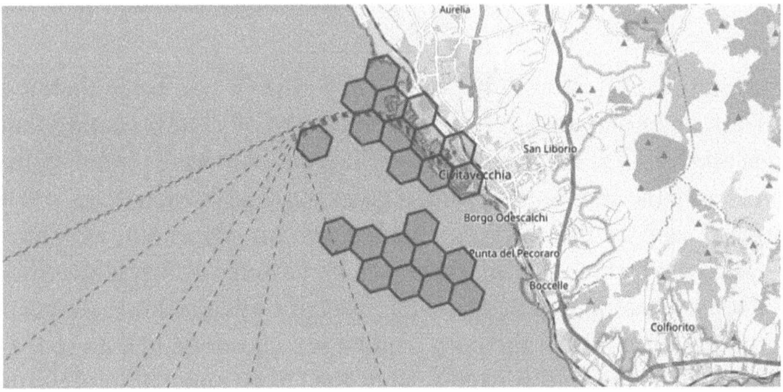

Fig. 12.6 H3 hexagons identifying a port area. (Source: authors' own elaboration using an OpenStreetMap basemap © OpenStreetMap contributors, https://www.openstreetmap.org/copyright)

its geographical coordinates and the first ring of hexagons surrounding the main hexagon, and we added them to the set representing its area. As a second step, we collected AIS data for 6 months in a reference year (e.g., 2022). We filtered the data by specific ship types, namely, cargo, tanker, and passenger, which were stationary in the port areas (i.e., in the main hexagons containing the geographical coordinates of the ports). Finally, we added the H3 hexagons of resolution eight containing the stationary ship positions to the port area (Fig. 12.6). Visualizing a visit to a port is made easier by defining each port area as a set of hexagons.

Furthermore, two other critical issues, namely, incorrect records and missing data, affect the AIS data. A record is erroneous if one or more variables assume a wrong value, creating a false port call within the vessel's timeline. For example, in Fig. 12.7 (on the left), the AIS observation $T3$ is erroneous as, even if the vessel appears with speed 0, it generates a false port call in *ITRAN*. In this case, we find two trips for the vessel, *ITGOA-ITRAN* and *ITRAN-ITNAP*, instead of *ITGOA-ITNAP*. Missing data refers to a period with no AIS observations for a vessel. In Fig. 12.7 (on

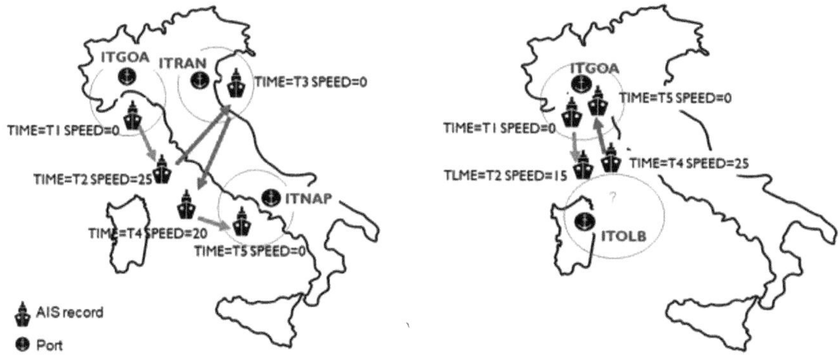

Fig. 12.7 Example of incorrect record (on the left) and lack of data (on the right)

the right), the observations are regular until time $T2$. After no data for several hours, the vessel reappears in the position in the figure at time $T4$. If, between $T2$ and $T4$, the vessel enters *ITOLB*, we lose both the two voyages *ITGOA-ITOLB* and *ITOLB-ITGOA*. To face these problems, it is necessary to use also the AIS observations of vessel movements that we initially thought were not useful (records $T2$, $T3$, and $T4$ in the example of Fig. 12.7). Each record is compared with its predecessor in the timeline by measuring the difference in the time value (Δ_T), the difference in the distance of the coordinate values (Δ_P), and the resulting speed:

$$resulting\_speed = \frac{\Delta_P}{\Delta_T}. \tag{12.1}$$

Finally, the algorithm will calculate the correct trips: *ITGOA-ITOLB* and *ITOLB-ITGOA*. To perform this step, we chose the following thresholds:

$$\Delta_T > 60\,\text{min},\ \Delta_P > 0,\ resulting\_speed < 15\text{km/h}. \tag{12.2}$$

Port Imputation Method: Heuristics and Deep Learning

The voyage dataset generated by the voyages calculation may contain some records where the departure or arrival ports are not specified. This occurs in the case of a lost port call due to AIS missing data or ambiguous value of the *destination* field. In this case, the missing value must be imputed. To do this, we use the last 3 months' voyages. To determine the port of arrival, we use the most frequent arrival port in the vessel's history, given the port of departure. Similarly, to determine the port of departure, we use the most frequent departure port, given the port of arrival. We limit this rule to "scheduled" routes to prevent making uncertain imputations.

This is only the first step. We are working on sequence classification models to classify arrival ports from AIS trajectories. Preliminary results for this analysis are presented in Pappagallo et al. (2024). We are now working on assessing the performance of deep learning models on real-world unlabeled ship trajectories,

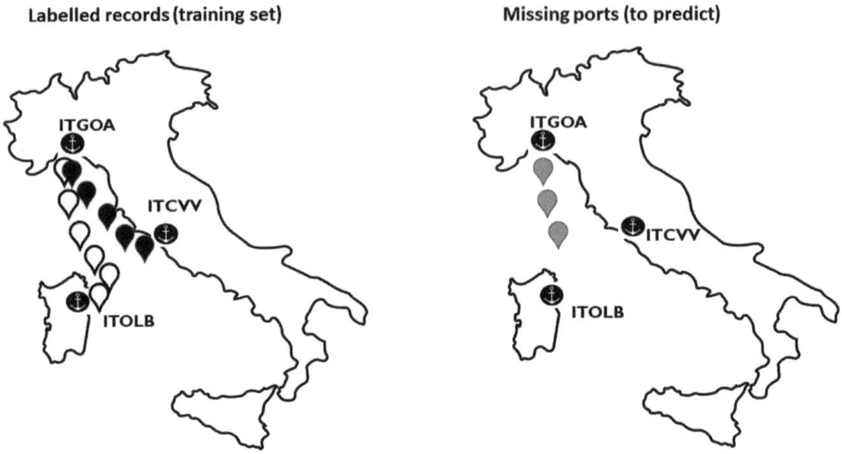

Fig. 12.8 Example of records used by the "Port Classifier" neural network. On the left, the labeled voyages are used to train the model. On the right, the voyage with the missing arrival port to predict. The model must use the incomplete route to distinguish the correct arrival port *ITOLB*

often characterized by noise and/or missing data points. At this moment, the best-performing model is an long short-term memory (LSTM) network, with an accuracy of 0.82 and a macro F1 score of 0.56 on the real-world test set[19] over 93 prediction classes. The key idea is to train the network using the dataset of complete voyages (i.e., voyages without missing AIS data) and use it to predict unspecified arrival ports. The model uses a dataset consisting of vessel routes (latitude, longitude, speed over ground, course over ground) as input and arrival ports as output.

This is a typical case of a supervised learning approach. The model is trained using labeled voyages where both departure and arrival ports are known and applied to predict cases where the arrival port is unknown. Figure 12.8 gives an example of how the model has to work, showing on the left the voyages used for training. The vessels departing from the port of *ITGOA* (Genova) might arrive at *ITCVV* (Civitavecchia), sailing on the black route, or arrive at *ITOLB* (Olbia), sailing on the white route. On the right is a voyage with a missing port of arrival. In this case, the model should recognize the pattern and predict *ITOLB* as the correct port of arrival.

[19] The test set mimics the characteristic of unlabeled problematic routes, i.e., incomplete trajectories and missing observations near ports.

12.3.1.2 Results

In previous sections, we presented and discussed the first results of the AIS data processing pipeline. In particular, we compared the number of arrivals produced by the proposed procedure with the official statistics for Eurostat (F2 Table). As a first experiment, we calculated the voyages for the fourth quarter of 2021. We could use the 2021 F2 Table as a benchmark. The resulting dataset contained 86,545 voyages. Among these, 18,088 had a missing port (in 9042, the arrival port was missing; in 9046, the port of departure was missing), 3274 of their records were imputed with the current imputation algorithm, and 14,814 remained unknown. The total number of arrivals in Italian ports calculated from AIS during the period was 57,659 instead of the 85,075 arrivals reported in the F2 Table. In practice, the procedure detected only 68% of the arrivals, and the remaining 32% were lost, but this varied among ports. First of all, the following are a few observations:

1. The F2 Table includes all vessels with a gross tonnage of more than 100 tons, but only vessels with a gross tonnage of more than 300 tons are required to have an AIS tracking system onboard. Therefore, some vessels in F2 may not be in the AIS dataset.
2. For this first run, we did not have accurate historical records of AIS voyages, and we used only the historical records of PMIS, which does not include all Italian ports. Also, imputation is not always possible for routes monitored by PMIS.
3. In case of missing data for an extended period, several port calls may be lost, but the algorithm will have only one prediction. This problem is particularly noticeable for shorter routes.

The Italian F2 Table consists of 56 ports. We compared the statistics for each port and obtained very different results. We calculated two types of indicators from AIS voyages: the number of arrivals at ports where the port of departure is known (column *AIS from*) and the number of all arrivals at ports (column *AIS to*) including cases where the port of departure is unknown.

Tables 12.1, 12.2, 12.3, and 12.4 provide a detailed comparison of ports divided into five classes:

1. Ports where the number of arrivals in F2 falls within the range of "AIS from" and "AIS to": This includes 14 major ports, such as Ancona, Bari, and Gioia Tauro.
2. Ports with over-coverage: This list contains only four ports. Milazzo and Eolie Islands ports have greater over-coverage and define a route not monitored by PMIS, where under-coverage of F2 is possible.
3. Ports with small under-coverage: List of eight ports with under-coverage limited to 10%. This includes several big ports like Civitavecchia, Livorno, Napoli, Palermo, and Trieste.
4. Ports with medium under-coverage: List of 17 ports with varying degrees of coverage. Some ports are in short routes like Elba and Piombino, Messina and Reggio Calabria, and Procida. Unfortunately, two important ports, Genova (−25%) and Ravenna (−37%), are also in this class.

Table 12.1 Result of the comparison between the F2 Table and the statistics regarding voyages produced from AIS data. Ports where F2 is *in the range*

UNLO code	Port name	AIS from	AIS to	F2
ITAOI	Ancona	401	562	426
ITAUG	Augusta	239	525	523
ITBRI	Bari	434	503	494
ITCHI	Chioggia	57	105	81
ITFAL	Falconara Marittima	42	63	43
ITGIT	Gioia Tauro	335	520	509
ITMDC	Marina di Carrara	120	137	133
ITMNF	Monfalcone	86	122	108
ITOTN	Ortona	43	85	70
ITPFX	Porto Foxi (Sarroch)	137	297	205
ITPNG	Porto Nogaro	46	77	76
ITRRO	Sorrento	972	1,094	1,032
ITSIR	Siracusa	69	140	109
ITVCE	Venezia	434	757	630

Table 12.2 Result of the comparison between the F2 Table and the statistics regarding voyages produced from AIS data. Ports with *over-coverage* or *small under-coverage*

UNLO code	Port name	AIS from	AIS to	F2	Difference
IT004	Eolie Islands	1,147	1,292	679	+68.9%
ITMLZ	Milazzo	1,155	1,297	827	+39.7%
ITPRJ	Capri	2,154	2,182	2,072	+4.0%
ITSPE	La Spezia	245	290	217	+12.9%
IT001	Ischia Island	3,334	3,408	3,686	–7.5%
ITCVV	Civitavecchia	505	550	588	–6.5%
ITLIV	Livorno	1,210	1,319	1,428	–7.6%
ITNAP	Napoli	6,434	6,563	7,215	–9.0%
ITPMO	Palermo	857	911	1,000	–8.9%
ITPZL	Pozzallo	183	219	241	–9.1%
ITSVN	Savona	327	391	430	–9.1%
ITTRS	Trieste	422	478	535	–10.7%

5. Ports with large under-coverage: This group contains only small ports or ports located in short routes.

Future work planned for this project includes implementing port imputation by machine learning methods. The first experiments achieved accuracies exceeding 80%. A more sensitive algorithm will also be developed to process only short tracks. These changes are expected to significantly reduce the current under-coverage of trips, making AIS a reliable source for generating improved statistics on maritime traffic.

Table 12.3 Result of the comparison between the F2 Table and the statistics regarding voyages produced from AIS data. Ports with *medium under-coverage*

UNLO code	Port name	AIS from	AIS to	F2	Difference
IT002	Elba Island	1,763	1,767	2,489	–29.0%
ITBDS	Brindisi	226	273	321	–15.0%
ITCAG	Cagliari	372	404	488	–17.2%
ITCTA	Catania	276	294	368	–20.1%
ITGAE	Gaeta	36	60	96	–37.5%
ITGAI	Golfo Aranci	79	81	95	–14.7%
ITGOA	Genova	911	995	1, 332	–25.3%
ITMSN	Messina	9,264	9,296	13,449	–30.9%
ITOLB	Olbia	387	387	479	–19.2%
ITPIO	Piombino	1,768	1,932	2,231	–13.4%
ITPRO	Procida	3,032	3,035	4,338	–30.0%
ITRAN	Ravenna	212	538	853	–36.9%
ITREG	Reggio Calabria	9,025	9,035	13,263	–31.9%
ITSAL	Salerno	440	504	588	–14.3%
ITTAR	Taranto	110	205	265	–22.6%

Table 12.4 Result of the comparison between the F2 Table and the statistics regarding voyages produced from AIS data. Ports with *large under-coverage*

UNLO code	Port name	AIS from	AIS to	F2	Difference
IT003	Egadi Islands	952	954	3,391	−71.9%
IT88P	Off-shore platforms	0	2	167	−98.8%
ITCLF	Carloforte	1,139	1,188	2,638	−55.0%
ITCLS	Calasetta	650	652	1,353	−51.8%
ITFCO	Fiumicino	0	0	15	−100.0%
ITGEA	Gela	4	4	45	−91.1%
ITIDG	Isola del Giglio	1	17	509	−96.7%
ITMDA	La Maddalena	14	41	3,402	−98.8%
ITPAU	Palau	19	38	3,607	−98.9%
ITPNZ	Ponza	137	154	495	−68.9%
ITPSS	Porto Santo Stefano	2	4	512	−99.2%
ITPTO	Porto Torres	175	184	324	−43.2%
ITPVE	Portoscuso (Porto Vesme)	528	554	1,329	−58.3%
ITQOS	Oristano	18	52	96	−45.8%
ITTPS	Trapani	1,023	1,122	3,180	−64.7%

12.3.2 *Satellite Images to Quantify Urban Green Areas*

Measuring urban green cover is crucial for analyzing and developing various indicators related to different aspects of city life (Jabbar et al. 2022). For instance, the "quality of life" often refers directly to citizens' ability to access public and private green spaces (such as parks, gardens, historic estates, and sports facilities). Furthermore, environmental quality depends on the presence and health of vegetation cover in each area.

The proposed statistical analysis could be a fundamental tool for further exploration of the dynamics governing large cities. Numerous studies have utilized remote sensing for vegetation analysis, focusing on how chlorophyll absorbs light radiation at different wavelengths. The literature presents various indicators for vegetation classification (Xue and Su 2017; Pristeri et al. 2021). Following several experiments, we opted to use the NDVI (Normalized Difference Vegetation Index), which relies on the behavior of *chlorophyll a* and *b*. Its standard formula is

$$NDVI = \frac{(NIR - RED)}{(NIR + RED)} \tag{12.3}$$

where RED is the red wavelength used for absorption and NIR is the near infrared for reflection.

We employ high-resolution remote sensing images (AGEA orthophotos with 20 and 50 cm pixel resolution), which span the entire Italian territory over 3 years. Starting in 2012, these images have been provided to Istat in four spectral bands. From this data, we produce continuous NDVI images (in float format) that require reclassification to identify pixels corresponding to vegetation specifically. This process involves a detailed analysis of the NDVI image histogram to determine a threshold value above which pixels are likely to represent "green" areas. Figure 12.9 demonstrates the clustering of vegetation index pixels within the image histogram.

12.3.2.1 Methodology

Typically, we expect four NDVI classes corresponding to distinct landscape features: water, built-up areas, bare soil, and vegetated areas, with a standard threshold

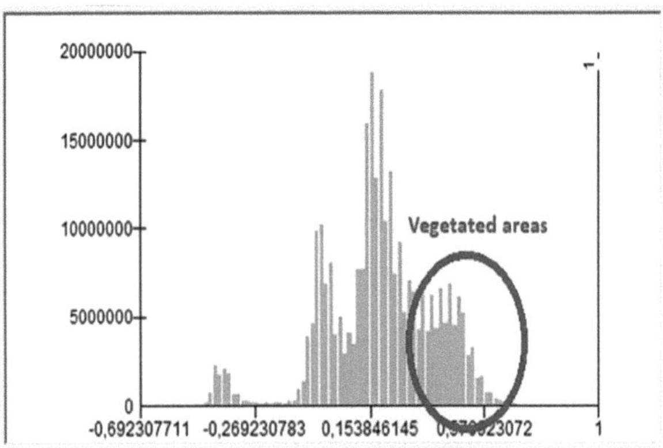

Fig. 12.9 NDVI histogram for determining the threshold for vegetated ('green') areas, based on a true-color satellite image of an urban area in Rome. (Source: authors' own elaboration)

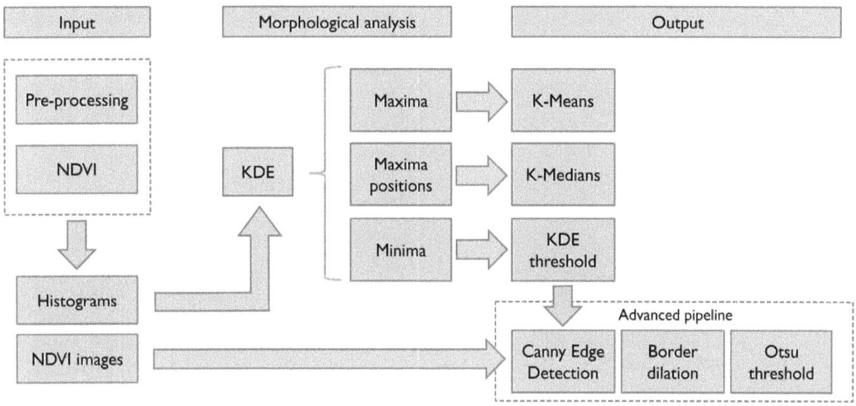

Fig. 12.10 Data processing pipeline

value of approximately 0.2 (Aryal et al. 2022) used to discern green vegetation. The NDVI threshold of 0.2 cannot always be used as a reference for any orthophoto since the histograms show considerable shifts in the vegetation cluster even on orthophotos taken of the same area in different years. Although there are strong theoretical arguments for expecting these four distinct regions in the NDVI distribution, these are rarely observed in actual data. Sometimes, one of the classes overshadows the one next to it, and in some extreme cases, we observe distributions that exhibit only two maxima or even just one. Consequently, we explore various techniques to determine the threshold automatically. The overall methodological approach is structured into three primary stages, as shown in Fig. 12.10: data input and pre-processing, morphological analysis of the histogram, and evaluation of the outcomes.

At first, a pre-processing data step is performed:

- Ortho-images, in ecw format, are converted into GeoTiff format.
- Images are cropped as per the shape file to select the urban areas correctly.
- NDVI is evaluated; note that the input images are multi-band, while the output is single band.
- Multiple images are combined to form a mosaic: a single image of the full city.

Then, three groups of methods are studied. One is based on kernel density estimation (KDE), the second is based on clustering algorithms, and the third is referred to as a refined version of green detection.

A common feature of NDVI histograms is that they are usually quite noisy (fluctuating), with frequencies often obscuring the meaningful maxima and minima; this makes it difficult to discern clear patterns. We address this using the first method: kernel density estimation, which estimates the distribution value for each point as the one-weighted sum of the contributions of a centered Gaussian kernel. KDE acts as a smoothing mechanism for our data. The smoothed curve represents

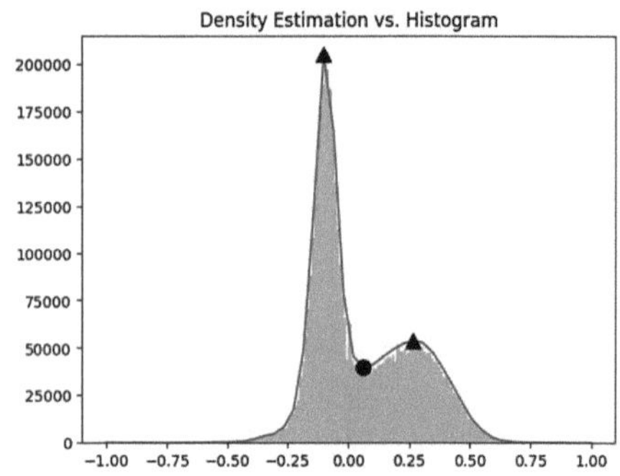

Fig. 12.11 Kernel density estimation KDE (line) vs. histogram. Maxima (triangles) and minima (dots) are shown

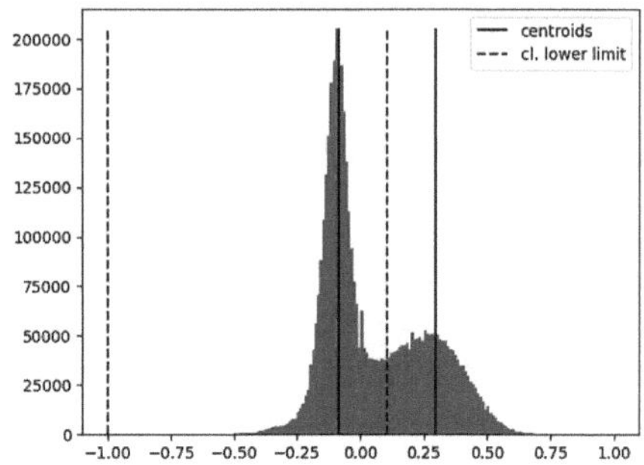

Fig. 12.12 Clusters centroids (black lines) and clusters lower limits (dashed lines) found by K-means

the frequency distribution function of our data, as seen in Fig. 12.11; it is crucial in our analysis since it makes it possible to estimate local maxima and minima. The local maxima, highlighted by KDE, serve a dual purpose: they indicate the likely number of clusters and provide the starting positions for the centroids in subsequent clustering algorithms. Both parameters are crucial for correctly applying the subsequent clustering algorithms.

This approach lets us identify a first-order estimate of the green areas: the threshold is the minimum rightmost in the distribution. This estimate requires no additional parameter, i.e., the number of expected clusters.

In the second approach, we used the K-means and K-medians clustering algorithms. The number of clusters is obtained by the number of local maxima identified by KDE; these values are also set as the algorithms' initial centroids. The urban green threshold is then given by the lower limit of the rightmost cluster, i.e., the cluster related to the centroid with the highest NDVI value; see Fig. 12.12. We also explored the K-medians clustering approach. K-means is efficient for segmenting one-dimensional data but can be sensitive to outliers. K-medians effectively address this issue.

Incorporating K-medians into our methodology allowed us to validate and enhance the clustering results obtained from K-means. This complementary approach added a layer of robustness to our analysis, improving the overall accuracy and reliability of our conclusions about vegetation distribution in urban environments.

The third approach refines the thresholds evaluated via the clustering methods. We focus mainly on two challenging aspects commonly encountered in this field. First, we encountered the challenge of varying green nuances across different images. One significant factor contributing to this variation is seasonal effects, which can alter the type of green detected in each image. For instance, the lush vegetation in spring presents a different shade of green compared to those of a later period (Eastman et al. 2013). We adopt the approach of Donchyts et al. (2016). This was initially applied in water-index detection and offers valuable insights for our context. We customized the methodology for vegetation analysis.

The method consists of the following complete pipeline:

- The KDE threshold is set as a preliminary threshold for green pixels in the image. Pixels below the KDE threshold are masked, while those above are kept unchanged. This masking enables the following steps to focus on green areas.
- The Canny edge detection algorithm is applied on the masked image. It segments the input image based on the variation of the NDVI values. This generates segments along the KDE threshold since there is a steep variation between green and non-green areas, and in all green regions above the KDE threshold, faint-green pixels can be separated by strong-green pixels.
- A buffering technique is applied around the detected edges. This step narrows the focus to the immediate areas surrounding the vegetation, providing a more targeted region for analysis. A buffered image is used as a mask on the original image to resample pixels in the regions of high index variation. This reduces noise from the irrelevant areas.
- Otsu clustering is applied to the resampled histogram to determine an optimal threshold for binary separation. It works on single-tone images (such as the NDVI images) and splits the values into two classes with a threshold. It efficiently distinguishes non-green from green areas.

12.3.2.2 Results

We applied the methods to Ravenna, a municipality in Northern Italy. Table 12.5 shows the estimated threshold, the number of green pixels, and the green area in terms of hectares and percentage of covered territory. There is a general agreement

Table 12.5 Results of the different approaches for determining the urban green threshold. Ortho-images for Ravenna and satellite (Advanced pipeline S2)

Approach	Threshold	Green pixels	Green area (ha)
KDE	0.06	182,322,979	729.29
K-Means	0.11	165,830,738	663.32
K-Medians	0.08	174,698,659	698.79
Advanced pipeline	0.17	139,599,919	558.40

on the first three approaches: KDE, K-means, and K-medians. KDE and K-medians are closer, while K-means is slightly different due to the algorithm's less robustness to the outliers.

The advanced pipeline remarks a different trend. In this case, there is a strong segmentation inside the green area, meaning a great variety of vegetal structures with highly varying NDVI, and the Otsu clustering determines a shift in the threshold to discard areas of less intense green vegetation.

The research findings suggest that orthophotos are highly effective in extracting and quantifying green spaces within urban areas. Additionally, machine learning techniques present a promising approach to overcoming challenges, such as accurately determining threshold levels in an automated and objective manner. The positive and promising results indicate that future efforts will be focused on refining methodologies, expanding the study area, enhancing the precision of urban green space quantification, and categorizing these spaces into established land cover types. The success of a project like this depends on methodological and thematic expertise, and the authors of this section who conducted this research are specialists in both fields.

12.3.3 Web Intelligence: Automated Analyses of Enterprises' Websites

Among the vast category of big data, the Internet can be considered a data source that may be used in substitution (with the aim of reducing respondent burden) or in combination with data collected using traditional statistical survey instruments to increase the accuracy of estimates.

One of the approaches for big data-based data collection on the Internet is web scraping, which is a process that permits the extraction of information from websites. There are two different kinds of web scraping: specific web scraping and generic web scraping.

Specific web scraping refers to the case when both the structure and content of websites to be scraped are almost perfectly known in advance. Generic web scraping, instead, assumes that no a priori knowledge on the content is available; in this case, the whole website has to be scraped and subsequently processed in order to infer information of interest.

Starting in 2013, Istat began exploring the potential of web scraping techniques within both national (Virgillito et al. 2017; Barcaroli et al. 2015a,b, 2016) and European (Stateva et al. 2018; Kühnemann et al. 2022) contexts, integrated during the estimation phase with text and data mining algorithms, with the goal of replacing traditional data collection methods to reduce respondent burden, enhancing estimation processes, or combining these approaches within an integrated framework. Some notable use cases that have been covered are:

- Use Case 1: Solve the "URL retrieval problem" of identifying the official enterprise website starting from the enterprise's name and administrative information.
- Use Case 2: Establish whether an enterprise does e-commerce.
- Use Case 3: Establish if an enterprise exposes job vacancies on its website.

This experience represented a truly extensive application of generic web scraping, characterized by (i) the need to scrape data from several thousand websites, highlighting its massive scale, and (ii) the lack of prior knowledge or reproducibility of the websites' structures, reflecting their inherent heterogeneity and the generic nature of the approach.

12.3.3.1 Use Case 1: URL Retrieval

Since the main input of a web scraper is an initial list of URLs to scrape, it is necessary to have them available before starting the scraping phase of the job. Unfortunately, this was not the case for most enterprises belonging to the datasets of interest. Hence, in this use case, the task was to obtain the official URL of their websites, if available. An automated procedure has been implemented to obtain an enterprise website (if it exists) starting from a search of the enterprise name in a search engine. The procedure involves several steps with a mix of techniques, ranging from scraping and crawling techniques to machine learning ones. The results show the feasibility of addressing this problem with an automated solution that gets good results both in terms of quality and efficiency.

More in detail as for Istat, it was used as a case study in the "Survey on ICT usage and e-Commerce in Enterprises" (shortly ICT survey), whose population of interest is composed of enterprises with at least ten employees and operating in different branches of industry and services (the population size is around 200,000). By the ICT survey estimates, it is known that about 70% of these enterprises own a website for different purposes. Unfortunately, only a subset of the enterprises' URLs is available (derived from the ICT survey answers integrated with administrative data).

Therefore, the objective is to determine the official website for the remaining enterprises. In this regard, the basic idea is to search a search engine for the name of each enterprise, collect and submit the search results to web scraping, and apply ML algorithms to the scraped web content in order to make the prediction. The first choice to make is to identify the search engine to rely on.

Search Engine Selection

When you think about search engines, Google is probably the first thing that comes to mind because it dominates the market by a long margin. But there are also several alternatives that it is possible to consider, especially if you are more interested

Table 12.6 Number of correct domains caught by link position

	Google	Bing	Yahoo	DuckDuckGo
No domain match	25	35	36	42
Domain matched by link 1	61	55	50	41
Domain matched by link 2	6	5	6	7
Domain matched by link 3	2	0	2	2
Domain matched by link 4	1	1	0	3
Domain matched by link 5	0	1	2	2
Domain matched by link 6	0	1	0	0
Domain matched by link 7	1	1	1	0
Domain matched by link 8	1	0	1	0
Domain matched by link 9	2	0	1	1
Domain matched by link 10	0	0	0	1
Total cases	99	99	99	99
Total matched domains	74	64	63	57
% of matched domains	74.75	64.65	63.64	57.58

in privacy-related features. However, in this specific use case, the most important feature is the ability to identify the URL of a searched enterprise's website (by providing the enterprise name).

To establish which search engine has the best "enterprise URL finding" performance, NSIs have tested the four search engines (Google, Bing, Yahoo, DuckDuckGo) already used in previous projects. A random sample of 99 Italian enterprises with the corresponding URLs has been downloaded from the publicly available website https://www.downloadaziende.it, and the name of each enterprise was used as a search parameter in each search engine.

Table 12.6 illustrates the number of correct domains identified at each link position in the search engine results page (SERP) by each search engine. Based on this test (acknowledging that its robustness could be improved by including a larger sample of enterprises), Google emerged as the most effective option and was therefore chosen as the search engine for the use case.

It is possible to query Google in two alternative ways: (i) by using their API or (ii) by scraping the SERP. In order to establish the best way to adopt in terms of "enterprise URL finding" performance, the enterprise sample was also searched by using the Google API, and the results were compared to the results obtained by using the SERP scraping strategy. The output contained in Table 12.7 shows that the SERP strategy leads to better results and was therefore adopted.

Technological Solution

The technological solution implemented for Use Case 1 is shown in Fig. 12.13. The starting input is a file containing each considered enterprise's ID code and name; each step uses the output of the preceding program as input in a pipeline fashion.

Table 12.7 Comparison between SERP and API strategies

	Google SERP	Google API
No domain match	25	33
Domain matched by link 1	61	52
Domain matched by link 2	6	7
Domain matched by link 3	2	4
Domain matched by link 4	1	2
Domain matched by link 5	0	0
Domain matched by link 6	0	0
Domain matched by link 7	1	1
Domain matched by link 8	1	0
Domain matched by link 9	2	0
Domain matched by link 10	0	0
Total cases	99	99
Total matched domains	74	66
% of matched domains	74.75	66.67

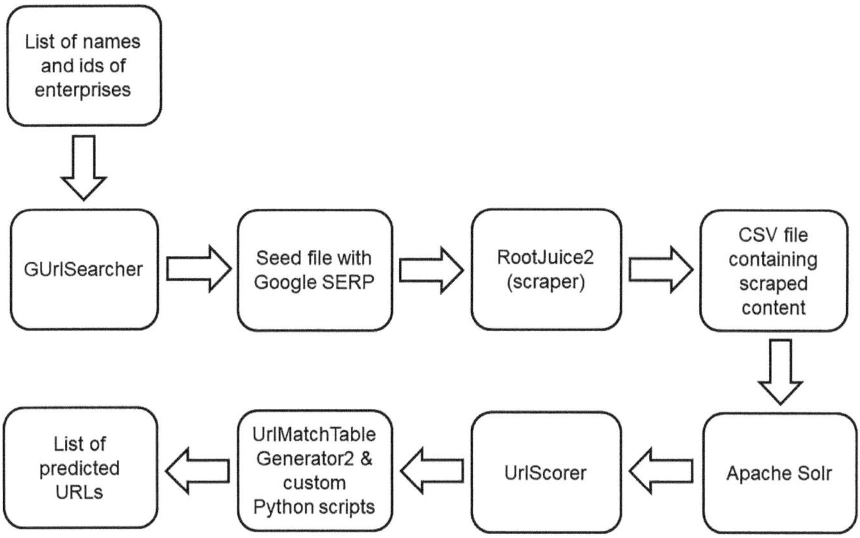

Fig. 12.13 URL retrieval software pipeline

The developed software pipeline is mainly made by the following custom software programs:

- GURLSearcher[20] is a custom Python application that takes as input a list of enterprises names and identification codes and, for each of them, performs a query on the Google search engine. The output is a text file containing the first

[20] https://github.com/SummaIstat/GUrlSearcher

ten URLs the search engine returns for each enterprise. This program was used in order to collect a list of web links for a given enterprise name. The underlying assumption is that if an enterprise has an official website, this should be found within the first ten results provided by Google.

- RootJuice2[21] is a custom Python application based on the Scrapy project[22] that takes as input a list of URLs and, on the basis of some configurable parameters, retrieves the textual content of that website plus the textual content of binary files contained in websites and loads it into a platform named Solr (see next).
- Apache Solr[23] is a NoSQL database. It parses, indexes, stores, and allows for searching of scraped content. Providing distributed search and index replication, Solr is highly scalable and, for this reason, suitable to be used in big data context.
- UrlScorer[24] is a custom Java program that reads one by one all the documents related to a specific enterprise contained in a specified Solr collection and calculates for each one the value of binary indicators, for instance: the URL contains the enterprise denomination (Yes/No); the scraped website contains geographical information coincident with enterprise information already available in the Register (Yes/No); the scraped website contains the same fiscal code in the Register (Yes/No); the scraped website contains the same telephone number in the Register (Yes/No); etc.
- UrlMatchTableGenerator2[25] is a custom Python application that is responsible to create the training dataset for the machine learning algorithms used in the analysis phase of the work.
- Custom Python scripts are used in the analysis phase, which turns to be the last phase of the process.

Analysis

Once the programs mentioned above retrieved a list of possible URLs associated with each enterprise name, a machine learning task was set up. Survey data and administrative data were integrated as ground truth to predict the correct association of a URL from the list with an enterprise's name.

The available dataset was split into three parts: training (70% of observations), validation (20% of observations), and test (10% of observations). Several supervised machine learning algorithms were trained on the training set, and the related performances have been evaluated and compared on the validation set by using the metrics normally adopted in classification tasks (see Table 12.8). Based on the calculated metrics, the neural network model was chosen as the champion model because it simultaneously presents the best accuracy, the best recall, the best F1 score, and the best AUC. After this initial model comparison, the neural network

[21] https://github.com/Summalstat/RootJuice2

[22] https://scrapy.org

[23] http://lucene.apache.org/solr/

[24] https://github.com/Summalstat/UrlScorer

[25] https://github.com/Summalstat/UrlMatchTableGenerator2

Table 12.8 UrlRetrieval performance metrics for different models

Model	Accuracy	Precision	Recall	F1 score	AUC
Decision tree	0.864292	0.773571	0.791667	0.782514	0.844171
Logistic regr.	0.862038	0.771552	0.785088	0.778261	0.840719
Gradient boosting	0.866546	0.767956	0.812865	0.789773	0.851674
SVM	0.866997	0.767538	0.815789	0.790928	0.852810
Neural network	0.867899	0.764189	0.826754	0.794242	0.856500
Random forest	0.864743	0.759109	0.822368	0.789474	0.853003
Naive Bayes	0.859558	0.756720	0.802632	0.779000	0.843786
K neighbors	0.813345	0.728426	0.629386	0.675294	0.762379

Table 12.9 Neural network tuned model performance

Dataset	Accuracy	Precision	Recall	Specificity	F1 score	AUC
Training	0.8743	0.7801	0.8432	0.8888	0.8104	0.8660
Validation	0.8684	0.7652	0.8268	0.8869	0.7948	0.8568
Test	0.8698	0.7671	0.8388	0.8839	0.8014	0.8614

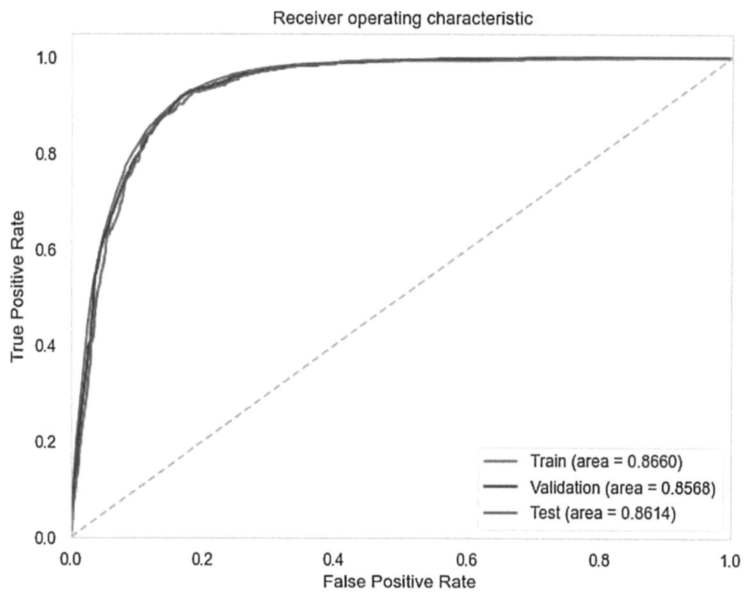

Fig. 12.14 ROC curve for the neural network tuned model

model was tuned in order to maximize its predictive capabilities, and the best cut-off value for the ROC curve was computed. As the last step, the tuned model was applied to the test set. In Table 12.9, there are the obtained metrics on the three datasets for the tuned model, while Fig. 12.14 represents the ROC curves. The neural network tuned model was then used to make predictions.

Performing some manual checks on a small random sample of misclassified cases in which the target variable was 0 (the URL is not correct) and the predicted value was 1 (the URL is correct), it emerged that in many cases, the model correctly identifies the real website, although this is different from what should be confirmed as the company provided it.

This can happen for several reasons, including (i) the company provided an incorrect web address (typos); (ii) the company provided a web address that is no longer valid (replaced by the one found by the software); (iii) the company has several valid domains (e.g., "rossi.it" and "rossi.com") and provided just one of the two, while the software pipeline found the other; and (iv) redirect phenomena (e.g., the provided domain "rossi.it" redirects to "rossi.com" found by the software).

On the other hand, most of the cases in which the target is 1 and the prediction is 0 can be explained by the fact that the enterprise incorrectly provided an official website: (i) a URL that points to a web directory website such as yellow pages; (ii) the URL of a third-party website that sells the products of the reference enterprise; (iii) the URL of the mother company in case of franchising enterprises; and (iv) a social network webpage.

In all of these cases, the procedure correctly states 0 (not the official website), but they are formally considered errors because the value of the prediction (even if it is actually correct) differs from the value of the target variable supposed to be true.

12.3.3.2 Use Cases 2 and 3: E-Commerce and Online Job Advertisements (OJA)

The objective is to determine whether each enterprise's website has e-commerce features (Use Case 2) and job advertisements (Use Case 3). Once the complete or nearly complete list of URLs is available (combining the a priori known URLs and those obtained through Use Case 1), it is used as input for the scraper software program. The scraped content is indexed and loaded into Solr, which will be available for any use case that relies on it to produce outputs. For Use Cases 2 and 3, the next step is executing the program responsible for creating the TDM (term-document matrix), which represents the analysis's starting point.

Technological Solution

The technological solution implemented for Use Cases 2 and 3 is shown in Fig. 12.15. The starting input is the URL list; as in Use Case 1, each software program uses the output of the preceding program as input in a pipeline fashion. The developed software pipeline is mainly made by the following custom software programs:

- RootJuice2 (see Use Case 1)
- Apache Solr (see Use Case 1)

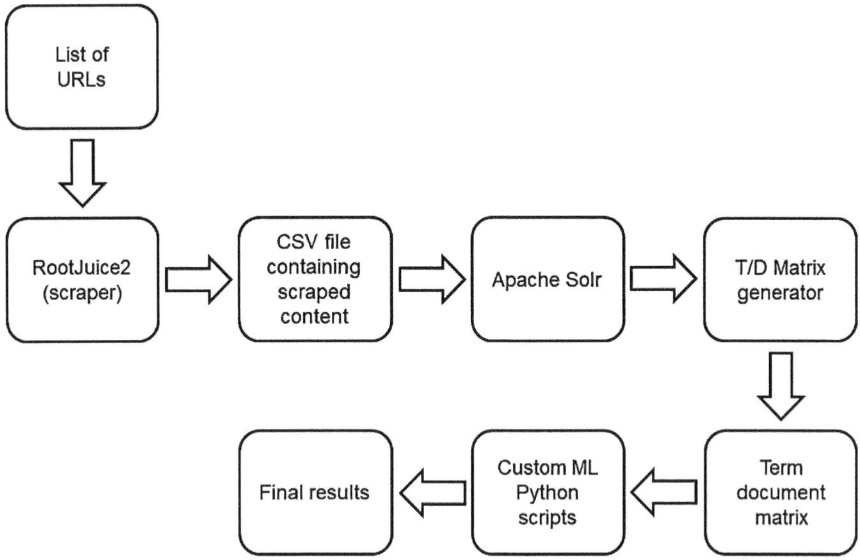

Fig. 12.15 Web scraping pipeline

Table 12.10 Proportion of websites with and without e-commerce facilities

	Yes	No
E-commerce (declared)	19.24%	80.76%

- FirmsDocTermMatrixGenerator[26] is a custom Java program that reads all the documents contained in a specified Solr collection, extracts and stems all the words from them, and generates a matrix having one word for each column and one entity (enterprise in this case) for each row; the integer values contained in the cells represent the number of occurrences of each word in each enterprise website.
- Custom Python scripts are used in the analysis phase.

Analyses

Even though Use Case 2 (establish whether an enterprise does e-commerce or not) and Use Case 3 (establish whether an enterprise exposes job vacancies on its website) have quite different objectives, the adopted strategy to determine the target value is the same. In both cases, the starting labeled datasets were unbalanced (see Tables 12.10 and 12.11); hence, as a first step, it was necessary to balance them by using the random oversampling strategy. The resulting dataset was split into three parts: training (70% of observations), validation (20% of observations), and test (10% of observations).

[26] https://github.com/SummaIstat/FirmsDocTermMatrixGenerator

Table 12.11 Proportion of websites with and without OJA

	Yes	No
Online job advertisement (declared)	28.17%	71.83%

Table 12.12 E-commerce performance metrics for different models

Model	Accuracy	Precision	Recall	F1 score	AUC
Random forest TFIDF	0.947964	0.956869	0.941331	0.949036	0.948165
Random forest NOSEL	0.946886	0.954835	0.941331	0.948035	0.947054
Random forest	0.926395	0.947974	0.906757	0.926908	0.926990
Neural network	0.909679	0.920856	0.902043	0.911352	0.909910
Neural network NOSEL	0.908061	0.888119	0.939759	0.913209	0.907102
Logistic NOSEL	0.869507	0.837198	0.926663	0.879662	0.867776
SVM	0.822324	0.890625	0.746464	0.812197	0.824621
K neighbors	0.818280	0.817808	0.832373	0.825026	0.817853
K neighbors NOSEL	0.813966	0.804902	0.842850	0.823439	0.813091
Gradient boosting TFIDF	0.771637	0.835651	0.692509	0.757376	0.774032
Gradient boosting	0.721758	0.784925	0.632792	0.700696	0.724452
Logistic	0.701806	0.759871	0.614982	0.679792	0.704435
Decision tree NOSEL	0.678889	0.724656	0.606600	0.660393	0.681078
Decision tree	0.678889	0.752107	0.561027	0.642664	0.682458
Naive Bayes	0.669992	0.732835	0.564694	0.637870	0.673180
Naive Bayes TFIDF	0.661364	0.718395	0.562598	0.631022	0.664355
Naive Bayes NOSEL	0.661364	0.718395	0.562598	0.631022	0.664355

In order to obtain the best possible predictive performance, three different strategies have been applied to the dataset. First, it was used almost as it was; the values contained in the term-document matrix produced by the FirmsDocTermMatrixGenerator software (numbers representing word occurrences) have been converted to Boolean values (presence/absence). The algorithms trained and evaluated on this dataset have been reported in Tables 12.12 and 12.13 with the "NOSEL" suffix.

Afterward, the dataset was reduced by selecting the words with the highest predictive power for each of the two classes by means of a chi-square test. The algorithms trained and evaluated on this dataset have been reported in Tables 12.12 and 12.13 without any additional suffix.

Finally, the TF-IDF strategy was tested and justified by the fact that occurrence count is a good starting point but contains an issue: longer documents will have higher average count values than shorter documents; to avoid these potential discrepancies, it suffices to divide the number of occurrences of each word in a document by the total number of words in the document: these new features are called tf for Term Frequencies. Another refinement on top of tf is to downscale weights for words that occur in many documents in the corpus and are, therefore, less informative than those that occur only in a smaller portion of the corpus. This downscaling is called tf-idf for "Term Frequency times Inverse Document

Table 12.13 OJA performance metrics for different models

Model	Accuracy	Precision	Recall	F1 score	AUC
Random forest TFIDF	0.880570	0.872289	0.888344	0.880243	0.880661
Random forest NOSEL	0.868445	0.853846	0.885276	0.869277	0.868642
Random forest	0.863292	0.848198	0.880982	0.864279	0.863499
Neural network	0.858442	0.854789	0.859509	0.857143	0.858454
Neural network NOSEL	0.840557	0.797735	0.907362	0.849024	0.841338
Logistic NOSEL	0.832980	0.809879	0.865031	0.836547	0.833354
SVM	0.807517	0.824104	0.776074	0.799368	0.807150
Gradient boosting TFIDF	0.793271	0.804236	0.768712	0.786073	0.792984
K neighbors	0.777508	0.806430	0.723313	0.762613	0.776875
K neighbors NOSEL	0.766293	0.804826	0.695706	0.746298	0.765468
Gradient boosting	0.766293	0.787291	0.722086	0.753280	0.765776
Logistic	0.751440	0.782033	0.688957	0.732551	0.750710
Decision tree	0.731737	0.781557	0.634356	0.700305	0.730599
Decision tree NOSEL	0.722037	0.800337	0.582822	0.674476	0.720410
Naive Bayes	0.665353	0.719167	0.529448	0.609894	0.663765
Naive Bayes NOSEL	0.654441	0.704849	0.517178	0.596603	0.652837
Naive Bayes TFIDF	0.654138	0.704261	0.517178	0.596392	0.652537

Table 12.14 Tuned random forest model performance on e-commerce use case

Dataset	Accuracy	Precision	Recall	Specificity	F1 score	AUC
Training	0.978425	0.990587	0.965630	0.990990	0.977950	0.978310
Validation	0.951469	0.975261	0.929282	0.975000	0.951717	0.952141
Test set	0.948248	0.975000	0.920601	0.976165	0.947020	0.948383

Frequency." The algorithms trained and evaluated on this dataset have been reported in Tables 12.12 and 12.13 with the "TFIDF" suffix.

For both use cases, on the basis of the metrics and the ROC curves (see Figs. 12.16 and 12.17), the champion model was chosen. In both cases, the choice was pretty straightforward because the random forest algorithm applied to the "tf-idf dataset" obtained the best performance in all the metrics.

As in Use Case 1, the champion models have been tuned in order to maximize their predictive capabilities, and the best cut-off value for the ROC curve was computed. The tuned models were then applied to the test set (see Tables 12.14 and 12.15) and used to make predictions.

For both use cases, some manual checks on a small random sample of misclassified cases have been performed.

In particular, for the Use Case 2 (e-commerce), the focus was on the cases in which the target variable was 0 (no e-commerce) and the predicted value was 1. It emerged that in many cases, the model correctly identifies the "real truth," although this is different from what should be true as the company declared it. This can happen for several reasons, including (i) the company provided an incorrect

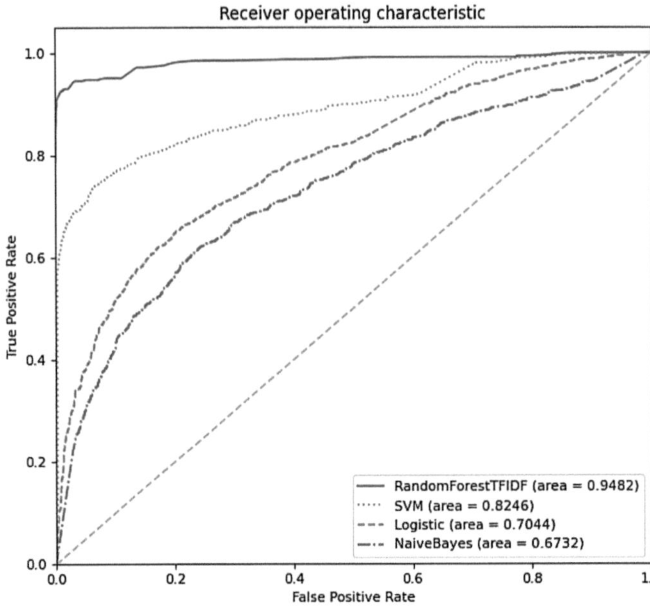

Fig. 12.16 E-commerce ROC curve comparison

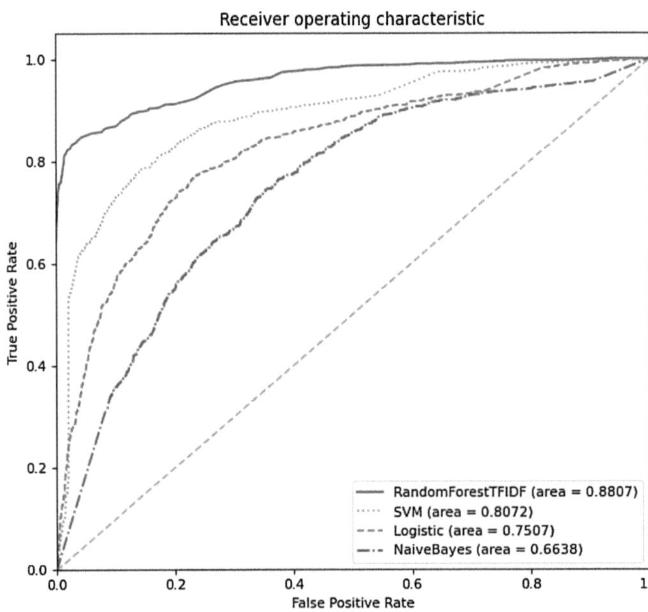

Fig. 12.17 OJA ROC curve comparison

Table 12.15 Tuned random forest model performance on OJA use case

Dataset	Accuracy	Precision	Recall	Specificity	F1 score	AUC
Training	0.979728	0.999462	0.960234	0.999477	0.979456	0.979855
Validation	0.895423	0.918567	0.865031	0.925105	0.890995	0.895068
Test	0.900000	0.925730	0.864932	0.933571	0.894298	0.899251

answer by mistake; (ii) the company provided an answer that was changed during the time; and (iii) the company voluntarily provided the wrong answer in order to avoid the subsequent series of related questions since the e-commerce question is a filter question in the questionnaire.

For Use Case 3 (online job advertisements), it emerged that when the target variable was 0 (no OJAs) and the predicted value was 1, the incorrect prediction is often caused by the presence of a "work with us" section of the website that does not contain job advertisements for specific positions but only invites to send resumes claiming that they are always looking for candidates. On the contrary, when the target variable was one and the predicted value was 0, in many cases, the model correctly identifies the "real truth," although this is different from what should be true as the company declared it. This can happen for several reasons, including (i) the company provided an incorrect answer by mistake; (ii) the company provided an answer that was changed during the time; (iii) the company provided the wrong answer because it did not understand correctly the question asking for the presence of job advertisements limited to their website (hence not including third-party job posting websites).

12.3.4 Input Privacy

As stated in the previous sections of this chapter, NSIs actively participated in research projects to harness the potential of new data sources for producing official statistics. They began their research journey many years ago and have recently focused on transferring the knowledge acquired from research projects into the statistical production processes. NSIs have embarked on a path affecting technological, methodological, organizational, and legislative aspects (Ricciato et al. 2019). They aim to enable a new official statistics production system that can integrate traditional data sources with new ones, adhering to the principles of official statistics and maintaining the quality levels of statistical products that NSIs have always upheld.

In the traditional paradigm of statistical production, NSIs addressed privacy concerns during the dissemination phase of statistical results. However, integrating new big data sources, when acquired from public or private entities external to the NSIs, requires implementing privacy-preserving techniques in the input phase.

Thus, significant privacy preservation considerations are introduced during the data collection and processing phases.

Private entities external to NSIs often collect and hold data autonomously. Institutions have also initiated several collaborations aimed at sharing data for the mutual goal of producing joint statistics. In both scenarios, whether accessing external data held by private entities or engaging in data-sharing collaborations among multiple entities, the sensitivity of the processed data often necessitates ensuring their confidentiality. To address the challenges of privacy protection and data confidentiality, researchers started to investigate new techniques, known as *input privacy techniques*, that have recently emerged from the intersection of multiple disciplines such as cryptography, computer science, and distributed processing. At the heart of these techniques is the basic concept of transforming data to fulfill confidentiality requirements. The data transformation must ensure at least two conditions: (i) the transformed data cannot be restored to the original data and (ii) the transformed data must allow the calculation of the correct output. How the processing phase transforms the original data distinguishes one technique from another. In the following paragraphs, we will illustrate some of the main input privacy techniques. The advent of big data and the availability of huge amounts of open data have also required increased privacy protection in traditional outputs. Traditional output privacy techniques, such as statistical disclosure controls, need enhancement to address the growing risk of privacy breaches. In this context, a new term has emerged: *privacy-enhancing technologies* (PET), which encompass techniques for preserving privacy on both the input and output sides (as sketched in Fig. 12.18).

12.3.4.1 Techniques

In this section, we will describe the most relevant techniques for ensuring input privacy in data processing. These methods enable secure data handling by minimizing the exposure of sensitive information while preserving analytical value. We will explore approaches such as federated learning, secure multiparty computation (SMPC), homomorphic encryption (HE), differential privacy, and synthetic data

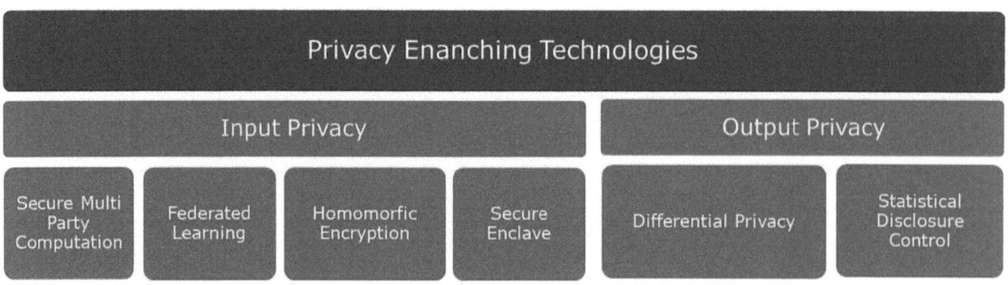

Fig. 12.18 Privacy-enhancing technologies: input privacy techniques (left) and output privacy techniques (right)

generation, highlighting their applications, advantages, and limitations in various domains, including official statistics.

Federated Learning

Recent advancements in input privacy include distributed training methods such as federated learning (Stock et al. 2023). This technique allows training machine learning models on datasets held by multiple entities (clients) without sharing them, enhancing data confidentiality. Clients share a neural network model, while a central authority oversees the training process and guides the model training without accessing sensitive data directly. At the end of each training round, the central authority applies a strategy to aggregate the model parameters received from clients. The federated learning protocol can be combined with homomorphic encryption to increase the security level. In official statistics, federated learning can be applied in experimental statistics and developing trusted smart statistics, such as those that need to exploit smartphone data input.

Secure Multiparty Computation

Secure multiparty computation (SMPC) is a cryptographic technique that enables a group of parties to compute the functions on their private data collaboratively (United Nations 2023b). It is a technique falling in the field of input privacy, and it is mainly thought to prevent collisions among a subset of involved parties aimed at controlling results. SMPC protocols are suited to balance privacy and utility needs but display disadvantages, like the high communication cost between parties. To minimize risks and simplify communication, data can be stored in various private domains, and an SMPC technique is used to perform a calculation using them all for the benefit of all parties involved. Overall, SMPC provides superior privacy protection compared to data obfuscation methods. It only reveals the final result, enhancing data privacy and security in collaborative computations. SMPC finds applications in secure data sharing, privacy-preserving analytics, and confidential computation.

Private set intersection (PSI) (De Cristofaro and Tsudik 2010) is a particular case of SMPC and an important cryptographic tool. PSI has received much attention from the crypto community since it represents a building block for many more complex functionalities. PSI is applied in cases where two parties (public or private) are interested in linking their respective datasets while disclosing information relative only to the data included in the dataset intersection to the other party. In the PSI framework, we assume a scenario where both parties are honest but curious, meaning that they respect the protocol but still try to learn information about the other party's dataset.

There are several possible implementations of PSI (Kamara et al. 2014), namely:

- Three-party solution: A reliable third party is introduced in the process, and it is responsible for carrying out the linkage between the datasets of the two parties.
- Two-party solution for supposed dishonest parties: More complex than the previous one.

- Solution with data transfer: It is similar to the three-party solution with the difference that the third party sends some microdata to one of the two parties. This solution requires perfect linkage.

Various scenarios have been defined to support the linkage of sensitive information in a manner that preserves privacy. They differ in the performed join, the shared information, and the type of analysis to perform. The PSI is applicable in numerous instances within official statistics, as in the case of the integration of external data sources with those gathered directly by NSIs.

Homomorphic Encryption

Homomorphic encryption (HE) is a specific type of cryptography that allows calculations to be performed directly on encrypted data without the need to perform the decryption step. The HE concept was first introduced in 1978, and it remained largely theoretical until 2009, when Craig Gentry presented the first fully homomorphic encryption scheme. Gentry's work, based on lattice cryptography, revolutionized the field and triggered a surge in research aimed at optimizing HE for practical applications.

HE scheme operates through the following phases:

- Encryption phase: Data is encrypted using complex algorithms, often involving vectors and matrices, especially in lattice-based systems.
- Computation phase: Encrypted data undergoes arithmetic operations, e.g., addition and multiplication.
- Decryption phase: The calculated encrypted result is converted back into the original value. This phase must effectively counteract the noise introduced during the computation phase to ensure accurate decryption.

The applications of HE vary from cloud computing to statistical analysis, where private data are computed without exposing individual information. Homomorphic encryption also promotes secure data sharing and collaborative practices, allowing collaborative processing of the data where the privacy of a single-party dataset is preserved. Homomorphic encryption represents a revolutionary innovation in cryptographic technology, offering a pathway for secure and privacy-respecting data processing. Although current implementations face significant computational complexity, efficiency, and scalability challenges, the field is rapidly evolving, and ongoing research is progressing toward more practical, high-performance, and robust systems.

Differential Privacy

Differential privacy aims to ensure that the output of a system or algorithm cannot be used to infer information about single individuals. This is assured if the response to a query or the release of a computational result is substantially the same in both cases when a particular individual is or is not included in the dataset. There are various techniques for implementing differential privacy, and one of the most common involves introducing statistically controlled noise into the dataset or in the algorithm output. The noise insertion makes it challenging to infer whether a particular individual is included in the dataset, and the difficulty varies according to the amount of noise. The adversary's inability to determine the presence of a record in the database is measured in terms of similarity between the probability distributions on outputs when the record is present or missing in the database. This similarity measure is numerically parameterized, with smaller values of these parameters representing stronger privacy protection. Although these values have an exact statistical interpretation, there is no general, application-independent recipe to set appropriate values. This represents one of the current limitations to the usability of this technique. Differential privacy works well when a single query is sent to a database, or the output of a computation is evaluated. Still, it can break down when the issuer repeatedly asks several queries to the database. The privacy protection effect due to the noise diminishes as the number of observed samples increases. Hence, the noise level required to adhere to a differential privacy level of security is higher in the case of multiple queries. A technical component in the protocol, called "privacy accountant," keeps track of previously sent queries and calculates the information gained from the combination of query results. However, increasing the amount of the inserted noise also results in the reduction of output data usability. Noise degrades the quality and utility of the output data. In cases where a large amount of noise is needed for privacy requirements, the possibility of extracting useful information from the data is jeopardized. Therefore, the differential privacy technique relies on the fundamental trade-off between privacy preservation and output utility. When the data output is too distorted, it can become difficult or impossible for researchers, analysts, or machine learning algorithms to extract meaningful insights or make accurate predictions based on that data (Drechsler and Bailie 2024).

Differential privacy was formally introduced in Dwork (2006). Since then, it has been the subject of extensive research, along with developing several protocols for its implementation. These include high-level mechanisms like SQL engines intercepting SQL queries and adding appropriate noise to return differential privacy-compliant output (Johnson et al. 2018). Other protocols focus on training machine learning models in a differential privacy mode (Ruan et al. 2023).

Synthetic Data

Generating synthetic datasets for statistical purposes allows the production of datasets containing different information than real data but with the same statistical characteristics. The utility of using synthetic rather than real data is to protect against privacy attacks. The advantage of synthetic data is that, depending on the user's purpose, it offers a trade-off between the analytical value of the dataset and the risk of disclosure. Every NSI seeks to use and share data securely. Synthetic data are increasingly considered an alternative to the privacy-preserving exchange of sensitive data. There are various levels of synthetic data, each with a different trade-off between analytical value and disclosure risk. Depending on this trade-off and the intended use of the dataset, synthetic data can be used for the following applications (United Nations 2023a):

- To release synthetic microdata to the public: Traditional output disclosure measures used by NSOs can limit users' access to high-quality microdata. NSOs are exploring generating synthetic data as a new output disclosure option.
- To test analysis: Some NSOs provide limited access to microdata to trusted entities, either through remote access or at physical research data centers. However, security checks, audits, and approvals can slow down research and analysis projects. Synthetic data could enable researchers to develop and test their models, algorithms, or analyses and conduct preliminary data analyses as they wait to access the original real data.
- Education: High-quality data is needed for students, academics, and users to learn new concepts and methods; by providing data for education purposes, an NSO may preserve only the distributions related to the studied dimensions and synthetically generate the others.
- Testing technologies: Dummy data are often used when testing new software. The file layout and error rates represent real data, but there is no analytical value. However, as machine learning becomes prevalent, the testing will require more analytically realistic data. In these cases, synthetic data with some inferential validity can be beneficial

There are many methods for generating synthetic data, such as sequential modeling, statistical obfuscation, and innovative deep learning methods, such as generative adversarial networks.

12.3.4.2 Challenges

From a technical perspective, privacy-enhancing technologies are in a mature development phase and are ready for use within NSIs. However, their application in the official statistics has not yet been realized due to a series of challenging issues that remain unsolved. Among those issues are legal concerns: every process using sensitive information must comply with the GDPR. Also, outside the European Union, there are other privacy regulations and legislation, and a standard is lacking. From an organizational perspective, the application of PETs has impacts both in terms of the costs required for their implementation and in terms of integration of PETs into current production processes. Furthermore, the adoption of PET requires researchers and analysts to change their working procedures. A crucial aspect regarding the trust model to adopt is the centralized versus distributed trust model and trust models with or without the presence of a neutral third party. It's essential to assess each use case individually to determine the most appropriate trust model. Finally, another critical issue concerns choosing between custom and general-purpose solutions. To address the abovementioned open issue, the UN Statistical Division currently promotes experimental projects on PETs involving several national statistical institutes, Istat included. The main case studies developed are collected in the UN case study repository.[27]

12.3.5 *Analyzing Trade Data with Network Analysis Techniques: The Experimental Statistics TERRA*

Istat found a valuable resource in the COMEXT data source provided by Eurostat, which contains information on international trade in goods. COMEXT covers the value and quantity of goods traded between EU Member States (intra-EU trade) and between EU Member States and non-EU countries (extra-EU trade). Trade data is disseminated monthly and at the most detailed level of the following product nomenclatures: the Combined Nomenclature (CN),[28] the Standard International Trade Classification (SITC),[29] the Broad Economic Categories classification (BEC),[30] the Classification of Products by Activity (CPA),[31] and the Standard Goods Classification for Transport Statistics/Revised (NSTR).[32] In addition, trade flows are classified by mode of transport, providing helpful information for transportation policy, monitoring international transport routes, and evaluating the impact of trade on the environment.

As international trade represents a major part of the world economy, statistics on trade in goods are an instrument of primary importance for numerous users, including public and private sector decision-makers. For example, statistics regarding international trade in goods are valuable to:

- Inform on recent and long-term developments in trade and economy
- Help EU businesses conduct market research and define their commercial strategy
- Enable EU authorities to prepare multilateral and bilateral negotiations under the common commercial policy

[27] https://unstats.un.org/wiki/display/UGTTOPPT/Case+study+repository

[28] Official Journal of the European Union, Commission implementing regulation (EU) 2024/2522, Council Regulation (EEC) No 2658/87 https://eur-lex.europa.eu/legal-content/EN/TXT/PDF/?uri=OJ:L_202402522.

[29] Standard International Trade Classification: revision 4 (ISBN: 9211614937) https://unstats.un.org/unsd/trade/sitcrev4.htm.

[30] United Nations Statistics Division, Classification by Broad Economic Categories, Defined in terms of SITC, Rev.3, ST/ESA/STAT/SER.M/53/Rev.3, E.89.XVII.4.

[31] Official Journal of the European Union, Commission delegated regulation (EU) 2024/3103, amending Regulation (EC) No 451/2008 of the European Parliament and of the Council as regards updating the classification of products by activity (CPA) https://eur-lex.europa.eu/legal-content/EN/TXT/PDF/?uri=OJ:L_202403103&qid=1736955449829.

[32] United Nations, Report of the working party on transport statistics on its fifty-ninth session, ECE/TRANS/WP.6/155/Add.1 https://unece.org/fileadmin/DAM/trans/doc/2008/wp6/ECE-TRANS-WP6-155a1e.pdf.

- Provide an essential source of information for other statistical domains, such as Balance of Payment statistics or national accounts.

International trade provides a rich data source for users to explore different aspects of global commerce. However, navigating and understanding the data can be challenging with so much information available. An innovative visualization tool that displays trade data in a suitable graphical form can illustrate the evolution of trade flows and show the trade volume and the composition of the basket of traded goods. TERRA (imporT ExpoRt netwoRk Analysis) is a data visualization tool designed at Istat. TERRA is an experimental statistics,[33] publicly available from November 2023. TERRA offers researchers and policymakers the opportunity to explore the dynamics of trade flows, with the possibility of focusing on specific products and transport modes, tracing the critical phases of recent years. It also makes it possible to simulate flow disruptions or closures of specific logistics hubs or transport routes, allowing the outline of possible scenarios of modification or relocation of global chains capable of mitigating the risk of shock transmission (changes in bilateral relations between countries, logistics, or transport investments, increased foreign investment, etc.).

European Big Data Hackathon
The European Big Data Hackathon is an international competition (hackathon) organized by Eurostat as part of the New Techniques and Technologies for Official Statistics (NTTS) conference. Teams of experts (data scientists, researchers, methodologists, domain experts, etc.) from different European statistical institutes participate in the competition to implement innovative products that integrate traditional data sources and big data sources. As part of the competition, teams must produce a product that responds to a policy question related to a pressing issue in the European context (e.g., COVID-19) and a statistical problem (integration of big data sources, use of new data collection tools, increasing the quality of outputs of specific production processes, etc.). The team representing Istat (Team-Istat) implemented TERRA's prototype during the virtual competition from February 26 to March 4, 2021. Team-Istat won first place in the competition.

12.3.5.1 Data Source and Data Analysis Tools

TERRA monthly processes about one billion foreign trade records produced by the member countries according to harmonized methodologies publicly available on Eurostat's COMEXT database. The information base provides official estimates of trade flows in monetary value and physical quantities at the highest granularity in temporal resolution (monthly frequency), characteristics of traded products, trading partner countries, and mode of transport. COMEXT's bulk download function lets

[33] Detailed information on TERRA is available in Istat experimental statistics website: https://www.istat.it/en/experimental-statistic/terra-import-export-network-analysis-3/.

users download large volumes of data in .dat format, facilitating easy import for R, SAS, or Python analysis. The datasets include metadata such as classifications, data availability details, and methodological notes. COMEXT data are organized in different folders. TERRA is fed from the folders:

- PRODUCTS: Containing EU countries' monthly and annual trade interchange with each partner country in value (Euro) and quantity (kg and any additional units) for products classified according to Combined Nomenclature, SITC, and CPA.
- TRANSPORT: Containing data on monthly trade interchange of EU countries with non-EU partner countries in value (Euro) and quantity (kg) by means of transport. Products are detailed according to NSTR classification.

The main functionalities implemented in TERRA allow for the analysis of the impact of shocks in the means of transportation and the effects of disruptions in cross-country trade relations for specific products with social network analysis techniques implemented in Python, offering a set of indicators proper to graph analysis.

The data analysis features offered by TERRA are:

- Interactive map: This section provides a map showing for each country the total population, some macroeconomic indicators, the main imported and exported products, and major trading partner countries. In addition, a time-lapse feature is provided to depict the changes in monthly trends in international trade for the past three years.
- Graph EU–extra-EU: This section displays graphs representing international trade by product (NSTR classification, see above) and mode of transport from COMEXT source, along with relevant global and local measures detailing the structure of trade relations between EU and extra-EU countries. The panel is interactive and allows for the application of various filters. In addition, an animation shows the evolution of the international trade graph over time.
- Graph EU–World: This section displays graphs representing international trade by product (CPA 2.1 classification, see above) from COMEXT source, along with relevant global and local measures detailing the structure of trade relations recorded by EU countries to/from each trading partner country.
- Time series: COMEXT data time series visualization tool. This section allows for easy and understandable reading despite the underlying data volume. Series are provided in value and quantity, and download is available in various formats.
- Basket of traded products: This section provides, for each EU Member State, a monthly time series of trend changes in the composition of the basket of exported and imported goods, classified according to the divisions of the CPA 2.1 classification.

12.3.5.2 Methodology

International trade relationships can be represented in interactive graphs and analyzed using network analysis techniques (Wasserman and Faust 1994). These networks represent countries as nodes and trade flows between them as directed weighted edges. The weights indicate the value of a product exchanged (Euro) as a percentage of the total trade volume for a specific time frame. Network metrics provide quantitative insight into the trade structure and the roles of individual countries. Density, a global metric, measures how interconnected the network is, ranging from 0 (no connections) to 1 (fully connected). A dense network reflects more complex trade relationships, while a sparse network indicates fewer interactions. Other centrality measures quantify node-specific roles:

- **Degree centrality:** quantifies the number of connections a node has. It is calculated based on the count of connections (edges) a node possesses, with higher values indicating a more central position within the network.
- **In-degree centrality:** the number of import partners, reflecting the country's reliance on others. Low in-degree centrality indicates potential vulnerability due to dependence on a few suppliers, who may have greater bargaining power.
- **Out-degree centrality:** the number of export partners, indicating trade outreach.
- **Closeness centrality:** measures how close a country is to others, calculated as the inverse of the sum of the shortest paths to all nodes:

$$CC(u) = \frac{n-1}{\sum_{v=1}^{n-1} d(v,u)}$$

  where $d(v, u)$ is the shortest-path distance between v and u, and $n - 1$ is the number of nodes reachable from u. High closeness indicates a hub role, minimizing trade distances (De Benedictis et al. 2014).
- **Betweenness centrality** of a node is the sum of the fraction of all-pairs shortest paths that pass through:

$$BC(u) = \sum_{s,t \epsilon V} \frac{\sigma(s,t|u)}{\sigma(s,t)}$$

  where V is the set of nodes, $\sigma(s, t)$ is the number of shortest (s, t)-paths, and $\sigma(s, t|u)$ is the number of paths passing through some node u other than (s, t). High values indicate control over trade flows, positioning a country as a bridge or bottleneck in the network.
- **Distinctiveness centrality:** highlights connections to peripheral nodes over hubs, emphasizing the importance of countries bridging the core and periphery (Fronzetti Colladon and Naldi 2020):

$$DC(u) = \sum_{j=1, j \neq u}^{n} w_{uj} log_{10} \frac{n-1}{g_j^{\alpha}}$$

where w_{uj} is the weight of the link between the node u and j (0 is there is no link between those nodes) and g_j^{α} is the degree of the node j elevated to $\alpha \geq 1$ and is used to allow a stronger penalization of connections with highly connected nodes. Unlike traditional measures, distinctiveness centrality penalizes redundant ties to dominant hubs, favoring strategically critical links to less-connected nodes.

COMEXT data gives a general overview of Europe but does not fully represent global trade. It relies solely on declarations from European countries and excludes interactions between non-European (extra-EU) countries. This limitation can distort density, closeness, and betweenness as the network remains incomplete. To mitigate this, extra-EU countries are grouped into a single node, balancing data completeness with detail. Future development plans include integrating additional data sources to address these gaps and build a more comprehensive trade network.

12.3.5.3 Architecture

TERRA implements a data processing pipeline to ensure data analysis is promptly available to stakeholders a few hours after Eurostat data is published. The application retrieves and processes monthly data spanning a 10-year time series. Additionally, TERRA offers the option to select two types of monthly time series—raw data and yearly variation.

To download and process a large volume of data, a specific batch program (shown in the left panel of Fig. 12.19), implemented in Python, runs automatically each month. This batch script is scheduled to start on the 24th day of every

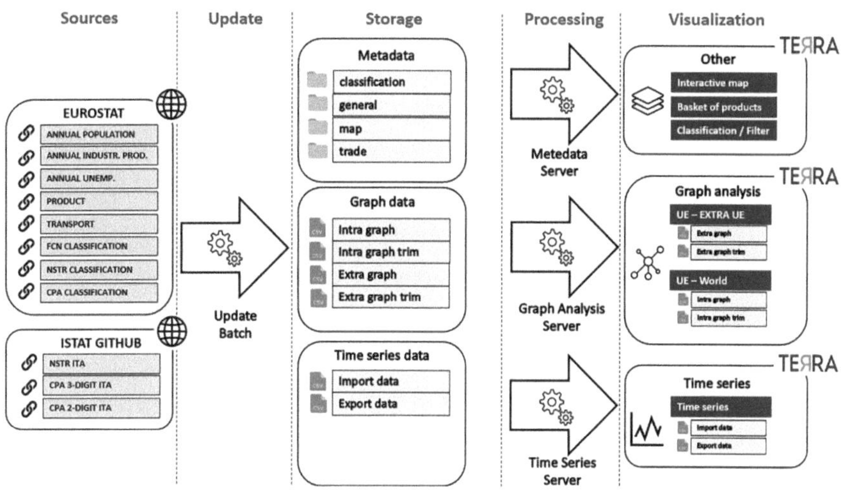

Fig. 12.19 TERRA main architectural layers and components

month. It downloads files, performs the necessary processing, produces outputs, and ultimately updates the data stored on the server.

The script runs on a cloud platform to minimize processing time using high-performance algorithms. It accesses the bulk download section of the Eurostat portal. The script initiates a parallel process to download 144 files containing monthly product data, two files for annual product data, 144 files for monthly transport data, and the associated classification files. Due to the varied and complex nature of the data analysis algorithms, we opted for a microservices architectural paradigm. This approach divides the application's functions into multiple services, allowing for individual responsibility; each service is implemented and deployed independently of the others.

Figure 12.19 sketches the main architectural components of TERRA.[34] Three different layers can be identified in the architecture: data storage, data processing, and data visualization.

1. **Data storage layer** contains raw data downloaded from the Eurostat portal. Batch processing stores the aggregated data in this layer, which is further processed by the microservices in the data processing layer.
2. **Data processing layer**, or backend, includes the three components: the first two aim to process Python scripts, and the third exposes static data, such as classifications and metadata. The frontend communicates with the backend through requests according to the HTTP protocol, exchanging messages in JSON format.
3. **Data visualization layer**, or frontend, includes the web component as the user interface. This application was implemented as a single-page application using open-source web frameworks. The use of these modern technologies makes the application responsive. For instance, the layout adapts to the size of the display, making it possible to access the dashboard from PCs and mobile devices.

12.3.5.4 Use Case

This section examines the trends in electricity supply when countries were primarily addressing the challenges of the COVID-19 pandemic.

Figure 12.20 illustrates the interactive map panel of TERRA, which facilitates the analysis of monthly percentage variations over the previous year for imported and exported data, over the last 5 years, expressed in Euro. It displays a color-coded heatmap based on the chosen variation index, with values ranging from −60% to +60%. In April 2020, European nations experienced a significant negative impact, as shown in the heatmap. This prompts further exploration of the effects on goods exchanges between countries.

[34] The source code of TERRA is open source and available in the following GitHub repository: https://github.com/istat-methodology/terra.

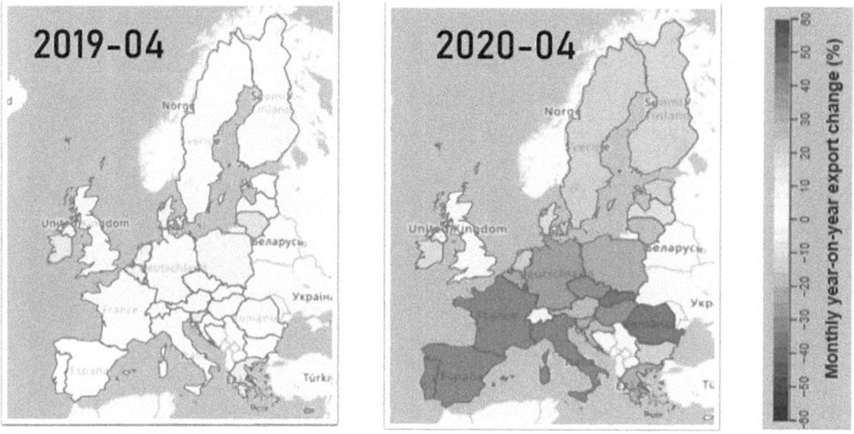

Fig. 12.20 Interactive map panel of EU exportation: comparison between April 2019 and April 2020. (Source: authors' own elaboration using an OpenStreetMap basemap © OpenStreetMap contributors, https://www.openstreetmap.org/copyright)

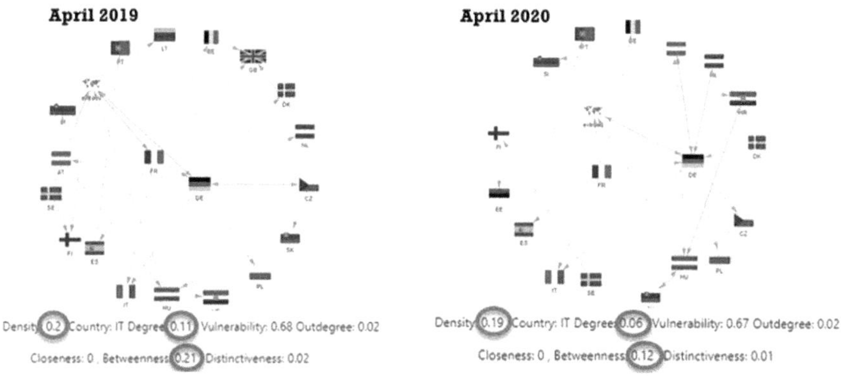

Fig. 12.21 Graph EU–extra-EU: prevalent trade network (60%) on "electricity, transmission and distribution services." Comparison between April 2019 and April 2020

The analysis focuses on the product category "electricity, transmission, and distribution services." By comparing the trade networks for this product in April 2020 with those from the same period the previous year—while combining extra-EU countries and averaging the flow, as illustrated in Fig. 12.21—we observe a slight decline in network density, which decreased from 0.20 to 0.19 (a 5% reduction). Despite this notable impact, the network remained relatively stable without significant disruptions.

However, looking into individual country metrics reveals more nuanced insights. For instance, Italy's betweenness centrality fell from 0.21 in April 2019 to 0.12 in April 2020, and its normalized degree metric dropped from 0.11 to 0.06. This

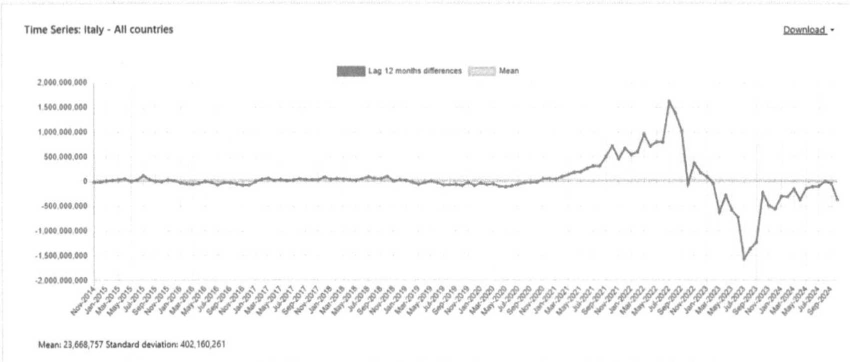

Fig. 12.22 Time series related to Italy's import of electricity between November 2014 and September 2024

indicates that while trade exchanges continued largely uninterrupted during the pandemic, dominant countries adjusted their strategies, leading to a loss of centrality for some nations.

To complete the analysis including a time series point of view, Fig. 12.22 shows Italy's electricity import data in Euro and includes these components:

- **Green line (lag 12-month differences):** This line shows the difference between data points lagged by 12 months.
- **Yellow line (mean):** A constant horizontal line representing the mean value of the data series. This provides a reference point for comparison against the fluctuations in the green line, which is approximately 23.7 million.
- **Standard deviation:** Approximately 402 million, suggesting high variability in the data.
- **Time-range:** The x-axis spans from November 2014 to September 2024.

The data remained relatively stable until March 2020, after which significant volatility was observed. There was a decline around mid-2020, followed by peaks in late 2021 and sharp decreases through mid-2022. In 2023, the data shows signs of recovery and stabilization, although fluctuations persist. This pattern is particularly noticeable during the COVID-19 pandemic, when electricity prices fell below the average. Additionally, a more pronounced effect can be seen in the subsequent years, where prices sharply increased in response to the Russia–Ukraine conflict, significantly exceeding the average value for the analyzed period, before decreasing a year after reaching their peak.

12.3.5.5 Results

TERRA demonstrates the potential of integrating advanced data visualization, analytical tools, and modern architectural solutions to disseminate and interpret complex trade data. Using Eurostat's COMEXT database, TERRA provides granularity and timeliness in monitoring international trade flows. Its innovative use of network analysis techniques offers policymakers and researchers with robust metrics to assess trade dynamics, identify vulnerabilities, and explore scenarios to mitigate disruptions in global supply chains. The dashboard includes interactive maps, trade network graphs, and time series data. The detailed case study on electricity trade during the COVID-19 pandemic showcases TERRA's ability to highlight critical shifts in trade patterns and economic impacts during crises, confirming its role as a robust tool for navigating an increasingly interconnected global economy. As TERRA evolves, the integration of additional data sources and the refinement of analytical capabilities promise to enhance its relevance and utility further, ensuring that it can meet future challenges and opportunities in the analysis of international trade.

12.3.6 Sentiment Analysis

Sentiment analysis identifies and categorizes opinions expressed as text to determine opinions toward a specific theme, i.e., positive, negative, and neutral. Indeed, sentiment analysis is a branch of natural language processing (NLP) that aims to identify a text's emotional tone. It allows for extrapolating opinions from any social media platform and determining users' feelings regarding specific topics. Sentiment analysis also allows for assessing public views on particular issues (Steinert-Threlkeld 2018). Social media are an excellent source of information for sentiment analysis to provide perceptions for various scopes, whether commercial or not. In recent years, the explosive growth of online media, such as social networking sites, has enabled individuals to express opinions on many potentially interesting subjects for statistical purposes and public policy evaluation. Moreover, evaluating real-time conversations and trends on social media allows for a timely representation of people's opinions, which cannot be obtained through traditional official statistics. However, sentiment indexes based on social media are not representative of the general population but only of the active platform users. For instance, Twitter users tend to have a lower average age than the real population. In addition, the users who discuss the topic of interest are a subset of the active users on the specific platform and may have different characteristics. This justifies why national statistical offices are interested in creating synthetic indexes to monitor the debate on topics relevant to public policy or for specific purposes for which traditional surveys can hardly collect data (such as cyber violence, prejudices, and homo-transphobia).

Sentiment analysis tasks, also known as opinion mining, consist of labeling people's opinions as different categories, such as positive and negative (or relevant and out of scope, subjective vs. objective) from a given piece of text (maybe a tweet or an article). Classifying these documents is an arduous task. In recent years, many methods, techniques, and enhancements have been proposed to solve the problem

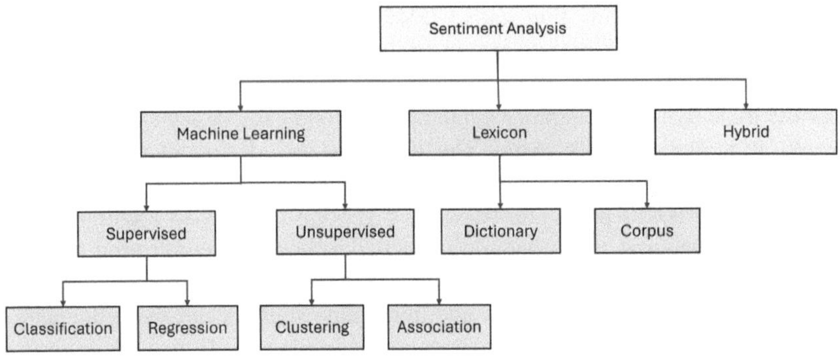

Fig. 12.23 Hierarchical structure of sentiment analysis techniques

of sentiment analysis in different tasks at different levels. Sentiment analysis can be measured using three main strategies:

1. Machine learning strategies (Pak et al. 2010)
2. Lexicon-based strategies (Taboada et al. 2011)
3. Hybrid strategies (Kolchyna et al. 2015), combining 1 and 2

These techniques can be split into several distinct methods, as shown in Fig. 12.23.

NLP analyzes language for its meaning; therefore, a "word" may have multiple meanings varying according to the context. It is a superior technique compared to lexicon-based approaches, which are essentially based on keyword processing: They assign positivity or negativity to individual words and calculate the overall percentage score for the post. Supervised machine learning techniques, however, rely on manually labeled data, i.e., manual processing: human interpretation of the sentiment must be accurate. In the machine learning approach, the supervised learning model can be easily trained, and the unsupervised model can easily categorize the data. The lexicon approaches are based on sentiment lexicons, namely, a list of lemmas (words or composite expressions) with pre-computed sentiment scores. The sentence scores of each sentence can easily be calculated as the mean of the sentiment scores of the matched words.

It must be stressed that most of these methods are developed for the English language. There are several quality problems in all approaches. The primary issue in all techniques is classification accuracy, and many efforts imply correctly classifying opinion and sarcasm, and it is well known that most of the text is often classified as neutral. In addition to this problem, most research on sentiment analysis focuses on text written in English. Indeed, the validation sets labeled in Italian were very few at the time (roughly less than 10,000). Even if an approach may provide a better accuracy on that set, when dealing with the specific problems of a daily index collecting billions of tweets with a specific domain, e.g., economy, it is not evident how this will affect the accuracy of the dynamics of a daily index of sentiment. Sentiment classification based on insufficient labeled data is still a challenging problem, and the amount needed for specific tasks is unclear even when a lexicon-based approach is validated. Semi-supervised or unsupervised methods need less human labor and may provide the same accuracy. However, lexicon-based approaches usually have limited word coverage and thus may fail to recognize emotional words (especially domain-specific words) and are still unable to deal with complex sentences, for instance, with mistakes in spellings, acronyms, or neologisms (see *Covid*, *lockdown*) and have static sentiment.

In addition, but outside the scope of this paragraph, sentiment analysis has various sub-streams such as bias analysis and emotion detection.

12.3.6.1 Social Mood on Economy Index (SMEI)

In recent years, the Italian National Institute of Statistics (Istat) has exploited social media messages to assess the mood of Italians about the country's economic situation. In October 2018, this effort led to the release of the Social Mood on Economy Index (SMEI) (Zardetto 2018; Catanese et al. 2022), an experimental high-frequency sentiment index based on Twitter data. Among the first experimental statistics published by Istat, the Social Mood on Economy Index is a daily index computed starting from the Italian Twitter's public stream aimed at representing the perception of the evolution of economic features.[35] Though computed daily, the index is published since October 2018 each quarter, with the daily time series provided as an attachment to the publication. The SMEI production pipeline is shown in Fig. 12.24.

The sentiment analysis procedure adopts an unsupervised lexicon-based approach (based on vocabularies whose lemmas are associated with pre-computed sentiment scores), as applying supervised methods requires large sets of manually

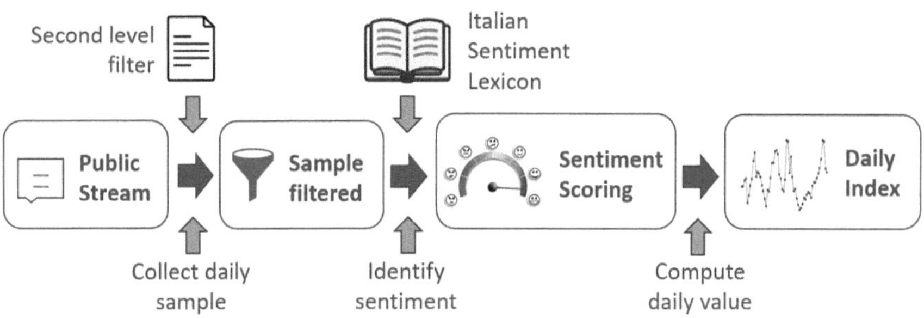

Fig. 12.24 Social Mood on Economy Index data processing pipeline

[35] Detailed information on SMEI is available from Istat's website, at the following link https://www.istat.it/wp-content/uploads/2024/05/Methodological_Note.pdf.

labeled texts. Briefly, the index calculation pipeline consists of four fundamental steps:

1. **Data collection:** Collection of a daily sample of public tweets matching a set of keywords (the filter) related to the economy.
2. **Text cleaning:** We only analyze the textual content of the tweets (no information concerning Twitter users is ever stored; the index only uses unlinked anonymized data). We perform standard natural language processing (NLP) pre-processing steps: (i) convert text to lowercase, (ii) tokenize the text into words, (iii) apply basic orthographic repairs, (iv) remove URLs, (v) remove nonalphabetic characters (e.g., # or @), and (vi) remove stop words and (vii) if needed, stem words to get rid of inflected forms.
3. **Score extraction:** Calculation of tweets' positive and negative sentiment scores, as weighted averages of the sentiment scores of each word in the tweet.
4. **Index computation:** Computation of the daily index value as an appropriate central tendency measure of the tweets' score distribution.

Subject matter experts, borrowing keywords from the Italian Consumer Confidence Survey (CCS) questionnaire, designed the first-level filter in 2018 in the following CCS filter. It consists of questions deemed most appropriate to assess the optimism/pessimism of consumers (namely, assessments and expectations on the Italian general economic situation, on unemployment, assessments on households' financial situation, opportunity and possibility of savings, current of durable goods purchases, assessments on the family budget). A typical social media pipeline consists of choosing a set of keywords and then filtering all the texts that contain at least one of these keywords. There is solid evidence that a significant portion of all the messages being exchanged on whatever social media platforms lack any relevant information (Morstatter et al. 2013). Ideally, filters should be able to capture relevant messages and eliminate off-topic messages from the beginning. For this purpose, the filter choice should be split into a top-down and bottom-up approach. First, experts choose a list of words according to the intended statistics. Then, the list should be validated by some data-driven analysis that ensures the relevance of the sampled texts. It is a well-known problem that words may have multiple meanings, but they also can be used in different contexts, and a priori, it is impossible to establish which ones. This data-driven analysis can be performed through machine learning methods for semantic analysis, such as word embedding or topic modeling, a frequently used text-mining tool for discovering hidden semantic structures in a text body. In 2021, a quality enhancement concerned the filter design, i.e., the relevance: (i) analyze the filter via data-driven techniques to enhance tweets' relevance, i.e., grasp appropriate information for economic statistical purposes, and (ii) address the "multiple meaning problem," i.e., a word in the filter may have different meanings in different contexts. To accomplish both tasks, an analysis through word embedding techniques was performed. This, in particular, led to redesigning the filter (Catanese et al. 2022) and further investigation of the coherence with other economic series as shown in Catanese et al. (2024a).

In particular, the change of sampled tweets, induced by the changes in the filter, enhanced the economic interpretability of observed peaks.

As supervised methods require large sets of manually labeled texts, sentiment scores are obtained in an unsupervised setting using an Italian sentiment lexicon, a vocabulary whose lemmas are associated with pre-computed sentiment scores. The texts of all tweets are compared with the lexicon. Based on the scores of the matched words, for each tweet, the positive (p) and negative (n) sentiment scores are obtained averages (more details can be found in the methodological note[36]). The calculation of the daily index (I) uses polar coordinate transformation and is a weighted average of the polarity in terms of intensity, as shown in Eq. 12.4:

$$I = \frac{\sum_t i_t \omega_t}{\sum_t i_t} \cdot 100 \tag{12.4}$$

where polarity $\omega = 1 - 4\pi/\theta$ so that $\omega \in [-1, 1]$ and $\theta = \arctan(n/p)$, intensity $i = \sqrt{p^2 + n^2}$, so that $i \in [0, 1]$.

There are two quantitative lexicons in Italian (Basile and Nissim 2013; Castellucci et al. 2015) for the corpus-based method on tweets. In 2021, the lexicon of SMEI has been revised mainly to increase the coherence with other economic time series. In Daas et al. (2012) and van den Brakel et al. (2017), several quality issues are posed that are relevant in evaluating SMEI's quality, like the selectivity concerning the intended target population and the bias estimation problem. A method for validating and interpreting SMEI is correlating the series with official statistics derived from monthly surveys, e.g., the Consumer Confidence Survey. However, there is no guarantee that the correlation is based on true causality and that the correlation will occur in the future. Indeed, the probability sampling for finite population inference is stronger than reliance on co-integration. Series obtained from social media are selected by maximizing the correlation with the series from the sample survey and do not necessarily measure the same concept as the survey. The 2021 validation of the DPL vocabulary and redesign of the filter were carried out by comparing the trend of SMEI with official economic statistics indicators (such as Industrial Production Index and Consumer Confidence Index). The new SMEI improved all analyzed correlations, increasing consistency with other short-term business statistics.

As described, the SMEI index has been revised by changing the filter and replacing the lexicon. The former change ensured a better filter relevance, while the second change improved the coherence of the index with other NSI economical time series. For all these reasons, the entire time series of SMEI has been reviewed and published since the last quarter of 2021 with the new methodology. A note to explain to the user the significant changes has also been published on the site. In the present

[36] SMEI's methodological note is available at the following link: https://www.istat.it/wp-content/uploads/2024/05/Methodological_Note.pdf.

study, we highlighted the enhancements of the SMEI achieved by redesigning the filter and changing the vocabulary.

With rare exceptions, the big data generation mechanism does not fall under the control of the statistician, and it is not known. Therefore, ensuring the accuracy and reliability of the statistical outputs derived from these sources is still a matter of research. For these outputs to be configured as Trusted Smart Statistics, they must be evaluated and validated ex post according to rigorous and shared methodologies. These methodologies do not fall within the traditional quality assurance framework of the European national statistical institutes and are often the subject of current research in the academic field. Integrating big data into the production processes of the national statistical institutes presents complex challenges in reviewing and adapting business processes and assuring and evaluating the quality of products. Supervised machine learning techniques have the main advantage of evaluating accuracy on the test set. However, they cannot compute the "bias" or "concept drift." This can be overcome by dynamically retraining them.

12.3.6.2 Istat Supervised Machine Learning Approaches for Text Classification

Since 2018, Istat has relied on lexicon-based sentiment analysis to calculate the SMEI, mainly due to the lack of labeled Italian datasets dealing specifically with the economy. In general, there are no significant amounts of Italian-labeled datasets. However, while relatively straightforward and efficient to implement, lexicon-based sentiment analysis presents shortcomings due to the grammar of a sentence. Indeed, each word in a sentence is assigned a positive, negative, and neutral score, and these scores are static and independent of context. The sentiment of a given text is then obtained summing all the scores. This implies that "negations," "adverbs," or adjectives may alter the sense of a sentence, and this approximation may induce biases. Some experimentation is carried out to include the use of valence shifters as shown in Catanese et al. (2024b).

Due to both advancements in natural language processing techniques (Vaswani et al. 2017) and available labeled datasets (Basile et al. 2014; Mencarini et al. 2019; Bianchi et al. 2021), Istat has explored the possibility of computing the SMEI using supervised machine learning techniques, following two main approaches: (1) training recurrent neural networks (RNNs) from scratch and (2) fine-tuning pre-trained encoder-only transformer-based models for sequence classification.

Recurrent Neural Networks (RNNs)

A popular way to deal with sequences of textual data is to use recurrent neural networks, i.e., neural networks designed to process sequential data by maintaining a "memory" of previous steps in the sequence. In particular, Istat experimented with RNNs like long short-term memory (LSTM) networks (Hochreiter 1997) and bidirectional LSTMs (Schuster and Paliwal 1997) to compute the SMEI (Bruno et al. 2024b). The results are promising and show an increased flexibility in the classification when the labeled utilized dataset contains an explicit reference to

COVID-19. This approach is conceptually equivalent to the classifier's dynamic re-training to overcome the concept drift problem. In addition, also pre-training an embedding layer on the entire prediction corpus using a FastText algorithm (Bojanowski et al. 2017), used as the first layer of the classification model, may provide some knowledge of COVID-related vector space. In general, this made the model more robust to the classification of sentences comprising words that are not seen in the training set but are seen in the corpus.

Pre-trained Transformers

After their introduction in Vaswani et al. (2017), transformer models have become state of the art in several natural language processing tasks, including text classification, such as hate speech recognition, racism, irony, and offensiveness. Only supervised methods can be utilized for these tasks.

In particular, pre-trained encoder-only transformers like *BERT* (Devlin et al. 2019) are particularly suited for fine-tuning for text classification tasks, as they are pre-trained on vast amounts of textual data, and thus have a general knowledge of how language and context work, and are easily fine-tuned to handle specific tasks, like those previously mentioned. The issue with models like *BERT* is that they are pre-trained on mainly English text and show sub-par performance in other languages. This was addressed with the introduction of *multilingual RoBERTa* (Conneau et al. 2020), pre-trained in over 100 languages. Istat used *multilingual RoBERTa* on labeled data from Basile et al. (2020) to experiment with the classification of hate speech in Italian tweets (Bruno et al. 2024a) and compared the results to an attention-based bidirectional LSTM (AT-BiLSTM), showing a significantly better performance of the former. It is, however, a matter of fact that supervised methods need a significant amount of labeled texts. For instance, for the task of hate speech detection against immigrants, Istat's training data comes from the EVALITA 2020 HaSpeeDe 2 task, which consists of roughly 7000 records. To improve the capability of the classifier, Istat labeled some of their most viral records. Texts likely to contain hateful tweets, i.e., those with offensive languages, such as *fate schifo* ("you suck") and *avete rotto i c****oni* ("you pi**ed us off"), were retrieved. This approach isolated 242,000 tweets, of which 67,000 were unique. Then, stratified sampling was carried out, an effective method for handling skewed distributions, using the total number of tweets as the target variable. The tweets were stratified into five classes based on the number of retweets, with the final class being a take-all stratum, resulting in 681 sampled texts, ensuring a coefficient of variation of 5%. These tweets were manually labeled, resulting in 225 hateful and 456 not hateful. Adding this dataset to the training led to more plausible results, at least in terms of average values, when the classifier was applied to the corpus of X posts.

12.4 Conclusion

Istat has successfully integrated big data processing pipelines into its statistical production system, marking a significant step toward modernizing official statistics. The maturity levels of these pipelines vary: some projects are still in the pilot stage, while others have progressed to the status of experimental statistics available on Istat's official website. A few projects have fully transitioned into operational production. At the core of this development is the institute's commitment to innovation, guaranteeing that new methodologies undergo thorough testing before they become operational.

Looking ahead, Istat is strategically investing in a fully developed production system based on nontraditional data sources. These investments will focus on strengthening cross-cutting capabilities, such as methodological frameworks, IT infrastructures, and governance models, as well as enhancing subject matter expertise in specific statistical domains. The ultimate goal is establishing robust and scalable processes capable of continuously integrating new data sources into statistical production.

Despite these advancements, several challenges remain. One of the key issues is the ongoing evaluation of accuracy and uncertainty in big data-driven statistics. Traditional statistical sources continue to play a crucial role, either by directly integrating with big data or serving as benchmarks for validation. While significant progress has been made in assessing the reliability of results across different projects, further refinement is still necessary. A potential path forward involves developing a generalized reference framework for evaluating big data-based statistical products. This framework would provide a standardized approach to ensuring methodological soundness and consistency across different applications.

In the coming years, Istat will continue to enhance its capabilities, address methodological challenges, and refine its production processes. Istat aims to reinforce its role in the European Statistical System by integrating big data and machine learning into official statistics. This will ensure that new data-driven insights contribute to informed policymaking and improve public understanding.

References

J. Aryal, C. Sitaula, S. Aryal, NDVI threshold-based urban green space mapping from sentinel-2A at the Local Governmental Area (LGA) level of victoria, Australia. Land **11**(3), 351 (2022)

G. Barcaroli, M. Scannapieco, M. Scarnò, D. Summa, Using internet as a data source for official statistics: a comparative analysis of web scraping technologies, in *Proceedings of Proceedings of the New Techniques and Technologies for Statistics Conference (NTTS)* (2015a)

G. Barcaroli, A. Nurra, S. Salamone, M. Scannapieco, M. Scarnò, D. Summa, Internet as data source in the Istat survey on ICT in enterprises. Austr. J. Stat. **44**(2), 31–43 (2015b)

G. Barcaroli, M. Scannapieco, D. Summa, On the use of internet as a data source for official statistics: a strategy for identifying enterprises on the web. Rivista Italiana di Economia Demografia e Statistica **70**(4), 25–41 (2016)

V. Basile, M. Nissim, Sentiment analysis on Italian tweets, in *Proceedings of the 4th Workshop on Computational Approaches to Subjectivity, Sentiment and Social Media Analysis* (2013), pp. 100–107

P. Basile, N. Novielli, et al., Uniba at evalita 2014-sentipolc task: predicting tweet sentiment polarity combining micro-blogging, lexicon and semantic features, in *Proceedings of EVALITA* (2014), pp. 58–63

V. Basile, M. Di Maro, D. Croce, L. Passaro, Evalita 2020: Overview of the 7th evaluation campaign of natural language processing and speech tools for Italian, in *CEUR Workshop Proceedings*, vol. 2765. CEUR-ws (2020)

F. Bianchi, D. Nozza, D. Hovy, FEEL-IT: Emotion and sentiment classification for the Italian language, in *Proceedings of the Eleventh Workshop on Computational Approaches to Subjectivity, Sentiment and Social Media Analysis*. Association for Computational Linguistics (2021)

P. Bojanowski, E. Grave, A. Joulin, T. Mikolov, Enriching word vectors with subword information. Trans. Assoc. Comput. Linguist. **5**, 135–146 (2017)

M. Bruno, E. Catanese, F. Ortame, Towards a hate speech index with attention-based LSTMs and XLM-RoBERTa, in *Proceedings of the Tenth Italian Conference on Computational Linguistics (CLiC-it 2024)* (2024a)

M. Bruno, E. Catanese, F. Ortame, F. Pugliese, Measuring social mood on economy during covid times: A BiLSTM neural network approach, in *International Conference on Learning and Intelligent Optimization* (Springer, Berlin, 2024b), pp. 305–317

G. Castellucci, D. Croce, R. Basili, Acquiring a large scale polarity lexicon through unsupervised distributional methods, in *Natural Language Processing and Information Systems: 20th International Conference on Applications of Natural Language to Information Systems, NLDB 2015, Passau, Germany, June 17–19, 2015, Proceedings 20* (Springer, Berlin, 2015), pp. 73–86

E. Catanese, M. Scannapieco, M. Bruno, L. Valentino, Natural language processing in official statistics: the social mood on economy index experience. Stat. J. IAOS **38**(4), 1451–1459 (2022)

E. Catanese, M. Bruno, L. Valentino, Quality enhancements in experimental statistics: The Italian social mood on economy index, in *New Frontiers in Textual Data Analysis*, ed. by G. Giordano, M. Misuraca (Springer, Berlin, 2024a), pp. 93–103

E. Catanese, G. Sacco, L. Valentino, A quantitative assessment of the impact of valence shifters and emoji in lexicon for Italian sentiment analysis. JADT 2024 Mots comptés, textes déchiffrés **1**, 179–188 (2024b)

S. Ceri, On the role of statistics in the era of big data: a computer science perspective. Stat. Probab. Lett. **136**, 68–72 (2018)

S. Chen, D. Haziza, Recent developments in dealing with item non-response in surveys: a critical review. Int. Stat. Rev. **87**, S192–S218 (2019)

S. Chen, D. Haziza, C. Léger, Z. Mashreghi, Pseudo-population bootstrap methods for imputed survey data. Biometrika **106**(2), 369–384 (2019)

A. Conneau, K. Khandelwal, N. Goyal, V. Chaudhary, G. Wenzek, F. Guzmán, E. Grave, M. Ott, L. Zettlemoyer, V. Stoyanov, Unsupervised cross-lingual representation learning at scale (2020). https://arxiv.org/abs/1911.02116

P.J. Daas, M. Roos, M. van de Ven, J. Neroni, *Twitter as a Potential Data Source for Statistics*. Statistics Netherlands (2012). http://www.pietdaas.nl/beta/pubs/pubs/DiscPaper_Twitter.pdf

M. Dagdoug, C. Goga, D. Haziza, Imputation procedures in surveys using nonparametric and machine learning methods: an empirical comparison. J. Surv. Stat. Methodol. **11**(1), 141–188 (2023)

L. De Benedictis, S. Nenci, G. Santoni, L. Tajoli, C. Vicarelli, Network analysis of world trade using the baci-cepii dataset. Global Econ. J. **14**(03n04), 287–343 (2014)

E. De Cristofaro, G. Tsudik, Practical private set intersection protocols with linear complexity, in *International Conference on Financial Cryptography and Data Security* (Springer, Berlin, 2010), pp. 143–159

F. De Fausti, M. Di Zio, R. Filippini, S. Toti, D. Zardetto, Multilayer perceptron models for the estimation of the attained level of education in the Italian permanent census. Stat. J. IAOS **38**(2), 637–646 (2022)

T. De Waal, WAID 4.1: a computer program for imputation of missing values. Res. Official Stat. **4**, 53–70 (2001)

J. Devlin, M.-W. Chang, K. Lee, K. Toutanova, BERT: Pre-training of Deep Bidirectional Transformers for Language Understanding (2019). https://arxiv.org/abs/1810.04805

M. Di Zio, M. Scanu, L. Coppola, O. Luzi, A. Ponti, Bayesian networks for imputation. J. R. Stat. Soc. Ser. A Stat. Soc. **167**(2), 309–322 (2004)

M. Di Zio, F. De Fausti, R. Filippini, S. Toti, D. Zardetto, The imputation of the "attained level of education" in the base register of individuals through neural networks using sampling weights, in *UNECE conference of European Statisticians, Expert Meeting on Statistical Data Editing* (2022). https://unece.org/statistics/events/SDE2022

G. Donchyts, J. Schellekens, H. Winsemius, E. Eisemann, N. Van de Giesen, A 30 m resolution surface water mask including estimation of positional and thematic differences using landsat 8, srtm and openstreetmap: a case study in the Murray-Darling Basin, Australia. Remote Sens. **8**(5), 386 (2016)

J. Drechsler, J. Bailie, The complexities of differential privacy for survey data (2024). https://arxiv.org/abs/2408.07006

C. Dwork, Differential privacy, in *International Colloquium on Automata, Languages, and Programming* (Springer, Berlin, 2006), pp. 1–12

J.R. Eastman, F. Sangermano, E.A. Machado, J. Rogan, A. Anyamba, Global trends in seasonality of normalized difference vegetation index (NDVI), 1982–2011. Remote Sens. **5**(10), 4799–4818 (2013)

B. Efron, Prediction, estimation, and attribution. J. Am. Stat. Assoc. **115**(530), 636–655 (2020)

A. Fronzetti Colladon, M. Naldi, Distinctiveness centrality in social networks. Plos one **15**(5), e0233276 (2020)

S. Hochreiter, Long short-term memory. *Neural Computation MIT-Press* (1997)

M. Jabbar, M.M. Yusoff, A. Shafie, Assessing the role of urban green spaces for human well-being: a systematic review. GeoJournal **87**, 1–19 (2022)

N. Johnson, J.P. Near, D. Song, Towards practical differential privacy for SQL queries. Proc. VLDB Endowment **11**(5), 526–539 (2018)

S. Kamara, P. Mohassel, M. Raykova, S. Sadeghian, Scaling private set intersection to billion-element sets, in *Financial Cryptography and Data Security: 18th International Conference, FC 2014, Christ Church, Barbados, March 3-7, 2014, Revised Selected Papers 18* (Springer, Berlin, 2014), pp. 195–215

H. Kühnemann, A. van Delden, D. Summa, J. Gussenbauer, A. Ils, K. Löytynoja, report: Url finding methodology (2022). https://cros.ec.europa.eu/system/files/2023-12/20220131_url_finding_methodology.pdf

P. Kilian, S. Ye, A. Kelava, Mixed effects in machine learning – a flexible mixedML framework to add random effects to supervised machine learning regression. *Transactions on Machine Learning Research* (2023). https://openreview.net/forum?id=MKZyHtmfwH

O. Kolchyna, T.T.P. Souza, P. Treleaven, T. Aste, Twitter sentiment analysis: Lexicon method, machine learning method and their combination (2015). https://arxiv.org/abs/1507.00955

K. Larbi, J. Tsang, D. Haziza, M. Dagdoug, On the use of machine learning methods for the treatment of unit nonresponse in surveys, in *Second Workshop on Methodologies for Official Statistics, Rome* (2023)

R.J. Little, Calibrated Bayes, an alternative inferential paradigm for official statistics. J. Official Stat. **28**(3), 309–334 (2012)

R.J. Little, Bayes, buttressed by design-based ideas, is the best overarching paradigm for sample survey inference. Surv. Methodol. **48**(2), 257–281 (2022)

L. Mencarini, D.I. Hernández-Farías, M. Lai, V. Patti, E. Sulis, D. Vignoli, Happy parents' tweets. Demograph. Res. **40**, 693–724 (2019)

C. Merrien, Worldwide list of seaports, version 2021 (2021). https://doi.org/10.12770/59ab5f6f-79ea-425d-830e-be5ecdb7bdbe

F. Morstatter, J. Pfeffer, H. Liu, K. Carley, Is the sample good enough? Comparing data from twitter's streaming API with twitter's firehose, in *Proceedings of the International AAAI Conference on Web and Social Media*, vol. 7 (2013), pp. 400–408

S. Mugnoli, A. Sabbi, F. De Fausti, G. Lancioni, F. Sisti, Quantification of urban green areas: An innovative remote sensing approach for official statistics, in *Second Workshop on Methodologies for Official Statistics, Rome* (2024)

S. Nordbotten, Neural network imputation applied to the Norwegian 1990 population census data. J. Official Stat. **12**, 385–402 (1996)

A. Pak, P. Paroubek, et al., Twitter as a corpus for sentiment analysis and opinion mining, in *LREc*, vol. 10 (2010), pp. 1320–1326

A. Pappagallo, F. Ortame, G. Massacci, F. Sisti, F. Pugliese, Deep learning for the classification of ports in maritime transport statistics via AIS data, in *International Conference on Learning and Intelligent Optimization* (Springer, Berlin, 2024), pp. 318–332

G. Pristeri, F. Peroni, S.E. Pappalardo, D. Codato, A. Masi, M. De Marchi, Whose urban green? mapping and classifying public and private green spaces in Padua for spatial planning policies. ISPRS Int. J. Geo-Inf. **10**(8), 538 (2021)

M. Puts, D. Salgado, P. Daas, Leveraging machine learning for official statistics, in *Foundations and Advances of Machine Learning in Official Statistics*, ed. by F. Dumpert, Chap. 2 (Springer, Berlin, 2025)

F. Ricciato, A. Wirthmann, K. Giannakouris, F. Reis, M. Skaliotis, Trusted smart statistics: Motivations and principles. Stat. J. IAOS **35**(4), 589–603 (2019)

W. Ruan, M. Xu, W. Fang, L. Wang, L. Wang, W. Han, Private, efficient, and accurate: Protecting models trained by multi-party learning with differential privacy, in *2023 IEEE Symposium on Security and Privacy (SP)*, IEEE (2023), pp. 1926–1943

H. Schulz-Kümpel, A.-L. Boulesteix, S. Fischer, R. Hornung, Challenges in resampling based performance estimation, in *Foundations and Advances of Machine Learning in Official Statistics*, ed. by F. Dumpert, Chap. 3 (Springer, Berlin, 2025)

M. Schuster, K.K. Paliwal, Bidirectional recurrent neural networks. IEEE Trans. Signal Process. **45**(11), 2673–2681 (1997)

G. Stateva, O. ten Bosch, D. Windmeijer, J. Maślankowski, G. Barcaroli, M. Scannapieco, D. Summa, M. Greenaway, I. Jansson, D. Wu, ESSnet Big Data deliverable 2.4: Final report on web scraping enterprise characteristics (2018)

Z. Steinert-Threlkeld, Twitter as data, in *Elements in Quantitative and Computational Methods for the Social Sciences* (Cambridge University Press, Cambridge, 2018)

J. Stock, O. Hauke, J. Weißmann, H. Federrath, The applicability of federated learning to official statistics, in *International Conference on Intelligent Data Engineering and Automated Learning* (Springer, Berlin, 2023), pp 70–81

M. Taboada, J. Brooke, M. Tofiloski, K. Voll, M. Stede, Lexicon-based methods for sentiment analysis. Comput. Linguist. **37**(2), 267–307 (2011)

United Nations, *Synthetic Data for Official Statistics: A Starter Guide*. United Nations (2023a). https://books.google.it/books?id=8jjTzwEACAAJ

United Nations, United Nations Guide on Privacy-Enhancing Technologies for Official Statistics (2023b). https://unstats.un.org/bigdata/task-teams/privacy/guide/2023_UN%20PET%20Guide.pdf

J. van den Brakel, E. Söhler, P. Daas, B. Buelens, Social media as a data source for official statistics; the Dutch consumer confidence index. Surv. Methodol. **43**(2), 183–210 (2017)

A. Vaswani, N. Shazeer, N. Parmar, J. Uszkoreit, L. Jones, A.N. Gomez, L.U. Kaiser, I. Polosukhin, Attention is all you need, in *Advances in Neural Information Processing Systems*, ed. by I. Guyon, U. von Luxburg, S. Bengio, H. Wallach, R. Fergus, S. Vishwanathan, R. Garnett, vol. 30 (Curran Associates, Red Hook, 2017)

A. Virgillito, D. Zardetto, M. Scannapieco, D. Summa, Istat's Reference Architecture for Internet as a Data Source for Official Statistics (2017). https://www.istat.it/wp-content/uploads/2018/09/Big-data-committee.pdf
V. Vovk, A. Gammerman, G. Shafer, *Algorithmic Learning in a Random World* (Springer, Berlin, 2005)
S. Wasserman, K. Faust, *Social Network Analysis: Methods and Applications* (Cambridge University Press, Cambridge, 1994)
J. Wieczorek, Design-based conformal prediction (2023). https://arxiv.org/abs/2303.01422
J. Wieczorek, C. Guerin, T. McMahon, K-fold cross-validation for complex sample surveys. Stat **11**(1), e454 (2022)
J. Xue, B. Su, Significant remote sensing vegetation indices: a review of developments and applications. J. Sens. **2017**(1), 1353691 (2017)
D. Zardetto, Using twitter data for the social mood on economy index, in *Atti della XIII Conferenza nazionale di statistica, Rome* (2018), pp. 4–6

Chapter 13
Streamlining Business Functions in Official Statistical Production with Machine Learning

Sandra Barragán, Adrián Pérez-Bote, Carlos Sáez, David Salgado, and Luis Sanguiao-Sande

13.1 The Production of Official Statistics, the New Data Ecosystem, Artificial Intelligence, and Quality

In the last two decades, the production of official statistics has increasingly been under stronger pressure to deliver "wider, deeper, quicker, better, cheaper" statistical products (Holt 2007; Brackstone 2007; Norwood 2007). This increasing pressure has grown along with the evolution of the economic, scientific, and technological progress of our society (see e.g. Radermacher 2020, Chapter 1).

Interestingly enough, the first electronic digital computer for civil usage produced in the United States was firstly devoted to the production of official statistics at the US Census Bureau (U.S. Census Bureau 2024). Nowadays, the computational power of statistical offices is far from that of, say, top IT companies. Before the Internet age, statistical offices used to concentrate comparatively vast amounts of data. In the last decade, we are living an increasing data deluge (Gleick 2011), and one of the most outstanding challenges for statistical offices is to access and use this data to reduce response burden, to gain efficiency, and to increase quality in the production of official statistics (DGINS 2013, 2018). More recently, the overwhelming deployment of applications based on artificial intelligence and machine learning (AI/ML) techniques clearly reveals the breed of experts excelling

The views expressed in this contribution are those of the authors and do not necessarily reflect the views of Statistics Spain (INE).

S. Barragán · A. Pérez-Bote · C. Sáez · D. Salgado (✉) · L. Sanguiao-Sande
Statistics Spain (INE), Madrid, Spain
e-mail: sandra.barragan.andres@ine.es; adrian.perez.bote@ine.es; carlos.saez.calvo@ine.es; david.salgado.fernandez@ine.es; luis.sanguiao.sande@ine.es

F. Dumpert (ed.), *Foundations and Advances of Machine Learning in Official Statistics*, Society, Environment and Statistics,
https://doi.org/10.1007/978-3-032-10004-7_13

in the use of data and statistical models outside the community of official statistics. The incorporation of data science, machine learning, and artificial intelligence into the routine production of official statistics is a current top priority in the modernisation of statistical offices (UNECE 2021).

In our view, the two main ingredients in this pressing environment are the new data ecosystem and the exponentially progressive success of artificial intelligence and machine learning, i.e. the use of new data sources and new statistical methods to increase quality in many of its dimensions. On the one hand, the new data ecosystem arises as a natural consequence of the digitisation driving the big data phenomenon and of the central role of data in the digital economy bringing the need of data governance, data management, and data stewardship in the construction of data spaces and similar data ecosystems. The integration of all kinds of digital transactional and administrative data together with survey data into the production of official statistics stands as a natural demand on official statistics (see e.g. Hand 2018).

On the other hand, the public explosion of artificial intelligence applications naturally poses the question on its use by statistical officers. This is currently an intense activity in official statistics understanding not only off-the-peg applications but especially the adaptation of statistical and deep learning models, in particular in the realm of finite population estimation problems (see contribution by Puts et al. 2025, in this same volume).

This combination of factors (data + AI/ML), in our view, constitutes a devilishly complex challenge for official statistics. Traditionally, the discipline of official statistics has been working with so-called microdata, i.e. basically data matrices of multiple variables per statistical unit (households, establishments, enterprises, etc.). The digitisation of the last decades is increasingly bringing into play so-called nanodata, i.e. transactional data with a much finer degree of entity breakdown. This is the data incrementally feeding more and more AI/ML systems already impinging on what and how statistical offices produce (or should now produce) and impacting on the role of official statistics in society. To take an example, if policymakers have nowadays access to this nanodata and data scientists can process and analyse it (see e.g. Guerrero and Margetts 2024), what is the role of official statistics produced by statistical offices?

Regarding the first factor, the adaptation to the new data ecosystem must be undertaken beyond doubt, with the subsequent transformation and modernisation of multiple data management aspects (see e.g. DAMA International 2017). However, data and statistics must not be confused (Reister 2023), and the main mission of official statistics, i.e. to provide a quantitative description of the connection between data and reality, including an uncertainty assessment (statistical inference), must prevail, all under a scrupulous fulfilment of legal regulations of statistical products (release calendar, accuracy, cost restrictions, territorial and sectorial breakdowns, data privacy and statistical confidentiality, etc.). In this renewed scenario, a revamped quality management and quality assurance must play a central role embracing old and new aspects providing knowledge and a standard reference for an increasingly datafied society (see e.g. Financial Times 2025, for far-reaching consequences regarding new data sources).

Regarding the second factor, by and large, we distinguish two broad approaches in the use of AI/ML techniques in the production of official statistics. Firstly, as in any other industry, any task or activity related to information processing should be subjected to an assessment for the potential adoption of an AI/ML tool increasing the cost efficiency, improving timeliness, and enhancing overall quality. This is the case, e.g. of statistical classification coding (with automatic coding systems), dissemination (e.g. with a chatbot), computer code production (with copilots and generative AI), etc. Secondly, however, as a specificity of the business of official statistical production, statistical inference must be paid due attention since it constitutes the critical core of the whole business. Regarding the first approach, the identification, adaptation, and adoption of AI/ML tools is just a matter of time and resource investment to come up with the best options. Regarding the second approach, which is indeed deeply connected to our conception of statistical quality (beyond data quality), it will require greater efforts: is design-based inference still the preferred choice for inference? How estimators should be improved with these new models? Do we need to change the inference paradigm? Are all underlying statistical assumptions in AI/ML valid for the inference problem in official statistics?

Here we present ongoing initiatives at Statistics Spain (INE) to use statistical learning models to improve the production of official statistics. Our approach is quality oriented, where we focus on critical aspects of statistical products such as timeliness, granularity, accuracy, efficiency, response burden, and frequency and seek improvements in different business functions using the versatile statistical learning models. For this chapter, we shall adopt the definition of business function provided by the Generic Statistical Information Model (GSIM) v2.0 (UNECE 2024): "Activities undertaken by a statistical organisation to achieve its objectives".

The chapter is organised as follows. In Sect. 13.2, we describe how to possibly improve core business functions such as estimation, editing, and classification coding in a more traditional fashion. In Sect. 13.3, we propose alternative uses of these models to execute novel business functions to improve different quality dimensions. In Sect. 13.4, we close with some conclusions.

13.2 Streamlining Traditional Business Functions

We propose improved business functions related to inference, editing, and classification coding.

13.2.1 Design-Based Predictive Inference

The official statistics approach to inference has been traditionally the design-based inference from probability samples (see e.g. Hansen 1987; Smith 1994; Kalton 2002; Rao 2005). This approach relies on a *known* sampling design and thus it is valid by construction. The usual alternative approach to inference is model-based inference, where the uncertainty of estimation is evaluated with respect to an *assumed* statistical model. Model-based inference often leads to more accurate results but might be invalid because of model misspecification.

In order to recover (at least part of) this loss of efficiency, a model-assisted approach has been proposed (Särndal et al. 1992; Wu and Sitter 2001; Breidt and Opsomer 2017), where an auxiliar model is explicitly formulated but inference remains design based. Model-assisted estimators are often not design unbiased, but design consistent asymptotically for a hypothetical sequence of populations of increasing sizes. However, design-unbiased model-assisted estimators have been also proposed (see Hartley and Ross 1954; Mickey 1959; Sanguiao-Sande and Zhang 2020).

In this section, we present a brief summary of the work by Zhang et al. (2025), where estimation is given by an arbitrary prediction algorithm, while the uncertainty measure (mean squared error) is design based. This separation between estimator and its properties is what makes this approach different from any model-assisted estimation.

Denote by $U = \{1, ..., N\}$ a given finite population of size N. Let $y_k, k \in U$, be the associated values of interest. Denote by $\mathbf{x}_k$, $k \in U$, the collection of feature vectors, where $\mathbf{x}_k$ is the vector associated with each unit $k \in U$. Given any sample of units from U, denoted by $s \subset U$, let $\mu(\mathbf{x}, s)$ denote the prediction given the feature vector $\mathbf{x}$ for a certain prediction algorithm trained on the sample s. Note that the algorithm can be model based or just any heuristic machine learning algorithm. When the target parameter is the population total $Y = \sum_{k \in U} y_k$, the prediction estimator of Y is given as

$$\hat{Y} = \sum_{k \in s} y_k + \sum_{k \in U \setminus s} \mu(\mathbf{x}_k, s). \tag{13.1}$$

Note that many design-based estimators in survey sampling can as well be given as prediction estimators. Of course, as μ is arbitrary, this prediction estimator is in general biased, so variance is not a good measure of its accuracy, and we need to estimate both bias and mean squared error. Unfortunately, it is not possible to measure the bias for the model trained on the full sample, since we lack out-of-sample units to compare with model predictions. This has already been noticed too for usual machine learning cross-validation (Bates et al. 2024).

Following the ideas by Sanguiao-Sande and Zhang (2020), denote by $s_1 \cup s_2 = s$ and $s_1 \cap s_2 = \emptyset$ a *training-test sample split*, where s_1 is selected by a *subsampling design*, denoted by $q(s_1 \mid s)$, and the sample s is selected according to a sampling design $p(s)$. Denote by $\mu(\mathbf{x}, s_1)$ the predictor obtained from the subsample s_1, in the same way as $\mu(\mathbf{x}, s)$ from s. Its error $\mu(\mathbf{x}_k, s_1) - y_k$ can be observed for any $k \in s_2$.

We shall refer to the sampling design that yields (s_1, s) as the *pq-design*, denote by

$$f_{pq}(s_1, s) = q(s_1 \mid s)p(s) = f(s \mid s_1)f(s_1) \tag{13.2}$$

where the last product indicates that, conditional on the training set s_1, one can view the test set s_2 as a probability sample from $U \setminus s_1$, according to which s can vary under the pq-design. In particular, for any $k \in U$, let

$$\pi_{2k} = \mathbb{P}(k \in s_2 \mid s_1) = \sum_{s \ni k, k \notin s_1} f(s \mid s_1) \tag{13.3}$$

be its conditional s_2-inclusion probability given s_1 under the pq-design.

Now, for a given subsample s_1 under the pq-design, the *subsample-trained* prediction estimator is

$$\hat{Y}_1^* = \sum_{k \in s} y_k + \sum_{k \in U \setminus s} \mu(\mathbf{x}_k, s_1)$$

Note that while μ is trained on the subsample, its predictions are used out of the full sample s. Now, applying Rao-Blackwellisation to $\hat{Y}_1^*$, we obtain the subsampling Rao-Blackwellised (SRB) prediction estimator:

$$\hat{Y}^{RB} = \sum_{k \in s} y_k + \sum_{k \in U \setminus s} \bar{\mu}(\mathbf{x}_k, s), \tag{13.4}$$

where

$$\bar{\mu}(\mathbf{x}_k, s) = \mathbb{E}_q \left[\mu(\mathbf{x}_k, s_1) \mid s\right] \tag{13.5}$$

As we point out below, this conditional expectation value in practice is computed using Monte Carlo approximations.

Note that this SRB prediction estimator is still a prediction estimator but for a slightly different algorithm $\bar{\mu}$. This new predictor is trained on the full sample. However, we cannot be sure whether it is more accurate or not than the estimator μ trained with the full sample. The advantage of using $\bar{\mu}$ over μ trained with the full sample is that it allows to get design-based estimators for both its bias and its MSE.

Let us denote $e_{1k} = \mu(\mathbf{x}_k, s_1) - y_k$ for any $k \notin s_1$; then

$$\hat{B}^{RB} = \mathbb{E}_q\Big(\sum_{k \in s_2} (\pi_{2k}^{-1} - 1)e_{1k}\Big) \tag{13.6}$$

is a (design-unbiased) estimator of the bias.

The construction of an MSE estimator is not so straightforward, but as shown by Zhang et al. (2025, Theorem 1), a design-unbiased estimator for the MSE of the SRB prediction estimator is

$$\text{mse}^{RB} = \mathbb{E}_q\{\hat{B}^2 - \hat{V}_s(\hat{B} \mid s_1) + \hat{V}_s\{B(s_2) \mid s_1\} \mid s\} - V_q(\hat{Y}_1^* \mid s) \tag{13.7}$$

where $\hat{B} = \sum_{k\in s_2}(\pi_{2k}^{-1} - 1)\{\mu(\mathbf{x}_k, s_1) - y_k\}$, and $\hat{V}_s(\hat{B} \mid s_1)$ is unbiased for

$$V_s(\hat{B} \mid s_1) = \sum_{k\notin s_1}\sum_{l\notin s_1}(\pi_{2kl} - \pi_{2k}\pi_{2l})\left(\frac{1}{\pi_{2k}} - 1\right)\left(\frac{1}{\pi_{2l}} - 1\right)e_{1k}e_{1l}$$

where $\pi_{2ij} = \mathbb{P}(i, j \in s_2 \mid s_1)$, and $\hat{V}_s\{B(s_2) \mid s_1\}$ is unbiased for

$$V_s\{B(s_2) \mid s_1\} = \sum_{k\notin s_1}\sum_{l\notin s_1}(\pi_{2kl} - \pi_{2k}\pi_{2l})e_{1k}e_{1l} \,.$$

Regarding the computation of the MSE estimator, note that:

1. The number of subsamples for exact Rao-Blackwellisation is $\binom{n}{n_1}$ where n_1 is the subsample size. Since each subsample requires to fit the model again, exact Rao-Blackwellisation is usually computationally intractable.
2. Thus, in practice, we have to replace exact RB with Monte Carlo RB by approximating conditional expectations by sample means. More concretely, we choose some positive integer $T \le \binom{n}{n_1}$ and approximate $\mathbb{E}_q\left[\mu(\mathbf{x}_k, s_1)\right]$ by $\frac{1}{T}\sum_{t=1}^{T}\mu(\mathbf{x}_k, s_1^{(t)})$ based on sample splits $(s_1^{(t)}, s_2^{(t)})$. We similarly approximate the conditional expectations appearing in mse^{RB} by a sample mean.
3. This increases the variance of both $\hat{Y}^{RB}$ and mse^{RB}.
4. While the Monte Carlo variance of $\hat{Y}^{RB}$ is usually small, it can be much bigger for mse^{RB}.
5. Decreasing n_1 often decreases the Monte Carlo variance of the MSE estimator. This is because the consequent increase in n_2 improves the accuracy of the (conditional on s_1) design-based MSE estimators, while it does not affect much the variance of mse^{RB} as it is based on the full sample.
6. The number of subsamples is usually fixed by the computational resources available, so it makes sense to tune n_1 to increase the efficiency of the MSE estimator. This might also affect the overall (not Monte Carlo) performance of $\hat{Y}^{RB}$, so n_1 should not be too small.

Let us see now the results from a simple example by Zhang et al. (2025). A population of size $N = 1000$ was generated by $y_k = \beta_1 x_{1k} + \beta_2 x_{2k} + \epsilon_k$ with IID $x_{1k} \sim \text{LogN}(1, 1)$, $x_{2k} \sim \text{Poisson}(5)$, and $\epsilon_k \sim N(0, \sigma^2/4)$, where σ^2 is the population variance of x_{1k}. Let s be given by simple random sampling without replacement from this fixed population, where $n = 100$. Let the mis-specified full-

Table 13.1 MSE estimation from 250 samples, $T = 1000$, $\mu(x, s)$ for $\hat{Y}$ and $\bar{\mu}(x, s)$ for $\hat{Y}^{RB}$, (training, test) set of size (n_1, n_2), RE against variance of the HT estimator

(n_1, n_2)	MSE($\hat{Y}$)	RE($\hat{Y}$)	MSE($\hat{Y}^{RB}$)	RE($\hat{Y}^{RB}$)	CV($\widetilde{\text{mse}}^{RB}$)
(98, 2)	386532.7	0.44	386632.4	0.44	3.48
(80, 20)	363613.9	0.41	363441.5	0.41	0.31
(70, 30)	362673.0	0.41	357146.9	0.41	0.21

sample predictor be $\mu(x_1, s) = a + x_1 b$, where (a, b) are the sample ordinary least square fit of y on x_1 only.

Table 13.1 shows the results of simulating MSE estimation based on 250 independent samples, given $T = 10^3$, where the subsampling q-design is a simple random sampling without replacement of s_1 from each realised sample given $n_2 = 2, 20, 30$. The MSE is simply the average squared error of either $\hat{Y}$ or $\hat{Y}^{RB}$ over the 250 samples, and the relative efficiency (RE) is the ratio between either MSE and the variance of the HT estimator. Notice that the three MSE($\hat{Y}$) here are all estimators of the same MSE, each using 250 independent samples, since $\hat{Y}$ depends only on s.

For n_2 up to 20 (or even 30), MSE($\hat{Y}^{RB}$) is practically equal to MSE($\hat{Y}$). The CV of the MC-MSE estimator $\widetilde{\text{mse}}^{RB}$ is drastically reduced by setting n_2 to 20 or 30 instead of 2. In comparison, the CV of the exact-RB MSE estimator mse^{RB} is 0.14 by simulation, whereas the CV of the HT variance estimator is 0.32. This confirms that setting n_2 to be 20 (or even 30) and using a larger but practical T would work satisfactorily for MSE estimation in this setup.

In terms of the choice of estimator, we notice that the mis-specified predictor $\mu(x_1, s) = a + x_1 b$ yields a design-based MSE that is less than half of the variance of the HT estimator, and the bias of $\hat{Y}$ or $\hat{Y}^{RB}$ is in this case a negligible part of the MSE. Finally, as mentioned before, there is no reason why one cannot adopt $\bar{\mu}(x, s)$ in (13.4), for which MSE estimation is unbiased, instead of using $\mu(x, s)$, as they show similar accuracy.

Some final remarks:

1. Design-based predictive inference from finite population probability sampling allows great flexibility in the choice of the estimator.
2. It provides design-based bias and MSE estimates that are valid independently from the assumptions of the predictor (if there were any).
3. So, actually, the model versus design controversy disappears.
4. Providing a design-based MSE estimator looks like a natural extension to traditional variance estimator.
5. While here we focused on total estimation, Zhang et al. (2025) also develop a design-based theory for estimations at the individual level.
6. It is a computationally demanding method, as it requires to reestimate the model multiple times.

The highly predictive power of statistical learning models paves again the way to explore the combination of statistical models with design-based inference for estimation in finite populations. As stated by Toth (2024), model-assisted estimation techniques allow us to reduce the dependence on modelling assumptions, can lower variance over traditional linear models (Särndal et al. 1992), and, as shown above, when combined with subsampling, emulate train-test splits to estimate and correct errors.

13.2.2 Selective and Macroediting

Statistical data editing is the production task devoted to detect and treat errors in order to gain quality for the estimation phase (de Waal et al. 2011). It may consume up to 20% of the production resources (see e.g. de Jonge and van der Loo 2018). The complexity of the design, development, and execution of so-called editing and imputation (E&I) strategies has been growing since the early years of clerical work (see e.g. UNECE 1994, 1997; EDIMBUS 2007; UNECE 2019). High-level editing business functions, i.e. editing modalities, are nowadays assembled to deal with the different types of error present in different statistics.

In this resource-intensive scenario, more efficient forms of data editing are continuously searched, and selective editing stands as a relevant choice for efficiency gains (Pannekoek et al. 2013). Selective editing (de Waal 2013) allows the statistician to concentrate high-valued resources on errors having a substantial influence on final estimates, thus increasing quality while being cost-efficient. The basics of this editing modality amounts to assigning a local (item) score $s_k^{(q)}$ to each variable $Y^{(q)}$, $q = 1, \ldots, Q$, under inspection, combining these into a global (unit) score $S_k = S(s_k^{(1)}, \ldots, s_k^{(Q)})$, computing a threshold t, and, finally, selecting units above the threshold t for interactive editing, i.e. selecting $\{k \in s \subset U : S_k \geq t\}$.

By and large, no accepted theory for selective editing exists. In fact, selective editing is an umbrella term for several methods to identify the errors that have a strong influence on final estimates. In our optimisation approach to selective editing (Arbués et al. 2013), local scores can be understood as conditional expectation values of measurement error models:

$$s_k = d_k \cdot \mathbb{E}_m \left[|Y_k^{\text{raw}} - Y_k^0| | \mathbf{Z}^{\text{aux}} \right], \tag{13.8}$$

where d_k denotes the design weight of unit k, Y_k^{raw} stands for the raw (unedited) value of target variable Y and Y_k^0 for its true value, $\mathbf{Z}^{\text{aux}}$ denotes the available auxiliary information, and m stands for the underlying assumed measurement error model.

This expression opens the door to using supervised statistical learning models provided a historic dataset with both raw and validated values for the same variable is available (see Forteza and García-Uribe 2025, for a different approach). Furthermore, the versatility of these models allows us to compute local score values for both categorical and continuous variables. In the former case, the identity (13.8) reduces to

$$s_k = d_k \cdot \mathbb{P}\left(Y_k^{\text{raw}} \neq Y_k^0 \middle| \mathbf{Z}^{\text{aux}}\right), \tag{13.9}$$

where $\mathbb{P}(Y_k^{\text{raw}} \neq Y_k^0|\mathbf{Z}^{\text{aux}})$ denotes the error probability for unit k in the variable under inspection. Thus, the problem reduces to a probability estimation problem, which we have approached using random forests. This choice is due to its good properties (versatility, performance, robustness, etc.); see e.g. Barreñada et al. (2024). Ultimately, model selection should be carried out investigating different choices (logistic regression, boosting, etc.). We have used this approach in the editing phase of the Spanish branch of the European Health Interview Survey for the so-called social class variable, which is indeed a subject matter aggregation of the ISCO occupation code. These error probabilities are estimated using a random forest on the set of variables used during the clerical revision of the questionnaires (age, sex, economic activity, educational degree, professional situation, job situation, etc.). The target variable in the random forest model is the measurement error indicator $I(y_k^{\text{raw}} \neq y_k^0)$, where the training data is taken from the historic dataset of raw and validated values in the survey. In actual production conditions, once a record is clerically revised, thus producing both a new pair of raw and validated values, the record enters into the training dataset for a new random forest updating to be applied to the new data collection batch. The (unrevised) questionnaires are then given an editing priority upon their collection every time a new data collection batch is processed so that clerical revision in the interactive editing modality is rationalised.

In Fig. 13.1, we include a visualisation of the prioritisation performance by the local score s_k in Eq. (13.9) for an arbitrary data collection batch. In the left figure,

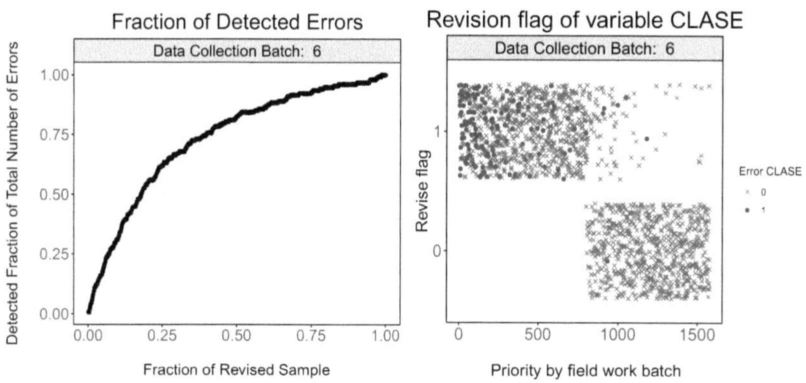

Fig. 13.1 (Left) Fraction of detected measurement errors in the categorical variable CLASE versus fraction of revised units according to the prioritisation provided by the local score (data collection batch no. 6). (Right) Revision flags (0 or 1) versus questionnaire (editing) priority from the score value s_k. Shape (cross or solid circle) indicates the absence or presence of error in each unit of data collection batch no. 6

we represent the fraction of detected errors as we revise questionnaires in the descending order derived from s_k in terms of the fraction of sampling units to revise. In the right figure, we represent those units marked for clerical revision (revision flag $= 1$) together with the absence/presence of errors upon inspection. The threshold was established by the subject matter experts after the first 5000 sampling units were systematically revised in this pilot experience based on the experimental finding that around the first 50% of the prioritised sample contained around 75% of the total number of erroneous units. The editing tasks were later completed during the macroediting phase.

This approach can also be adapted for continuous variables. Suppose the target variable Y represents the turnover of a statistical unit. Then, the measurement error $\epsilon_k = y_k^{\text{raw}} - y_k^0$ is indeed a semi-continuous variable, since it takes either the value 0 or a continuous value. Indeed, we can introduce (i) a binary variable $\delta_k^{(\epsilon)}$ taking the value 1 when $\epsilon_k \neq 0$ and the value 0 otherwise and (ii) a continuous variable $e_k^{(\epsilon)}$ with the non-null measurement error value when $\delta_k^{(\epsilon)} = 1$. Thus, we can write $\epsilon_k = \delta_k^{(\epsilon)} \cdot e_k^{(\epsilon)}$, and starting from Eq. (13.8), we can decompose

$$\begin{aligned} s_k &= d_k \cdot \mathbb{E}_m \left[\delta_k^{(\epsilon)} \cdot |e_k^{(\epsilon)}| \, |\mathbf{Z}^{\text{aux}} \right] \\ &= d_k \cdot \mathbb{E}_\delta \left[\mathbb{E}_e \left[\delta_k^{(\epsilon)} \cdot |e_k^{(\epsilon)}| \, |\delta_k^{(\epsilon)}, \mathbf{Z}^{\text{aux}} \right] \right] \\ &= d_k \cdot \mathbb{P}\left(\delta_k^{(\epsilon)} = 1 \middle| \mathbf{Z}^{\text{aux}} \right) \cdot \mathbb{E}_e \left[|e_k^{(\epsilon)}| \, |\delta_k^{(\epsilon)} = 1, \mathbf{Z}^{\text{aux}} \right]. \end{aligned} \tag{13.10}$$

We can then proceed in two steps. Again, assuming a historic dataset with both raw and validated values of this target variable is available for the same survey, firstly, we can estimate the error probability $\mathbb{P}(\delta_k^{(\epsilon)} = 1 | \mathbf{Z}^{\text{aux}})$. Then, in a second step, using those units with measurement error as training data, we can adjust a second regression model to predict $|e_k^{(\epsilon)}|$. As features for these models, we use both classification and other target survey variables and paradata as well as new variables derived thereof (Bohnensteffen 2020), i.e. all available information. The guiding principle is to consider the variables taking into account a golden-standard editing procedure.

In addition, the value y_k^{raw} can also be missing, which obliges us to propose a specific regression model for the measurement error $|\epsilon_k|$ of these units. Once the score values s_k are available for the entire sample, an editing priority in terms of their descending values can be assigned to each questionnaire.

In Fig. 13.2, we represent the reduction of the relative pseudo-bias in absolute value $ABS = \frac{|\hat{Y}(n_{(ed)}) - \hat{Y}^{(val)}|}{\hat{Y}^{(val)}}$, where $\hat{Y}(n_{(ed)}) = \sum_{k \in s_{ed}} d_k y_k$ and s_{ed} denotes the sample of units with n_{ed} revised units and $n - n_{ed}$ units with raw values and $\hat{Y}^{(val)} = \hat{Y}(n)$, in terms of the number of revised units in the priority order derived from the score values. These are the results of a pilot study with real data from the Spanish short-term business statistics called Service Sector Activity Indicators

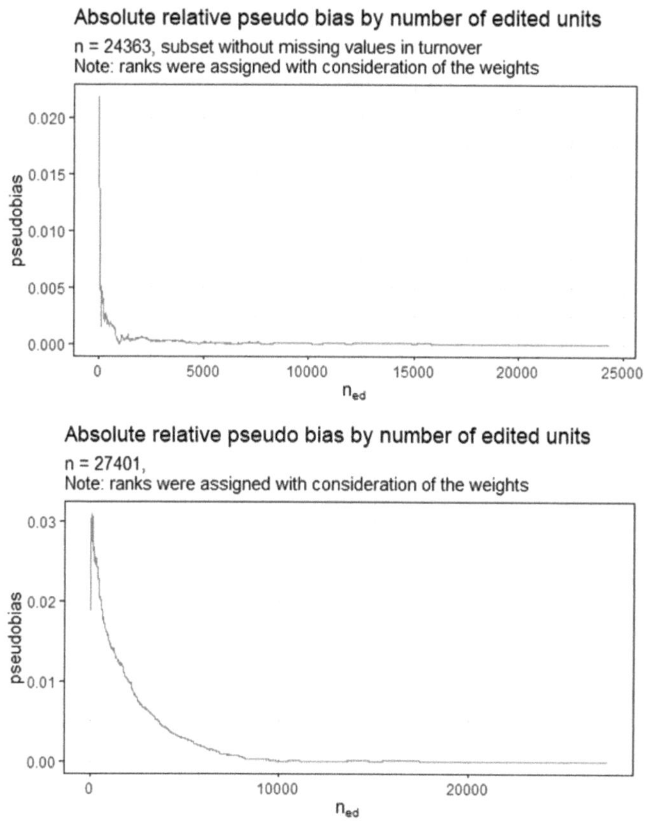

Fig. 13.2 (Top) Reduction of the relative pseudo-bias in absolute value in terms of the number of revised units. Units with missing values are not included in the list of units to revise. (Bottom) Reduction of the relative pseudo-bias in absolute value in terms of the number of revised units. Units with missing values are included in the list of units to revise. Results correspond to the consultations test set

Survey (Bohnensteffen 2020). You can observe that efficiency gains can be obtained by focusing on those units with large score values.

Some remarks:

- Historic datasets with raw and validated values are a rich source of information to predict measurement errors to gain efficiency during the editing phase. Both the data architecture and the process design and execution in a statistical office should be oriented towards their preservation and storage. Statistical learning models constitute a promising tool to dive into the data error patterns present in the process.
- Although hyperparameter optimisation and model selection are compulsory steps in statistical learning model adjustment, the ultimate goal in the production

process should drive the modelling exercise. In our case, random forests have provided enough quality to get some first efficiency gains using editing quality indicators as figures of merit (error detection rate, pseudo-bias, etc.).

- This proposal covers only the computation of local scores. Their combination to obtain global scores can be carried out as in the traditional approach. However, a more natural generalisation can be provided in the following way. For categorical variables, instead of estimating $\mathbb{P}(Y_k^{\mathrm{raw}} \neq Y_k^0 | \mathbf{Z}^{\mathrm{aux}})$ for a single variable Y, we can estimate $\mathbb{P}(Y_k^{(1)\mathrm{raw}} \neq Y_k^{(1),0} \vee \cdots \vee Y_k^{(Q)\mathrm{raw}} \neq Y_k^{(Q),0} | \mathbf{Z}^{\mathrm{aux}})$ for $q = 1, \ldots, Q$ variables, so that error detection in any of these variables is accomplished. For continuous variables, further research is needed.
- An indicator for the efficiency of the prioritisation in order to compare different methods is needed (e.g. related to the area under the curves in Fig. 13.2 see Arbués et al. 2013).
- Not only do efficiency gains impinge on the cost-efficiency dimension of production quality but also it allows the process to be more timely. We claim that the predictive power of machine learning techniques could in principle allow us to improve some editing business functions to the point that the editing process could a priori be redesigned to consider the production and release of some fast early estimates with influential errors under control (see also Sect. 13.3.1). Considering the Generic Statistical Data Editing Model (GSDEM) v2.0 (UNECE 2019) as the natural framework to design and develop editing and imputation strategies, any method involving a prediction will be improved by using these statistical learning models. Nonetheless, microdata quality in their final validated form is crucial for final estimation of the population quantity of interest (population total in this case), for training this kind of models, and for researchers and stakeholders accessing these data.

13.2.3 *Statistical Classification Coding*

A statistical classification is defined as a "hierarchically organised set of mutually exclusive and jointly exhaustive categories that share the same or similar characteristics, used for meaningfully grouping the objects or units in the population of interest" (UNECE 2024). It constitutes an extremely valuable statistical instrument extensively used by statistical offices as well as by other public organisations and even stakeholders for both statistical production and analysis.

They are commonly arranged in classification series, which in turn are grouped into classification families based on common concepts (UNECE 2024). The family of economic activity classifications probably conforms the most widely used example of statistical classification in the production of official statistics. We shall focus on this family in the Spanish context, which runs completely similar to many other countries.[1] The global reference is provided by the International Standard Industrial Classification of all economic activities (ISIC) (UNSD 2024), which is adapted to European needs in the *Nomenclature statistique des activités économiques dans la Communauté Européenne* (NACE) (Eurostat 2024), and it is further specialised for the Spanish context in the *Clasificación Nacional de Actividades Económicas* (CNAE) (Statistics Spain (INE) 2025a). This will be the focus of this section.

ISIC, NACE, and CNAE have undergone a thorough revision during the past years. This process has produced the ISIC Rev. 5, the NACE Rev. 2.1, and the CNAE-2025. A revision process of a statistical classification strongly impinges on many aspects of production and dissemination of official statistics. The core of this challenge is the need to assign economic activity codes in the new versions of the classification for all statistical units, hence also to aggregates and indices, in all produced and released statistics. Furthermore, in the Spanish context, the CNAE-2025 is intensively used for non-statistical purposes by other public bodies, especially the National Tax Agency and the Ministry of Social Security. Researchers and analysts also see themselves pushed to make use of the new classification. This fact, in turn, implies that citizens, researchers, and firms are compelled users of the classification.

The increasing use of administrative data for the production of official statistics, data thoroughly generated by administrative bodies, makes it advisable for statistical offices to focus on the quality of the generation process of economic activity codes. In this context, Statistics Spain has undergone two complementary strategic actions. On the one hand, the use of CNAE-2025 has been made legally compulsory to classify and code any economic activity variable in all public administrative registers (BOE 2025). This will ease the reuse of this data for official statistical purposes and, at the same time, will boost semantic interoperability in the Spanish Public Administration. On the other hand, Statistics Spain has developed and is maintaining and evolving an automatic coding tool named CodIA assisting in the selection of a CNAE-2025 activity code from a textual description of an economic activity (Statistics Spain (INE) 2025b).

CodIA has been built using Natural Language Processing (NLP) techniques. In the following, we shall describe the main aspects of technical and strategic interest of this process. Firstly, since the structure of statistical classifications is similar in different classification series (e.g. occupation and education), the coding tool should abstract the semantic content of the classification and benefit from its metainformation elements such as explanatory notes, introductory guidelines, and similar. In this sense, CodIA has been separately trained and tested using data regarding both the CNAE-2009 and the CNAE-2025 *mutatis mutandi*. In this sense, we shall always refer to CNAE embracing both versions, unless otherwise specified. This mindset allows us to face the challenge of building a tool which can be easily replicated for other statistical classifications and future versions of CNAE.

[1] See e.g. Beuter et al. (2025), Avouac et al. (2025), and Fiedler et al. (2025) in this book.

Secondly, the hierarchical structure in sections, divisions, groups, and classes can be approached either top-down or bottom-up. In the former, the task is to find the code for the section (one-digit code) and move downward providing then the code for the division (two-digit code), for the group (three-digit code), and finally for the class (four-digit code). In the latter, the task is to find the code for class and, making use of the nested hierarchy, to automatically provide the corresponding codes for the section, division, and group. After very short-ranged and preliminary tests suggesting so, for the interest of time, we decided to follow the bottom-up approach: CodIA provides CNAE codes for the class category of a given textual description of an economic activity.[2] These codes are provided with corresponding probability-like scores $s_c(text)$ for each returned class code c and an input text $text$. For nesting categories such as section, division, and group, the corresponding scores are elementarily computed by aggregation: $s_g(text) = \sum_{c \in g} s_c(text)$, where g denotes a group, and similarly for divisions and sections. Thus, results hereafter will focus on the class level, unless otherwise explicitly mentioned. More specifically, we shall focus on the CNAE-2025.

Thirdly, in the need to recode statistical units under the revised version of the classification, in the case of CNAE-2025, CodIA also accepts optionally the class activity code of CNAE-2009 as an input together with the textual description, thus assisting in recoding tasks.

At first sight, one might think that automatic coding is an already solved problem as a text classification application (see e.g. Kamath et al. 2019). We just need curated training sets and a state-of-the-art NLP model. One may even think that a large language model (LLM) may do it for us: "just ask ChatGPT". However, automatic coding of statistical classifications in the context of official statistical production presents some non-trivial issues:

- The task of classifying and coding descriptions of economic activities is hard even for a human expert for the following reasons:
  - The number of categories (classes) is very high (664 in the CNAE-2025).
  - The boundaries between classes are complex, subtle, and sometimes blurry.
  - Many descriptions are incomplete or ambiguous.
- Single classes (lowest hierarchical categories) contain indeed an unequal variety of economic activities.
- The coding tool must produce satisfactory results for different types of inputs depending on the needs of the user and/or the source of the textual description to be coded. In real production conditions, we may find descriptions with a single word or concept, while others may provide a long, detailed paragraph.

[2] See Beręsewicz et al. (2024) for results of a thorough comparison of both approaches in the context of the occupation classification family.

Table 13.2 Comparison of family of NLP models for automatic classification coding

Family of models	Main advantages	Main disadvantages
Bag-of-words + "classic" ML model	- Easy to implement - No confidentiality issues	- Performance far below state of the art
Fasttext (Joulin et al. 2017)	- Very easy to implement - Very short training time - No confidentiality issues	- Performance below state of the art
BERT-like (Liu et al. 2019)	- Almost state-of-the-art performance	- Training and deployment are time and resource demanding - Careful metaoptimisation is needed
Zero-shot LLM	- Very easy to implement	- Performance far below state of the art - Confidentiality issues
LLM + RAG	- Probably state-of-the-art performance	- Demanding in human effort, time, and computational resources
LLM + Fine-tuning	- State-of-the-art performance	- Training and deployment are time and hugely resource demanding

Thus, the first decision is to choose a family of NLP models. To this end, we carried out a comparison between the most popular options. A summary of the output of this analysis is presented in Table 13.2.

With this analysis in mind, taking into account (i) the scarcity of available computational resources, (ii) the time needed to collect and prepare enough training and test datasets, (iii) the deadline to deploy the solution, (iv) the robustness advised for technological solutions maintained in production conditions given the available resources, and (v) the inspiring experience and counselling by the French National Statistical Institute[3] (Faria and Seimandi 2023), we decided to use Fasttext. An additional consideration should be made here: BERT-like models represent a huge advancement over Fasttext, or other simpler and older models, because they incorporate the attention modules. However, we hypothesise that this advantage is reduced in our context because most texts are very short in the 1–10-word range.

Once the family of models was chosen, the next step was to build datasets for training and validation. By and large, pairs of text-code instances needed to be collected. Notice that differences must be taken into account between the CNAE-2009 and the CNAE-2025, because the latter is new and no data was available. For the CNAE-2009, data were collected from collection and editing paradata of existing official business statistics in Statistics Spain. Both structural and short-term business statistics were used. For the CNAE-2025, a twofold course of action was undertaken. On the one hand, taking advantage of the existing infrastructure

[3] We acknowledge the invaluable contact with Romain Lesur and INSEE's Data Science Lab. See https://ssplab.lab.sspcloud.fr.

for data collection, two large-sample structural business surveys were extended to include a specific questionnaire item requesting business units to provide their activity code according to the CNAE-2025. On the other hand, a specific ad hoc survey was implemented with a sample size of 110,000 business units with a 1:N correspondence between their CNAE-2009 and CNAE-2025 main activity codes. In this first step, in total, datasets amounting to around 420,000 instances were obtained. A test set with around 1000 instances was carefully revised by experts in order to ensure maximum coding quality. As a consequence, useless or incomplete descriptions were discarded.

At this point, textual descriptions were pre-processed; Fasttext metaparameters were optimised (validating the different combinations with the validation set), and a first Fasttext model was trained with the whole training and validation dataset. As an automatic coding tool providing just one single code for each textual description, CodIA reached a global accuracy of 0.63. A careful analysis of this performance, both quantitative and qualitative, showed that while the overall performance was not poorly bad so that this model could possibly work for manual coding assistance, it presented some weak points:

- The performance over minority classes was very poor. This did not heavily affect the overall metrics and use of the model because of the actual occurrence of these classes in the Spanish economy. However, it could erode the credibility of the coding tool.
- This model was unable to detect many concepts clearly related to one or few classes because they did not occur or occur very seldomly in the training set. Some of these concepts are, however, present in the titles and/or explanatory notes of the CNAE.

As a second step, thus, we decided to generate synthetic data to enrich the training dataset by applying three strategies:

- We added the titles and explanatory notes themselves, paired with the corresponding codes, as instances.
- We used LLMs to build a dictionary with word synonyms and we replicated the original textual descriptions by replacing some of its words by these synonyms.
- We asked LLMs via API to create synthetic economic activity descriptions. For this, we carefully tuned the prompt, including the title and explanatory notes of a given class as well as detailed instructions and examples to optimise the behaviour of the LLM.

Additionally, we trained the model initialised with pre-trained weights for the Spanish language, with the ambition that it would grasp semantic relations, including those not occurring in the dataset.

With all these initiatives, we got a dataset of about 2.5 million instances. After training the Fasttext model with the previous metaparameters over this larger dataset, the global accuracy as an automatic coder providing a single code was raised up to 0.69. A qualitative analysis showed that many of the problems listed above were softened, even despite the huge imbalance of the training dataset. The performance of the minority classes was still worse than the majority ones, and some unusual concepts were not treated correctly. However, it is worth noting that this model could be trained in a laptop in less than 10 minutes, which is very convenient for production conditions and model updating and retraining.

Finally, before deploying the tool into production, we undertook a stricter testing procedure. To this end, we collected user consultations received through Statistics Spain's corporate mailbox asking which code corresponded to a given provided description. It should be noted that these descriptions are much harder to classify and code than the average. These consultations, which are solved by classifications experts, are indeed motivated by this difficulty. The global accuracy over this combined dataset resulted in 0.53, while we got 0.41 when using the model trained only with real data. We could observe that the improvement of adding synthetic data is more acute in this case. We believe that the use of synthetic data for training language models is a strategic approach, which should be further pursued.

At this point, CodIA can be used as an automatic coder or as a coding assistant. As an automatic coder, it is configured to return the class code with the highest probability-like score. As a coding assistant, it is configured to return the class codes scoring above a chosen threshold. Notice that in either case, there will be textual descriptions possibly returning no class code because the model is configured to return class codes only if they are scoring above the threshold. With the choice of the threshold, the user can impose more or less stringent conditions on venturesome outputs.

To illustrate the performance of CodIA, we focus on four figures of merit:

- Precision, understood as the fraction of input instances resulting in the correct class code (automatic coder) or in a set of class codes containing the correct class code (coding assistant).
- Recall, understood as the fraction of input instances resulting in at least one class code. If recall equals 1, CodIA always returns at least one output class code.
- Threshold, selected by the classification expert to provide the minimum value of the probability-like scores of input instances producing an output class code. As an automatic coder, CodIA will produce either 1 or 0 output class code. As a coding assistant, it will produce 0 or more output class codes.
- Averaged number of output class codes, understood as the mean of the number of output class codes produced for each input instance. Notice that this figure of merit only makes sense when CodIA is used as a coding assistant.

The results on the combined dataset with real and synthetic data are illustrated in Fig. 13.3. The graph represents the performance of CodIA both as an automatic coder (lower curve) and as a coding assistant (upper curve). From left to right, results correspond to decreasing values of the threshold. Four threshold values are explicitly marked. Notice how CodIA diminishes its performance when forced to provide always a single output class code (lower curve). Notice also how it starts returning more than one output class code when the threshold is chosen below a given value. Indeed, as a coding assistant, the lower the threshold, the higher the number of output class codes. You can observe that with a very low threshold, it always returns

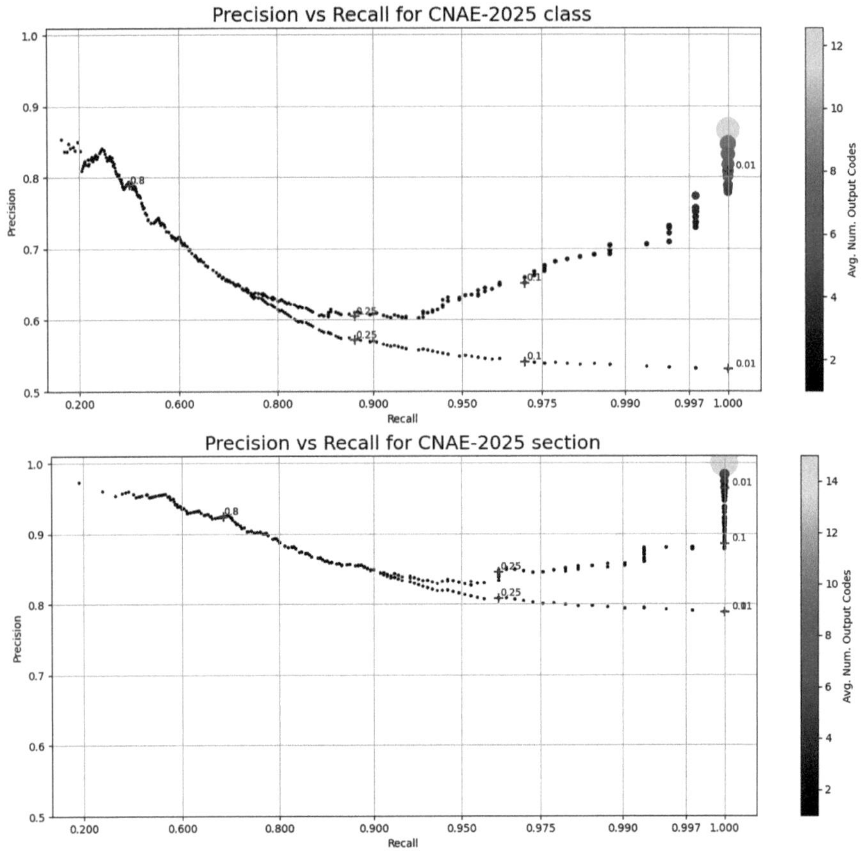

Fig. 13.3 Results of CodIA as an automatic coder and as a coding assistant (see text) for CNAE-2025 class codes (above) and section codes (below). Notice the logarithmic-like scale in the x-axis

output class codes (recall equals 1) providing 8 output class codes and 14 output section codes on average, out of which around 85% and 94%, respectively, contain the correct answer.

With these results, CodIA was deployed into production (Statistics Spain (INE) 2025a) on January 15, 2025. In the first 20 days, more than 20,000 requests were processed in Statistics Spain's website. A sample of them was manually labelled and the preceding four figures of merit were recomputed. Results are illustrated in Fig. 13.4.

To sum up, NLP techniques constitute a promising breadth of tools to handle statistical classifications, already producing results to improve classifying and coding activities. More research and analyses need to be carried out to disentangle and understand challenges as the construction of high-quality training sets or the blurry limits between textual descriptions of some economic activities with different

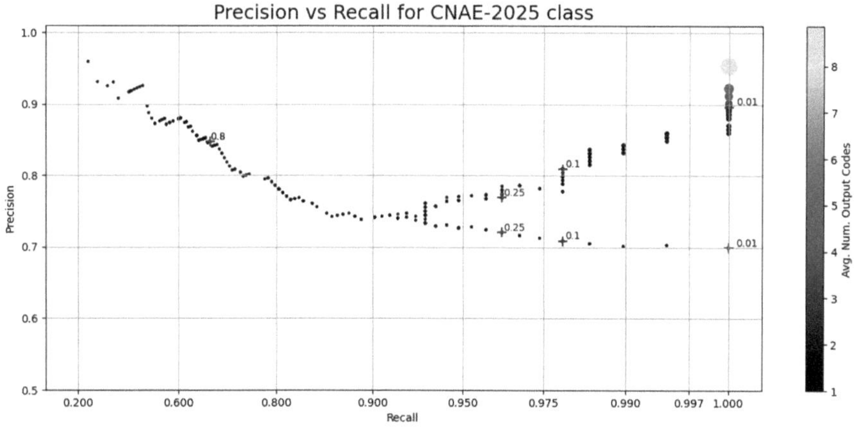

Fig. 13.4 Results of CodIA as an automatic coder and as a coding assistant in production for CNAE-2025 class codes. Notice the logarithmic-like scale in the x-axis

categories. Although the choice of language models is important, we detect a need to clearly quantify and unravel the quality of this type of data for statistical purposes.

13.3 New Business Functions for More Granular, Frequent, and Timely Statistics

Traditional demands on official statistics, underlined in crisis periods, e.g. in the last financial crisis and COVID pandemic, are the production and release of more granular, more frequent, and more timely statistics (Holt 2007; UNECE 2023). These demands are usually at the root of the proposals to use new data sources, especially big data and any other sort of digital data. This situation is especially visible and critical when producing survey statistics. The production of survey statistics requires in GSBPM's terminology to periodically collect, to process, to analyse, and to disseminate microdata and aggregates, where lower-level business functions are to be executed. To improve granularity directly, we would require larger sample sizes with a prohibitive cost. To increase frequency, we would need microdata to be more time disaggregated (sometimes not even possible because of the definition of the target variables themselves). To improve timeliness, we would need to execute these business functions at an extraordinary speed (e.g. could we request data providers to deliver their response to statistical offices in the first week of each month?).

In this section, we present proposals and proofs of concepts to improve these quality dimensions using survey data or a combination of survey and administrative data. The core of the proposals is the use of statistical learning models with survey and administrative data (already present and available in most statistical offices) to overcome the slow delivery of statistics based on them. This paves the way for their integration with new digital data sources, a priori more frequent and more granular.

13.3.1 Early Imputation

Short-term business statistics constitute a key set of indicators produced by statistical offices to monitor different aspects of the economic situation in a country. They are subjected to a high pressure to inform about the immediate state of the economy. When produced using survey data, the statistical process, broadly speaking, needs to execute data collection, data editing, estimation and index computation, and dissemination and communication. For monthly statistics, data collection takes 3–4 weeks at least, to be followed by editing and the computation of indices.

In this line of thought, we propose to make use of statistical patterns in the historical microdata to predictively construct every monthly sample along with the execution of the data collection and data editing processes so that we have a synthetic microdata (survey + predicted) set at every time together with an assessment of the uncertainty. In this way, we can compute an early estimate of all those indices to be released later when the whole production process finishes.

We apply this proposal to the Spanish Industrial Turnover Index, subjected to European Regulation in the European Statistical System (EC Council 1998). For ease of reading, Table 13.3 contains the list of symbols used in this section. Let z_k^{my} denote the value of the turnover of an industrial establishment $k \in s^{my}$ for the reference time period with month m and year y from sample s^{my}. Firstly, we need to introduce the time dependence $z_k^{my}(\tau)$ within the data collection period measured in number of days so that $\tau = d$ stands for the date $my + d$ days (d days after the reference time period is over). We notice the following:

1. The value $z_k^{my}(\tau)$ evolves along with the data collection and data editing from a missing value to a validated value going through different editing status (raw, edited during collection, edited interactively, etc.).
2. If we denote the press release date by d_{release}, when $\tau \geq d_{\text{release}}$, it turns out that we have non-response or the value could not have been possibly validated or the value is finally validated. In the first two cases, imputation is applied to obtain a synthetic value $\hat{z}_k^{my}(\tau)$ under a given imputation model.
3. We shall denote by $s^{my}(\tau)$ the set of sampling units having provided response up to time τ (response set) and $\hat{s}^{my}(\tau) = s^{my} - s^{my}(\tau)$.
4. In practice, three batches are made available to subject matter experts and processed by them during data collection at $\tau = 20, 29, 38$. The press release is published at $\tau_{\text{release}} = d_{\text{release}} = 51$ on average.
5. Occasionally, responses arrive after the release date. These data are also processed and validated for revised versions of the indices. We shall label values with this final state of validation by $\tau = \tau_f$.

Table 13.3 Notation

s^{my}	Sample of industrial establishment for the reference time period my with month m and year y
$s^{my}(\tau)$	Response subsample up to time τ for the reference time period my
$\hat{s}^{my}(\tau)$	Non-response subsample up to time τ for the reference time period my
$\bar{s}^{my}(\tau)$	Union sample of units $s^{my}(\tau)$ and $\widehat{s}^{my}(\tau)$
z_k^{my}	Value of the turnover of an industrial establishment $k \in s^{my}$
$z_k^{my}(\tau)$	Value of the turnover of an industrial establishment $k \in s^{my}$ at day τ after the reference time period my is over
$\widehat{z}_k^{my}(\tau)$	Predicted value of $z_k^{my}(\tau)$ according to a given imputation model
$\bar{z}_k^{my}(\tau)$	Synthetic value of $z_k^{my}(\tau)$ amounting to $z_k^{my}(\tau)$ or $\hat{z}_k^{my}(\tau)$ depending on the unit k
$\mathcal{F}_{\leftarrow}^{my}$	Set of validated turnover values for reference time periods prior to my
$\mathcal{F}_{\updownarrow}^{my}(\tau)$	Set of collected and partially edited turnover values for reference time period my up to day τ
$\mathcal{I}_{\leftarrow}^{my}$	Set of territorial and sectorial classification and collection paradata values for reference time periods prior to my
$\mathcal{I}_{\updownarrow}^{my}(\tau)$	Set of territorial and sectorial classification and collection paradata values for reference time period my up to day τ
$\mathcal{F}^{my}(\tau)$	Set of values of $\mathcal{F}_{\leftarrow}^{my}$, $\mathcal{F}_{\updownarrow}^{my}(\tau)$, $\mathcal{I}_{\leftarrow}^{my}$, and $\mathcal{I}_{\updownarrow}^{my}$
$\bar{\mathcal{F}}_{\leftarrow,\mathrm{unit}}^{my}$	Set of features for each unit $k \in s^{my}$ constructed individually with variables of each unit k
$\bar{\mathcal{F}}_{\leftarrow,\mathrm{aggr}}^{my}$	Set of features for each unit $k \in s^{my}$ constructed as an aggregated measure over a group containing of each unit k and using values prior to reference time period my
$\bar{\mathcal{F}}_{\updownarrow,\mathrm{aggr}}^{my}(\tau)$	Set of features for each unit $k \in s^{my}$ constructed as an aggregated measure over a group containing of each unit k and using values of reference time period my
$\bar{\mathcal{F}}^{my}$	Union set of features $\bar{\mathcal{F}}_{\leftarrow,\mathrm{unit}}^{my}$, $\bar{\mathcal{F}}_{\leftarrow,\mathrm{aggr}}^{my}$, $\bar{\mathcal{F}}_{\updownarrow,\mathrm{aggr}}^{my}(\tau)$
$\mathcal{D}^{my}(\tau)$	Union set of features $\bar{\mathcal{F}}^{my}$ and the validated turnover values for units of s^{my}
$\mathcal{D}_{train}^{my}(\tau)$	Subset of $\mathcal{D}^{my}(\tau)$ with values up to reference time period $(my) - 2$ (2 months prior to my)
$\mathcal{D}_{test}^{my}(\tau)$	Subset of $\mathcal{D}^{my}(\tau)$ with values of reference time period $(my) - 1$ (1 month prior to my)
$Z_{U_d}^{my}$	Turnover population domain total for domain d and reference time period my
$\widehat{Z}_{U_d}^{my}$	Estimator of $Z_{U_d}^{my}$

6. No consideration about the sample selection or the index computation has been done so far. We are focusing on the microdata level for the time being.

The core idea is to consider, for each reference time period my in turn, all past monthly validated microdatasets $\mathcal{F}_{\leftarrow}^{my} = \{z_k^{\bar{m}\bar{y}}(\tau_f)\}_{k\in s^{my}}^{\bar{m}\bar{y}<my}$, the collected and partially edited values $\mathcal{F}_{\updownarrow}^{my}(\tau) = \{z_k^{my}(\tau)\}_{k\in s^{my}(\tau)}$ at $\tau = 20, 29, 38$ and territorial and sectorial classification microdata and collection paradata of each statistical unit

$\mathcal{I}_{\leftarrow}^{my} = \{\mathbf{x}_k^{\bar{m}\bar{y}}\}_{k \in s^{my}}^{\bar{m}\bar{y}<my}$ and $\mathcal{I}_{\updownarrow}^{my} = \{\mathbf{x}_k^{my}\}_{k \in s^{my}(\tau)}$. These include continuous, semi-continuous, and categorical variables.

Next, we define $\mathcal{F}^{my}(\tau) = \mathcal{F}_{\leftarrow}^{my} \bigcup \mathcal{F}_{\updownarrow}^{my}(\tau) \bigcup \mathcal{I}_{\leftarrow}^{my} \bigcup \mathcal{I}_{\updownarrow}^{my}$. Notice that values from the past history are already validated (possibly not in their final form, but mostly) in preceding executions of the production process and values from the current time period correspond to responding units. Thus, no missing values are present in the dataset.

For each unit $k \in s^{my}$ and groups of units, we carry out different computations to obtain three kinds of derived variables so that the set of features $\bar{\mathcal{F}}^{my}(\tau)$ can be decomposed as $\bar{\mathcal{F}}^{my}(\tau) = \bar{\mathcal{F}}_{\leftarrow,\text{unit}}^{my} \bigcup \bar{\mathcal{F}}_{\leftarrow,\text{aggr}}^{my} \bigcup \bar{\mathcal{F}}_{\updownarrow,\text{aggr}}^{my}(\tau)$, where:

- $\bar{\mathcal{F}}_{\leftarrow,\text{unit}}^{my}$ contains features for each unit $k \in s^{my}$ constructed with variables of the same unit. For example, ${}_1\bar{x}_k^{my}$ can be the moving average for the past 3 months, thus only involving values from the past ($\leftarrow$) of the same unit k (unit).
- $\bar{\mathcal{F}}_{\leftarrow,\text{aggr}}^{my}$ contains features for each unit $k \in s^{my}$ constructed as an aggregated measure over a group which unit k belongs to and using values only from the past ($\leftarrow$). For example, ${}_2\bar{x}_k^{my}$ can be the $95th$ percentile of the moving averages ${}_1\bar{x}_k^{my}$ over the economic activity class (four-digit classification code) which unit k belongs to.
- $\bar{\mathcal{F}}_{\updownarrow,\text{aggr}}^{my}(\tau)$ contains features for each unit $k \in s^{my}$ constructed as an aggregated measure over a group which unit k belongs to and using values from the current time period ($\updownarrow$) at day τ. For example, ${}_3\bar{x}_k^{my}$ can be the $95th$ percentile of the turnover distribution of $z_k^{my}(\tau)$ over the economic activity class which unit k belongs to.

A detailed list of features and their coding for categorical variables is provided by Barragán et al. (2022). Notice that at time τ, features can be assigned for all units $k \in s^{my}$, since it does not require the values z_k^{my}. This set of features is completed with the target values $\{z_k^{my}(\tau_f)\}_{k \in s^{my}}$ to constitute the dataset $\mathcal{D}^{my}(\tau)$.

Next, we divide $\mathcal{D}^{my}(\tau)$ into the training data[4] $\mathcal{D}_{train}^{my}(\tau) = \bigcup_{\bar{m}\bar{y} \leq (my)-2} \mathcal{D}^{\bar{m}\bar{y}}(\tau)$ and test data $\mathcal{D}_{test}^{my}(\tau) = \mathcal{D}^{(my)-1}(\tau)$. We train a model on $\mathcal{D}_{train}^{my}(\tau)$, perform model selection on $\mathcal{D}_{test}^{my}(\tau)$, retrain the model on $\mathcal{D}_{train}^{my} \bigcup \mathcal{D}_{test}^{my}$, and apply it on $\mathcal{D}^{my}(\tau)$ to compute the predicted values $\{\hat{z}_k^{my}(\tau)\}_{k \in s^{my}}$. Details are given by Barragán et al. (2022).

The synthetic set $\bar{s}^{my}(\tau) = s^{my}(\tau) \bigcup \hat{s}^{my}(\tau)$ constitutes an anticipated version of the sample constructed using predicted values with the available information at time τ and data patterns from the past. This is the key idea: statistical offices potentially have enough microdata from short-term business statistics as to use statistical learning models to keep frequently updated anticipated versions of samples. These samples can now be used as in the traditional context to produce the corresponding aggregates.

[4] We overload the notation by writing $(my) - k$ to denote the reference time period k months before (my). Notice that this is easily translated into many programming languages following the object-oriented paradigm.

For the Spanish Industrial Turnover Index, the sample is selected by cut-off (Statistics Spain (INE) 2018), and the turnover total in population domain U_d (geographic, sectorial, etc.) is estimated by $\widehat{Z}_{U_d}^{my} = \sum_{k \in s_d^{my}} z_k^{my}$, where $s_d^{my} \subset U_d^{my}$ is the cut-off sample. At any time τ, this total can be early estimated by

$$\widehat{Z}_{U_d}^{my}(\tau) = \sum_{k \in \bar{s}_d^{my}} \bar{z}_k^{my}(\tau) = \sum_{k \in s_d^{my}(\tau)} z_k^{my}(\tau) + \sum_{k \in \hat{s}_d^{my}(\tau)} \hat{z}_k^{my}(\tau). \tag{13.11}$$

At any time τ, the total estimates (13.11) are ready to be used in the computation of econometric indices such as the fixed-base Laspeyres index used in the Spanish Industrial Turnover Index Survey (Statistics Spain (INE) 2018), producing thus early estimates of the same collection of indices and variation rates routinely produced and released monthly.

Our preliminary proof of concept focuses on a first implementation of the key idea above to construct the anticipated sample s^{my} for every month my over 5 years and compare the early estimates with the final released values and with the predicted values not using cross-sectional data $\mathcal{F}_{\updownarrow}(\tau)$ from each current month (initial estimate) (see Figs. 13.5 and 13.6).

Uncertainty in the predicted values for each unit and for the early aggregate estimates can be accounted for. Firstly, we focus on the conditional mean squared

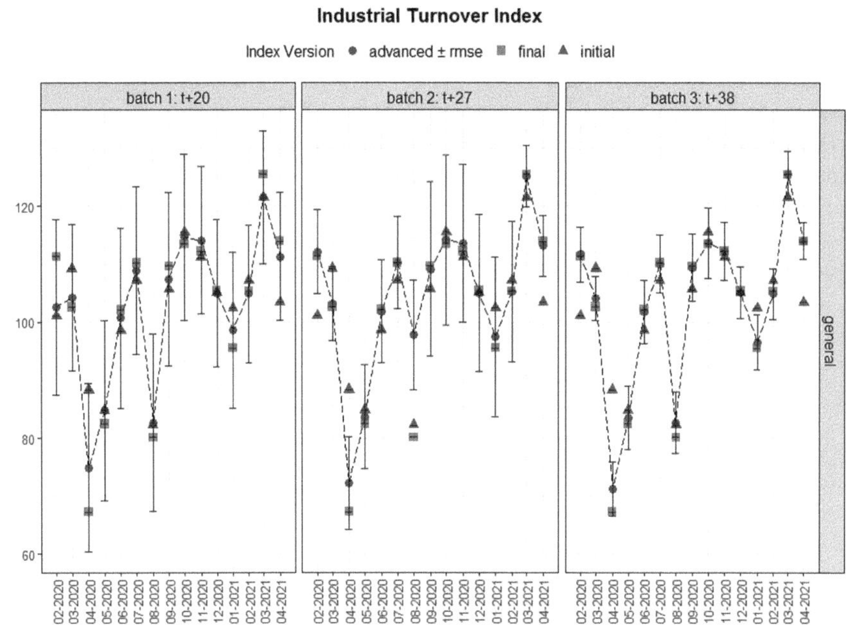

Fig. 13.5 Early estimates of the national Spanish Industrial Turnover Index from February 2020 to April 2021

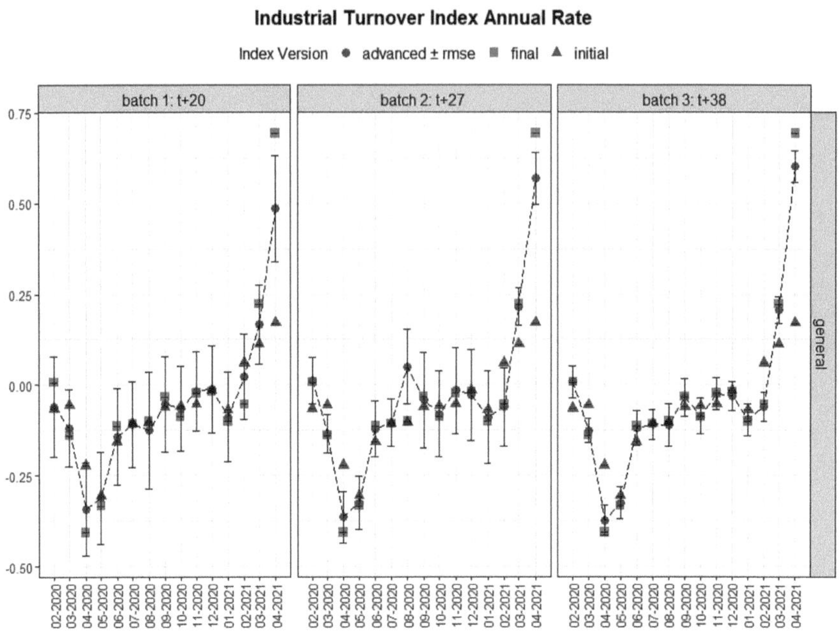

Fig. 13.6 Early estimates of the annual variation rates of the national Spanish Industrial Turnover Index from February 2020 to April 2021

error, i.e. on the uncertainty for a given sample $s^{my}(\tau)$ regarding the model prediction for these units and this time period. As a first immediate choice, neither do we assume independence and equal distribution nor exchangeability among the industrial establishments and estimate prediction errors at the current time period individually for each unit k using prediction errors from the past (see Barragán et al. 2022). The mean squared error for the total estimates (13.11) is computed by direct aggregation. We then use this aggregate as a first immediate figure of merit to account for the uncertainty in the early estimate (see Figs. 13.5 and 13.6 for a graphical representation of this aggregate).

We comment on methodological and strategic aspects regarding this proposal and its implementation for official statistical production:

1. The representation learning step is central in the prediction exercise. The construction of the features set $\mathcal{D}^{my}(\tau)$ from the original variables $\mathcal{F}^{my}(\tau)$ has been conducted in very close collaboration with domain experts to incorporate as much subject matter knowledge as possible (distribution tails, sectorial breakdowns, etc.) by defining and computing new features to incorporate into the prediction model. The possibility of using deep learning for this step remains to be explored.

2. No special attention at all has been paid to model selection and hyperparameter optimisation, which can and should be exhaustively improved in real production conditions. There is ample room for improvement in this course of action.
3. Uncertainty has been elementarily accounted for, only minimally avoiding the identical distribution or exchangeability assumptions. More sophisticated uncertainty assessments must be explored (e.g. conformal prediction).
4. In line with the avoidance of the identical distribution or exchangeability assumptions, outliers arise as an extraordinary challenge to predict, especially regarding large units, which strongly influence the indices. No special treatment has been explored and this needs to be addressed in real production conditions.
5. The traditional production process needs some adjustments, for example, by making collected and validated microdata as early available for model updating as possible, moving from a batch-based perspective to a continuous updating of the systems.
6. In this line, the microdata validation process is crucial for a high-quality algorithm training. Again, editing and validation business functions need to be accommodated to incorporate subject matter knowledge as automatically and early as possible. This will reduce the chance of model drift and model deterioration.
7. If we can provide reliable predicted values for each statistical unit, we can reformulate the sample selection problem as the problem to select those units which allow us to maintain the model quality so that not every single unit must be required to provide a response every single time period. The goal of the sample selection is not the final aggregate estimate but the quality of the prediction model. Therefore, we could reduce response burden.
8. Survey data do not play a central role in this exercise and a similar approach could be considered for administrative data. However, access to these data needs to be restated. Usually, statistical offices have access to these data after the reference period is over and after some pre-processing and revision tasks are concluded by their holders. Data access and use are usually approached in batches. Timeliness could be improved should offices have earlier and continuous access to administrative data as soon as they are generated and/or collected.

13.3.2 Imputation Beyond the Sample

One of the current demands from the users of official statistics is the need to improve the granularity of the statistics, providing more disaggregated information while maintaining a good quality. Moreover, there is also the need to reduce the response burden and costs of the operation. Clearly, there arises a trade-off. Most efforts to incorporate new data sources into the production of official statistics aim at seeking a solution for the trade-off. In order to achieve these goals, we need to leverage the large amount of administrative data available nowadays in order to provide better estimates with a smaller sample.

Our proposal consists in constructing the microdata for all the population. Acknowledging the highly predictive power of statistical learning models, especially in data-rich environments such as those arising from the access to administrative registers, we shall use machine learning techniques to impute the values of the variables of interest for the entire target population. While this construction of population microdata has several potential uses, in this section, we will focus on its use for the estimation of population totals. By using machine learning models, we hope to make use of the auxiliary information in a most accurate way to obtain good quality microdata for all the units in the population and hence to be able to provide high-quality estimates at a finer degree of granularity.

Let us describe the basic idea behind our proposal. Suppose we have some population U and for each unit $k \in U$ we have available some auxiliary data $\mathbf{x}_k$ (for instance, administrative data) and some variables of interest y_k, which we want to impute for all the population.[5] Suppose moreover that we have available a sample s obtained from U by probability sampling and that we know the value of y_k for $k \in s$. Then, we can proceed as follows. We use the sample s to select and train some model $\mu(\mathbf{x}, s)$ for the target variable y in terms of the auxiliary data $\mathbf{x}$. Then, for each unit k in the whole population, we can impute the value of y_k by $\hat{y}_k = \mu(\mathbf{x}_k, s)$. Finally, suppose we want to provide an estimate of the population total of the variable y. One can do it straightforwardly by using a prediction estimator, which just uses the imputed values outside the sample:

$$\hat{Y}^{pred} = \sum_{k \in s} y_k + \sum_{k \in U \setminus s} \mu(\mathbf{x}_k, s).$$

There are at least two issues that arise in this naive approach. First, there is the problem of how to choose an appropriate model for each target variable. This entails both deciding the kind of model to be used (linear models, tree-based models like random forest or XGBoost, neural networks, etc.) together with its hyperparameters and training the model. It is important to note that, in order to give the best possible estimate for the population total, we should select the model that minimises the squared total error:

$$E^2 = (Y - \hat{Y})^2,$$

which will not necessarily agree with the model that gives the more accurate individual predictions. Another important issue that needs to be addressed in this approach is that of uncertainty quantification. We want not just an estimate for the population total, but also a measure of the quality of that estimate (for instance, an estimate of the MSE of the total estimator).

[5] As explained below, imputed values for units in the sample with known value of y_k are used for better estimation performance.

Both problems are solved by using the algorithm-assisted[6] estimation paradigm described in Sect. 13.2.1. As described there, we can make use of the sample s in order to select and train the appropriate model. The same paradigm also provides us with an unbiased estimate of the MSE of the total estimator.

In some cases, we could even use prediction estimators based on pretrained models (for instance, models selected and trained in a sample of the previous year) to provide predictions of the population totals for the variables of interest. In this case, if $\mu(\mathbf{x})$ is our pretrained model for y, the prediction estimator would be given by

$$\hat{Y}^{pred} = \sum_{k \in U} \mu(\mathbf{x}_k).$$

This would allow us to obtain estimates for the population totals without using any sample. However, one has to keep in mind that this approach has several drawbacks:

1. We expect some model drift as time passes, meaning that the model performance degrades over time. By using the same model at different points in time, we are assuming the hypothesis that the model is not changing in a significant way. Moreover, if we have no available sample at the current time, we have no way to assess if this hypothesis is true.
2. If we have no sample, we have no way to obtain a reliable quality indicator of the estimates. Again, we must rely on the assumption that the quality is similar to that of the last time where we had some sample.
3. In the situation where we have an available sample, we can use the values of the sampled units in our estimation. This can improve substantially our estimation, especially in the presence of an exhaustive stratum, which contains the most influential units. On the other hand, when we have no sample, we have to rely on the imputed values for all the units of the population.

It is also important to remark that, while we are mostly ignoring this issue here, the quality of the auxiliary data is very important in order to guarantee good estimates. Therefore, some quality indicators of the auxiliary data should also be given (e.g. indicators about their completeness, coverage, validity, and timeliness).

As a proof of concept, we explain how we apply these ideas to the Spanish Structural Business Statistics (SBS). The goal of the SBS is to provide information on the main structural and economic characteristics of the enterprises, in the different sectors studied, through a wide range of variables relating to the personnel employed, turnover and other incomes, purchases and consumption, personnel expenditure, tax, and investment. This information is given separately for each of the following sectors: industry, trade, and services. The information is both provided for

[6] We use the term *algorithm-assisted* in the sense of generalising in a natural way the ideas of *model-assisted* estimation (see e.g. Särndal et al. 1992).

the totality of the sector and also disaggregated by NACE and by region. Currently, the sampling design used is stratified sampling, and the estimates of the variables are computed via Horvitz-Thompson (HT) estimators with calibration to the population totals of some auxiliary data. See Statistics Spain (INE) (2023) for details.

Our objective is to use imputation beyond the sample using ML methods in order to significantly reduce the sample size or, if possible, even eliminate the sample in some years while maintaining or improving the current quality of the statistics. More precisely, we want to estimate the population totals of 95 quantitative survey variables, using as auxiliary data 150 variables coming from administrative data of the Spanish and Statutory Tax Agencies. While the goal is to provide estimates at several levels of disaggregation, in this description, we will focus just on the aggregated totals for all the economy.

As the first step of the project, we have dealt with a worst-case scenario. We assume that we have no sample for the current year and that the only available sample is that of the previous year. Moreover, the year chosen for our study is 2020 (the COVID-19 pandemic year) using sample from the year 2019. Therefore, it is reasonable to expect that the model drifting effect will be more significant than in any other pair of consecutive years. As models for our target variables, we try a variety of machine learning models (linear regression with elastic net regularisation, random forests, and XGBoost) with hyperparameter tuning, and we make both model selection and training using exclusively the sample from the year 2019, therefore using the models in 2020 as pretrained prediction estimators. It is also important to remark that in training the models with the 2019 sample, we use the sampling weights, so that the different importance of each sampled unit can be taken into account in the model.

We have obtained estimates of the 95 variables of interest, both using the pretrained prediction estimators and the currently used HT estimators with the sample of 2020, and we have compared the accuracy of the two estimators using the sample for 2020. Let $\hat{Y}^{pred}$ be the pretrained prediction estimator of one variable and $\hat{Y}^{HT}$ the HT estimator of the same variable. In order to compare the accuracy of both estimators, we use the usual estimator of the variance for the HT estimator, $\hat{V}(\hat{Y}^{HT})$, and we estimate the bias of $\hat{Y}^{pred}$ by using a Horvitz-Thompson-like estimator upon the 2020 sample:

$$\hat{B}(\hat{Y}^{pred}) = \sum_{k \in s_{2020}} \frac{\hat{y}_k - y_k}{\pi_k},$$

where s_{2020} stands for the sample in the year 2020, $\hat{y}_k$ is the prediction of the pretrained model for the unit k, and π_k are the sampling weights for 2020. Observe that, since the pretrained estimators do not depend on the chosen sample for the year 2020, they have zero sampling variance but are generally biased estimators. On the other hand, HT estimators are unbiased but have nonzero sampling variance (see

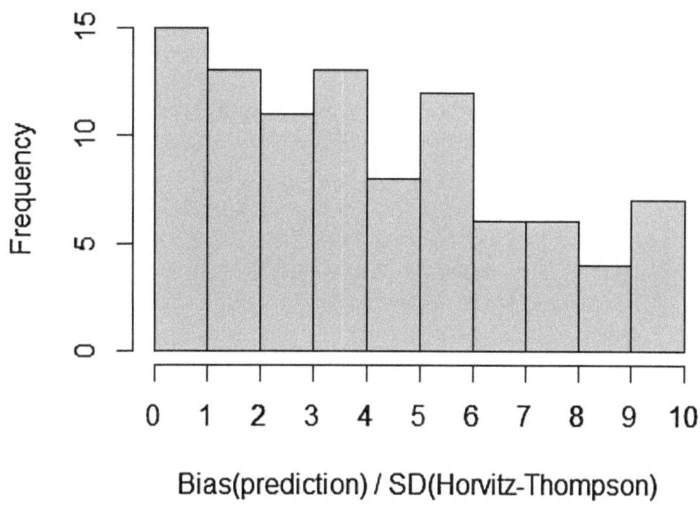

Fig. 13.7 Histogram of the relative efficiency of the prediction estimators with respect to the Horvitz-Thompson estimators for the 95 target variables

e.g. Särndal et al. 1992). Therefore,

$$\frac{\widehat{rmse}(\hat{Y}^{pred})}{\widehat{rmse}(\hat{Y}^{HT})} = \frac{\hat{B}(Y^{pred})}{\sqrt{\hat{V}(\hat{Y}^{HT})}}$$

provides an estimate of the relative efficiency of $\hat{Y}^{pred}$ with respect to $\hat{Y}^{HT}$. The prediction estimator is more efficient than the HT estimator when this quotient is less than 1. The histogram of this relative efficiency for the 95 target variables can be seen in Fig. 13.7.

We see that only for 15 of the 95 target variables the accuracy of the prediction estimator is better than that of the Horvitz-Thompson estimator. This means that, at least for 2020, the models pretrained with 2019 data in general do not provide enough accuracy to replace the HT estimators. However, it is noteworthy that even in this worst-case scenario, there are some variables for which the prediction estimator outperforms the HT estimator. From the results obtained thus far, we can conclude that some sample will be necessary each year in order to deal with model drift. Moreover, a well-chosen sample will very likely improve substantially the quality of the estimates.

We make some additional comments and remarks.

1. For some units, there is essentially no auxiliary data available, so for those units, the imputed values using models are unreliable. On the other hand, there are some big units which have a significant contribution to the total, and knowing the value of those units (by surveying them each year) would improve the quality of the

total estimation. It is important to notice that even statistical learning methods fall short of providing a working solution at this point.
2. There are some target variables which are essentially independent from the available auxiliary data. Therefore, for these variables, it will be impossible to obtain good predictions using models based only on the auxiliary data. Again, we notice that even statistical learning methods fall short of providing a working solution at this point.
3. No regressor selection has been made thus far. With well-chosen regressors for each model, we expect to improve the quality of the predictions and hence the quality of the total estimates. This situation is similar to that in early imputation for constructing the features of the model. The representation learning step in the use of these models constitutes the next step. In this regard, the collaboration with the business experts in order to define the best regressors for each target variable is indispensable.
4. The model selection process has not been very exhaustive, mainly due to the fact that this is the most computationally expensive part of the process. A more exhaustive model selection process should improve the estimates. As in early estimation (and use cases in the next sections), we have made an educated choice of the models, which now should be successively improved, potentially with an exhaustive search, but perhaps more sustainably with a continuously evolving process.
5. As mentioned before, in this exercise, we have used the sample of 2019 for the model selection and to train the models, while we use the sample of 2020 just to compute an estimate of the accuracy of our total estimates. In a situation where we assume we have a sample in the same year we want to make the estimation, we would use the same sample for model selection, training, and estimation of the MSE. We would also use the known values for the units in the sample instead of the predictions in the estimation of the total.

To conclude, we note that the next step is the determination of a reduced sample which allows us to significantly reduce the sample size while maintaining the quality using algorithm-assisted estimation. Given the characteristics of the Structural Business Survey, it seems that the better approach would be to survey the bigger enterprises, which contribute substantially to the population totals, survey also the units with no auxiliary information, and finally select a well-chosen sample within the companies of intermediate size. In designing this sample, the expertise of business experts and sampling experts will also be essential. The sample thus selected will be aiming at excelling model performance even for highly granular information. In our view, even when machine learning methods are used, surveys are not recommended to be discarded in order to achieve accurate estimates.

13.3.3 *Integration of Administrative Data as a Primary Source in Business Statistics*

The proposal for the production framework using administrative data as the primary source begins from the hypothesis of maintaining the same objectives as in the case of survey data, i.e. the aim is to estimate a set of population aggregates in a finite population U, defined as $Y_{U_d} = \sum_{k \in U_d} f(\mathbf{y}; \mathbf{x})$ for a collection of population domains $U_d \subset U$ (publication cells) for various target variables $\mathbf{y}$ and auxiliary variables $\mathbf{x}$. Without loss of practical generality, we can focus on population totals in the form $Y_{U_d} = \sum_{k \in U_d} y_k$, as other more complex aggregates can be expressed as functions of these totals. We have the complete sample $s = \bigcup_d s_d$, where $s_d \subset U_d$.

To achieve this, similar to survey data, we will analyse the use of linear estimators of the form $\widehat{Y}_{U_d} = \sum_{k \in s_d} \omega_{ks}(\mathbf{x}) y_k^{\circ}$, $\omega_{ks}(\mathbf{x})$ are pseudo-sampling weights (or true weights if a sampling design is used), and y° denotes a synthetic value for variable y, which can either be a transformation of the corresponding administrative variable or a predicted value for the survey variable based on all available information (administrative and survey). Accuracy measures must also be produced.

In this context, the fundamental concepts of finite population and target variable remain to be central. In consequence, for quality assurance, the paradigm of the total survey error model (Groves and Lyberg 2010) is still valid, even under its consideration as the second phase of the two-phase life-cycle model by Zhang (2012). We shall focus on short-term business statistics incorporating tax register data as primary data source in combination with survey data under a given probabilistic sampling design. In particular, we shall provide the description of an ongoing pilot experience with the Service Sector Activity Indicators (SSAI) survey, which is beginning to use VAT data from the National Tax Agency to reduce response burden.

Let us denote by U^{adm} the set of business units contained in the tax register and by U the finite population of analysis, which is represented by the population frame U_F obtained from the complete business register in our office. Our first concern is about the coverage error, particularly on identifying those administrative units $k \in U^{adm}$ considered as statistical units $k \in U$. From the tax register U^{adm}, we shall consider statistical units $k \in U$ those units k also contained in the frame population, i.e. $k \in U^{adm} \cap U_F$. The target statistical variable for these units will be synthetised using the raw administrative value y^{adm} in a model. We shall denote $U^{mdl} = U^{adm} \cap U_F$.

A first natural course of action is to use the administrative values y^{adm} by mere substitution as the value of the statistical values y^{stat}. This would be a more cost-efficient and timely strategy *provided the statistical quality of the input administrative data is guaranteed.* In this sense, taking advantage of having both the administrative and statistical values for a subset of units $k \in s \cap U^{mld}$, we assess the quality of the input tax data for statistical purposes as follows.

Several proposals in official statistics aim to measure quality at various stages of the statistical production process. However, historically, more emphasis has been

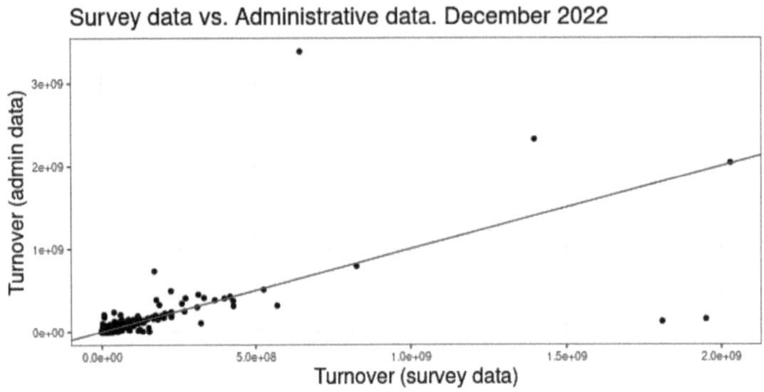

Fig. 13.8 Microdata comparison between survey and administrative data

placed on assessing the quality of final aggregates rather than on the input data, primarily due to the control inherent in the generation of survey data. With the growing incorporation of diverse data sources, there is an increasing need to evaluate their quality as well. Initiatives within the European Statistical System (ESS), such as the BLUE-ETS Project (Daas et al. 2011) and the ESSnet KOMUSO (see Ascari et al. 2020, and references therein), have addressed this need.

At Statistics Spain, motivated by the third round of Peer Reviews of the European Statistical System (ESS 2024), we have recently started to engage in the development of a diverse set of indicators aimed at evaluating the quality of data sources across multiple dimensions (Nieto et al. 2024). Furthermore, we are conducting retrospective analyses to juxtapose administrative data with survey data at the microdata level. This endeavor entails the creation of numerical indicators to facilitate a comprehensive and rigorous evaluation process for their usage as input data in official statistical production. Meanwhile, a graphical representation of this comparison can be seen in Fig. 13.8 where the turnover variable is represented for the units in common between administrative and survey data for reference time period of December 2022.

This rule-of-thumb comparison has been complemented focusing on the distribution of both datasets (admin and survey) for the target variable. This is relevant to understand where the differences lie. Graphical representations depicting the quantiles corresponding to each dataset have been generated. Figure 13.9 represents the sample quantiles of the survey data against those of the administrative data for the January 2022 period. This makes clearly explicit the issues with both distribution tails.

The comparison can be conducted also upon the population-level estimates. In particular, in Fig. 13.10, we represent the estimated population quantiles for the same period, thus making visible the effect of the sampling weights. As these representations show, the differences between both sources are significant at microdata level. The discrepancy is not as pronounced concerning the estimated population distributions as evidenced by the comparison in Fig. 13.10. However,

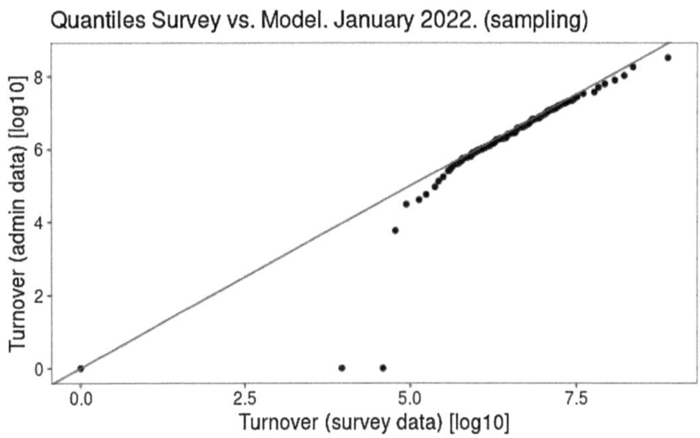

Fig. 13.9 Quantile comparison between survey and administrative data for January 2022 (sample level)

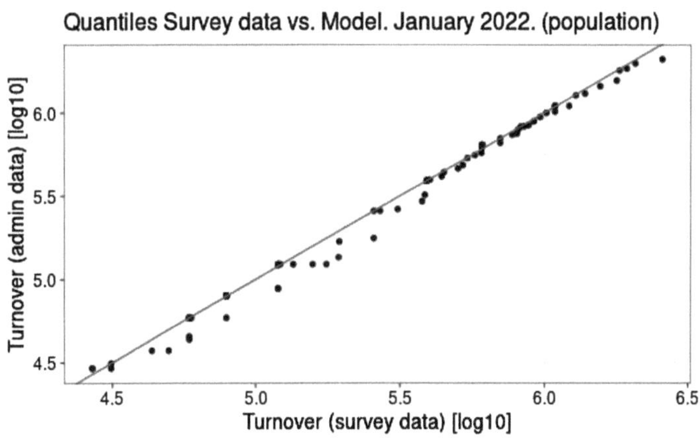

Fig. 13.10 Quantile comparison between survey and administrative data for January 2022 (population level)

these differences become greater as we focus on more disaggregated population domains U_d, thus impinging negatively on granularity.

As a consequence, our next concerns are the validity and measurement errors. It is well known (Ascari et al. 2020) that administrative data can severely differ from survey data since the latter are defined and collected for statistical purposes. In this sense, in the particular case of the SSAI survey, the administrative total sales value y^{adm} declared for tax purposes may differ from the statistical total turnover value y^{stat} traditionally collected in questionnaires. These differences cautiously discourage the use of y^{adm} by mere substitution as the value of y^{stat}.

Our proposal aims at a combined use of statistical learning models and data validation techniques to control this difference (validity error, but also measurement error since we use validated microdata as training data). Longitudinal information is of special relevance as auxiliary information. Consider several datasets for reference time periods t_{-1}, t_{-2}, t_{-3}, and so on, where past periods t_{-i} will be used for model training. The proposal focuses on predicting and validating successively each dataset t_0, t_1, t_2, etc. with their past datasets.

Firstly, to initialise the recurrent modelling exercise in successive time periods, training sets for t_0 are identified as those units in the probabilistic samples and the tax register, i.e. $k \in s_{t_{-i}}^{mdl} = s_{t_{-i}} \cap U^{mdl}$. Their corresponding synthetic target variable values y_k° are the validated values entering the computation of the indices, i.e. $y_{kt_{-i}}^{\circ} = y_{kt_{-i}}^{stat}$. Then, a statistical learning model $y^{\circ} = f(y^{adm}; \mathbf{x}) + \epsilon$ is adjusted using explicitly the value of the administrative variable as a feature. Once the model is constructed, it is used to predict the values of the variable $\widehat{y}_{kt_0} = \widehat{f}(y_k^{adm}; \mathbf{x}_k)$. Notice that this is the predicted value of the validated total turnover in terms of the raw administrative value of the total sales variable (and other features). The rest of features $\mathbf{x}$ are constructed as in the early imputation section for the early estimates of the ITI (cf. Sect. 13.3.1). These predicted values are candidates to enter the index computation.

Next, a data validation strategy is applied. This involves designing and applying both error detection functions (edits) and their treatment, which likely requires a more specific imputation model. Ideally, this part should be automated to the highest possible degree. The result will be a new refined set of validated synthetic target values together with the validated survey data values s_{t_0} used for index computation.

The main objective of using administrative data as the primary source is to reduce the burden on respondents. To achieve this, questionnaires should be eliminated for those units with reliable administrative information. However, the selection of such units must be carried out carefully, as there may be a lack of information to properly train models that predict values for units not included in the survey sample.

Many scenarios based upon scoring to select units reporting survey data and units reporting with their administrative records have been tested. The one presented here has been yielding the best results so far. In this scenario, specific criteria have been defined to identify units exhibiting erratic behaviour for survey reporting, thereby ensuring the quality of their values through the traditional data collection and data editing processes. Survey-reporting units satisfy at least one of the following criteria.

Criterion 1 Units with a high impact on the aggregate

A first period-wise score for unit k in period t_{-i} is defined as $r_{kt_{-i}} = \frac{\omega_{kt_{-i}} y_{kt_{-i}}^{stat}}{\widehat{Y}_{dt_{-i}}}$, where $\widehat{Y}_{dt_{-i}}$ is the aggregate estimator for domain d in period t_{-i}. The periods corresponding to the first 9 months of the previous year to the reference year are used, and a first global score $r_k^{(1)}$ for each unit k is defined as the pth quantile $r_k^{(1)} = Q_p(r_{kt_{-1}}, \ldots, r_{kt_{-9}})$. We have used the median ($p = 0.5$). A threshold is computed using a conservative elbow criterion (Satopää et al. 2011), and thus, units above the threshold are selected for survey reporting.

Criterion 2 New units

All new units in the sample s of the previous year to the reference time period that have $\omega_{kt_{-i}} = 1$ or annual turnover in frame U_F for the preceding year greater than a chosen threshold t_F are selected. We have used $t_F = 10^7$. A second global score $r_k^{(2)}$ is thus defined as $r_k^{(2)} = I\{\omega_{kt_{-i}} = 1 \vee a_F > 1e7\}$. Notice that $\omega_{kt_{-i}} = \omega_{kt_{-j}}$ for all t_{-i} and t_{-j} in the same year since sampling designs change only annually. Units with $r_k^{(2)} = 1$ are selected for survey reporting.

Criterion 3 Units with high variability in the target variable

Another global score is defined as $r_k^{(3)} = \sqrt{Var(\omega_{kt_{-1}} y_{kt_{-1}}, \ldots, \omega_{kt_{-9}} y_{kt_{-9}})}$. Again an elbow-based threshold is used to select those units for survey reporting.

Criterion 4 High difference between the survey and administrative values

A time-wise score for unit k in period t_{-i} is defined as $r_{kt_{-i}}^{(4)} = \frac{\omega_{kt_{-i}} |y_{kt_{-i}}^{adm} - y_{kt_{-i}}^{stat}|}{\widehat{Y}_{dt_{-i}}}$. A new global score is defined as $r_k^{(4)} = \sqrt{Var(r_{kt_{-1}}^{(4)}, \ldots, r_{kt_{-9}}^{(4)})}$. Again, an elbow-based threshold is used to select those units for survey reporting.

Criterion 5 High absolute differences between the survey value and the administrative record value

Using the same time-wise score for unit k in period t_{-i} as in Criterion 4, a new global score is defined as the pth quantile so that $r_k^{(5)} = Q_p(r_{kt_{-1}}^{(4)}, \ldots, r_{kt_{-9}}^{(4)})$. We have selected $p = 0.5$. Again, an elbow-based threshold is used to select those units for survey reporting.

Criterion 6 Zero values

All units with any administrative record value equal to zero in the periods under consideration are selected. The global score is defined as $r_k^{(6)} = I\{y_{kt_{-1}}^{adm} = 0 \vee y_{kt_{-2}}^{adm} = 0 \vee \cdots \vee y_{kt_{-9}}^{adm} = 0\}$. Units with $r_k^{(6)} = 1$ are selected for survey reporting.

Units not selected according to these scoring system will be modelled. These criteria are conservative with respect to reduction of response burden with the idea of keeping all challenging units in the survey. For example, for year 2021, there were 5281 units in the administrative dataset, and using these criteria, still 2320 would be needed to collect in the survey and 2961 could be dropped.

As preliminary results, again we can report both at the statistical unit level and at the aggregate level. Firstly, in Fig. 13.11, we represent the predicted values for the total turnover values in comparison with the survey value, which is equivalent to Fig. 13.8. In this case, the comparison is obtained after implementing the proposed model. A substantial improvement is now evident in how closely the model values resemble those of the survey, thus partially reproducing the data editing tasks to account for validity and/or measurement errors.

At the aggregate level, Fig. 13.12 presents a comparison of the SSAI index obtained through direct substitution of administrative data for those units intersecting with the sample (dotted line) and the index obtained using model-predicted and validated values (dashed line). The graph illustrates the differences between each of

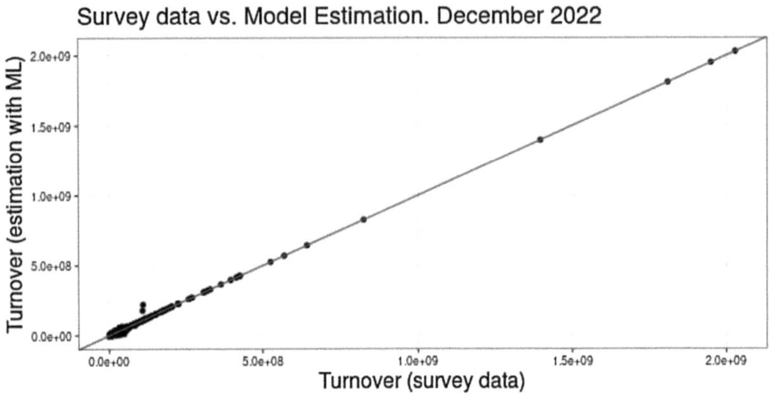

Fig. 13.11 Comparison between survey microdata and model-predicted microdata

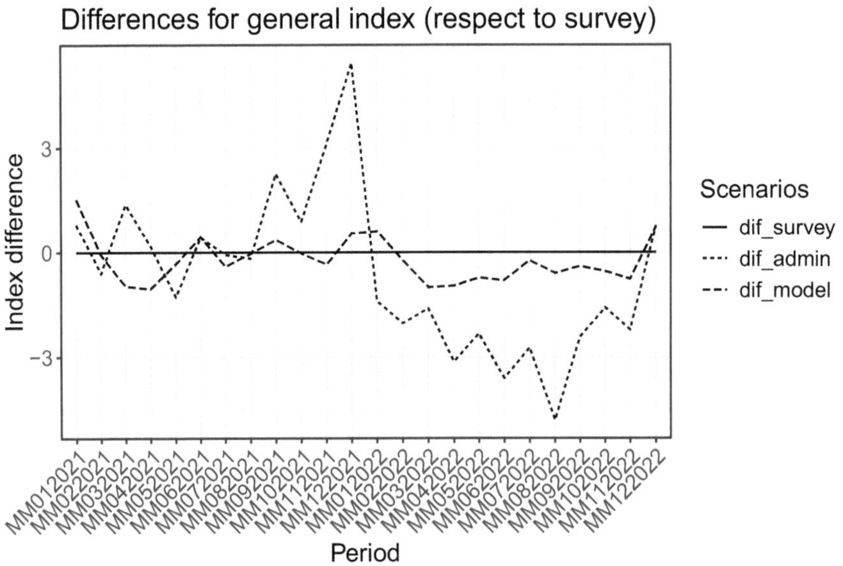

Fig. 13.12 Differences between the general index of SSAI obtained with administrative and the model respecting to only survey data

these indices and the index obtained exclusively from survey data (solid line). It can be observed how the differences are considerably smoothed out by using a model to account for validity and measurement errors in the statistical value based on the administrative value.

As main remarks, we state the following:

1. To account for validity and measurement errors in administrative data using statistical learning models aiming at response burden reduction, a selection of units maintaining the model quality is advised.
2. The selection problem is thus focused on the quality of the model, not on the quality of the final output, which is obtained by standard aggregation procedures.
3. More research is needed to find a trade-off between accuracy (lack of measurement errors) and response burden reduction. Our guess is that the higher the predictive power of the model, the higher the reduction.
4. Ongoing work is under development to estimate variances and mean squared errors arising from the combination of sampling designs and model predictions.

13.3.4 Time Disaggregation of Sampling Designs

Many official statistics collect data continuously (e.g. weekly) and process and aggregate them to produce and release monthly, quarterly, or even annual outputs. The whole production process follows the traditional survey methodology, with the selection of a probabilistic sample from a population frame, the use of well-known data collection modes integrating the interview administration, the data entry and the data editing during collection (e.g. using a CAPI data collection mode), the execution of a statistical data editing and imputation strategy, the calibration of sampling weights, the computation of the estimates and of coefficients of variation, and the preparation of the dissemination products (including statistical disclosure control), to name the most significant.

This process is usually deemed expensive and slow, negatively impinging on timeliness (see e.g. Eurostat 2017). Here we show how the introduction of novel methods in this process allows us to take advantage of the continuous data collection to produce more frequent estimates. We shall use the Spanish Labour Force Survey to illustrate our proposal. The starting point is the usual linear estimator formula for constructing the statistical outputs: $\hat{A}_d(C) = \sum_{k \in s_d} \omega_{ks}(\mathbf{x})\delta_k(C)$, where $A_d(C) = \sum_{k \in U_d} \delta_k(C)$ denotes the C-property class total in the population domain U_d, $\omega_{ks}(\mathbf{x})$ denotes the (usually calibrated with respect to marginal $\mathbf{X}$) sampling weight for unit k, and $\delta_k(C)$ is the binary indicator variable of property C for unit k (e.g. employed or unemployed).

We make the following considerations regarding time. Firstly, we observe two time scales in the survey. On the one hand, aggregates $A_d(C)$ and their estimators $\widehat{A}_d(C)$ are referred to a long time period such as quarters in a year (say, $\hat{A}_d^Q(C)$, with $Q = 1, 2, 3, 4$). On the other hand, collected target variables $\delta_k(C)$ are referred to specific weeks[7] in a year (say, $\delta_k^W(C)$, with $W = 1, \ldots, 52$). These two time scales

[7] Usually, the interview is administered 1 week after the reference week.

are connected through the sampling weights derived from the sampling design over the quarterly target population.[8] In this connection, the underlying assumption that the weekly collected value $\delta_k(C)$ is valid throughout the whole reference period Q is made. This is a kind of ergodic hypothesis assuming that a cross-sectional behaviour in a given week can be assigned longitudinally to the whole quarter. This hypothesis is made to make the sampling size valid for the whole reference time period (e.g. the quarter).

Instead, we avoid this ergodic hypothesis and use a random forest algorithm to disaggregate the sampling designs from the quarterly scale down to the weekly scale. The reasoning is as follows. Let us denote by π_k^Q the first-order inclusion probability for unit $k \in U$ according to the quarterly sampling design in production. As a production step in the whole process, every sampled unit $k \in s$ follows an assignment procedure in the quarter to administer the interview in the corresponding reference week W. This is accomplished semi-manually by experts accounting for a strict balance across the whole national territory for data collection fieldwork reasons. This week assignment is conducted taking into account frame and design variables such as regions, provinces, and strata. The quarterly sample s^Q can then be disjointly partitioned into 13 weekly samples $s^Q = \bigcup_{W=1}^{13} s^W$, $s^W \cap s^{W'} = \emptyset$, $W \neq W' \in \{1, \ldots, 13\}$. We define the weekly first-order inclusion probabilities as $\pi_k^W = \mathbb{P}\left(s^W \ni k\right)$ and compute them using compound probability properties:

$$
\begin{aligned}
\pi_k^W &= \mathbb{P}\left(s^W \ni k\right) \\
&= \mathbb{P}\left((s^Q \ni k) \wedge (k \rightsquigarrow W)\right) \\
&= \mathbb{P}\left((k \rightsquigarrow W)|(s^Q \ni k)\right) \cdot \mathbb{P}\left(s^Q \ni k\right) \\
&= \mathbb{P}\left((k \rightsquigarrow W)|(s^Q \ni k)\right) \cdot \pi_k^Q,
\end{aligned}
$$

where $(k \rightsquigarrow W)$ denotes the event "unit k is assigned to week W according to the assignment procedure".

To compute the weekly assignment probability $\mathbb{P}\left((k \rightsquigarrow W)|(s^Q \ni k)\right)$, we compose a dataset with the frame and design variables $x_k^{(q)}$ used for the assignment for each unit k as well as the weekly assignment w_k. Since this survey has a rotating sampling design where each unit k retains the assigned week w_k when it is firstly selected in the sample, we only include records corresponding to this first selection. In this way, we have a dataset with the empirical results of the allocation procedure. Only units from the past six time reference periods are included in this dataset.[9]

---

[8] In household surveys, this is usually the population at the middle date of the time span period coming from the Population Register.

[9] Six is the number of periods a statistical unit is kept in the rotating sample.

Next, using a random forest routine for this classification problem, we get the resulting probabilities computed by the algorithm for each unit k over the whole dataset. These probabilities are indeed the probabilities $\mathbb{P}\left((k \rightsquigarrow W)|(s^Q \ni k)\right)$ which we need. Notice that we do not need to split the dataset into train and test, since we are not going to predict any output. We are just *measuring* the probabilities $\mathbb{P}\left((k \rightsquigarrow W)|(s^Q \ni k)\right)$ from the empirical data. Overfitting is indeed beneficial because we want to measure the probabilities for this specific dataset.

Now, having both the weekly design weights $d_k^W = 1/\pi_k^W$ and the weekly target variable values $\delta_k(C)$ for the weekly respondent sample $r^W \subset s^W$, we proceed exactly the same way as in the quarterly time scale. We compute the Horvitz-Thompson estimate $\widehat{A}_d^{W,HT}(C) = \sum_{k \in s_d^W} d_k^W \delta_k(C)$ and apply the two-step calibration procedure (Särndal and Lundström 2005) to adjust for non-response and to reduce variance (Statistics Spain (INE) 2009), so that finally

$$\widehat{A}_d^W(C) = \sum_{k \in s_d^W} \omega_{ks}^W(\mathbf{x})\delta_k(C),$$

where $\omega_{ks}^W(\mathbf{x})$ stands for the two-step calibrated weekly sampling weight. As marginal totals for calibration, we use the monthly version of the original marginal population totals. This is indeed a decision depending on the production environment: should the weekly marginal population totals be available (nowadays uncommon in official statistics), we would use them; instead, the corresponding monthly population figures are used.

Figure 13.13 depicts an illustrative example for the 52nd week of 2022. We rapidly observe that (i) as expected, there roughly exists a factor 13 to go from the quarterly to the weekly scale and (ii) there exist outliers. Further insight is needed here to understand the model performance (e.g. with the Brier score) and its interrelation with the calibration process and the outlying values.

The weekly estimates $\widehat{A}_d^W(C)$ can now be combined to produce monthly estimates:

$$\widehat{A}_d^M(C) = \frac{1}{n_M} \sum_{W \subset M} \widehat{A}_d^W(C), \tag{13.12}$$

where n_M denotes the number of weeks W in the month M. Notice that this is possible not only for calendrical months M.

Alternatively, rescuing the aforementioned ergodic hypothesis for month M, and computing monthly inclusion probabilities as $\pi_k^M = \sum_{W \subset M} \pi_k^W$, we can also produce monthly estimates as

$$\widehat{A}_d^M(C) = \sum_{k \in M} \omega_{ks}^M(\mathbf{x})\delta_k^M(C),$$

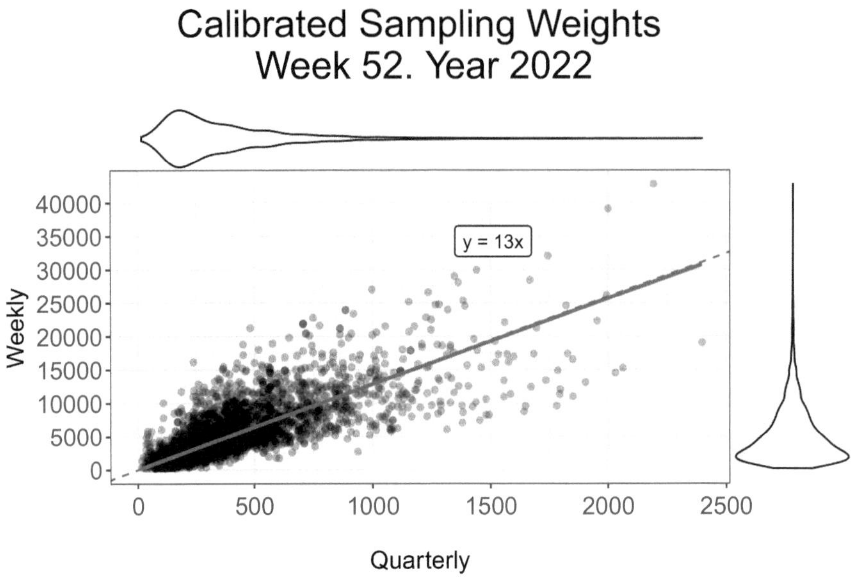

Fig. 13.13 Weekly calibrated sampling weights $\omega_{ks}^{W}(\mathbf{x})$ versus quarterly calibrated sampling weights $\omega_{ks}^{Q}(\mathbf{x})$. The regression line with no intercept (dashed) has an estimated slope $\hat{\beta}_1 = 12.857$ with $\hat{\sigma} = 0.096$

where $\omega_{ks}^{M}(\mathbf{x})$ denotes the calibrated sampling weight with initial design weight $d_k^M = \frac{1}{\pi_k^M}$ and $\delta_k^M(C)$ stands for the target variable value assumed for the whole month M from the weekly collected value $\delta_k(C)$.

In either form, both the calibrated weekly estimates $\widehat{A}_d^W(C)$ and the monthly estimates $\widehat{A}_d^M(C)$ show a great variability for the same quarter (in the same flavour as multiple Horvitz-Thompson estimates from the same population, especially for the weekly disaggregation). To reduce this variability, we apply a filter to the time series so constructed. This is beyond the methodological scope of this chapter. In Fig. 13.14, we show an example of weekly and monthly disaggregations compared to the original quarterly time series before applying any time-series filtering technique. The variability is visible, especially in the weekly scale.

This proposal is currently under deeper investigation (Barragán et al. 2025) to assess different issues potentially impinging on the quality of the final estimates:

1. Once higher frequency is gained, accuracy must be our first-order concern. Regarding bias, we remind that probability estimation by random forests is consistent (Biau et al. 2008), so that $\widehat{A}_d^{W,HT}$ is asymptotically unbiased considering both the sampling design and the estimation model. Sampling weight calibration only introduces an asymptotically negligible bias. The size of the training sets (the past six quarters in our case) needs to find a trade-off between the distance

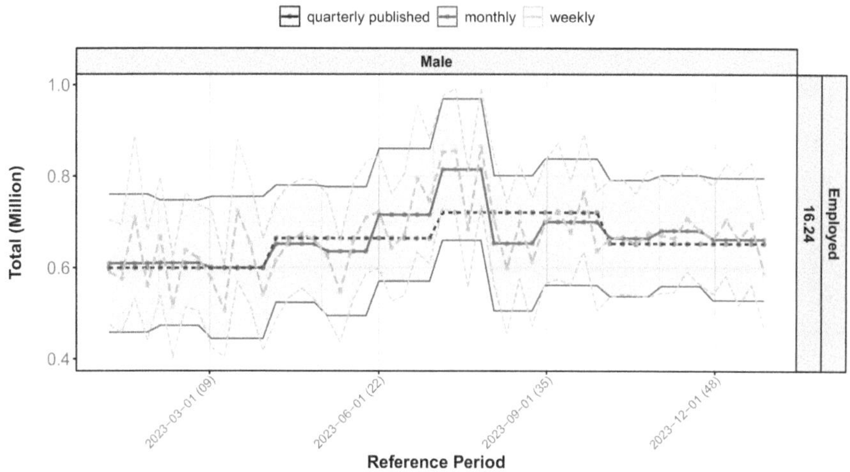

Fig. 13.14 Quarterly, monthly, and weekly estimates for the total number of employed people in the age group 16–24 during 2023. No time-series filtering technique has been applied at this point: sampling variability is clearly visible

in time in the past (so that weekly assignments actually are the result of the current practice, not that of the far past) and the consistency rate.

2. In the same line, regarding the variance, firstly the estimated variance for the weekly estimator $\widehat{\mathbb{V}}[\widehat{A}_d^W(C)]$ is computed with the same jackknife methodology as the quarterly time scale (Statistics Spain (INE) 2009). Then, for the monthly scale, we conceive the monthly estimator (13.12) as the result of the sampling design $p(\cdot)$ and a combination of weekly estimates w so that we can write

$$\mathbb{V}[\widehat{A}_d^M(C)] = \mathbb{V}_w[\mathbb{E}_p[\widehat{A}^M|s^M]] + \mathbb{E}_w[\mathbb{V}[\widehat{A}_d^M(C)|s^M]]. \tag{13.13}$$

This variance can be estimated by

$$\widehat{\mathbb{V}}[\widehat{A}_d^M(C)] = \widehat{\mathbb{V}}_{JK}[\widehat{A}_d^M(C)] + \frac{1}{n_M}\sum_{W\subset M}\widehat{\mathbb{V}}[\widehat{A}_d^W(C)], \tag{13.14}$$

where $\widehat{\mathbb{V}}_{JK}[\widehat{A}_d^M(C)]$ stands for a jackknife estimator of the monthly estimator $\widehat{A}_d^M(C)$ deleting 1 week at a time. This rough estimation of the variance needs to be refined taking into account the filtering procedure.

3. We have observed that dropping out the aforementioned ergodic hypothesis on the target variable values during the whole quarter occasionally drives us to three monthly estimates systematically below or above the original quarterly estimates. This needs further investigation and probably the introduction of a benchmarking step to reach full coherence between the original estimates and the time-disaggregated estimates.

4. So far, our focus has been placed on the estimation of population totals. Employment and unemployment rates, thereof derived, must be further estimated and analysed.

13.4 Some Conclusions

This collection of proposals and pilot experiences allows us to gather some conclusions regarding the use of statistical learning models for the production of official statistics.

Our main conclusion is that statistical learning models constitute a versatile tool enabling the improvement of the design, execution, and monitoring of statistical business functions, both traditional functions already in production and novel modifications impinging on different quality dimensions. Thus, in the same spirit as survey methodology has been successfully deployed in production in statistical offices, an adaptation of these organisations to integrate these new methods and technologies must be put in place. This involves different aspects, mainly data, technology, and skills.

Machine learning is a highly data-intensive activity. Therefore, data governance, data management, and data architectures are crucial to implement these methods at scale. The amount of pre-processing tasks in our proofs of concept to prepare data and compute regressors (feature engineering) would clearly benefit if a data architecture with fully fledged metadata is put in place and shared among all surveys. This would reduce the cost for the discovery and development of new ML-based applications.

Indeed, in this same line, a repository of regressors or features would be highly advised with a continuously updating process in place incorporating subject matter knowledge. According to our experience, representation learning or feature learning is a crucial step for the performance of the models. The subject matter knowledge needs to be transferred to the models, and the identification and construction of features from the available data constitute a fundamental task in this direction.

Data quality is another critical aspect for the performance of machine learning models. This does not only imply the curation of existing data but also (and even more strategically) the decision about what data to collect and to use for training algorithms. In particular, in relation with the necessary reduction of response burden generated by survey data collection, in our opinion, this high-quality data source should be strategically considered. When planned to be integrated with administrative data and digital transactional data, a change of focus should be undertaken: we need to pursue the quality of the trained models and not the quality of final statistical outputs themselves. The problem of sample selection in statistical offices should progressively focus on a selection problem to assure model quality (either assisted- or dependent-like) not direct estimator quality. In our view, this is already suggesting the role of statistical offices in the new data ecosystems, where statistical knowledge and quality assurance frameworks (e.g. for calibration) will be more relevant than statistical products (to be also released in the new future by other actors with more data, higher computational power, and more complex statistical tools—probably the policymakers themselves; see e.g. Guerrero and Margetts 2024).

Data intensivity requires also computational capacity. This is needed to continuously train the models with new collected and validated data. Technological platforms such as MLOps solutions providing these new functionalities are necessary. This computational environment should be provided with complementary and necessary functionality such as data security, software version control, continuous development and continuous integration tools, and model versioning tools.

Skills related to statistical learning need to be generalised in the same way as sample survey methodology is general knowledge among production staff in a statistical office. In other words, statistical methodology capabilities should be extended to comprise also statistical learning. For example, while the difference between a Horvitz-Thompson estimator, a ratio estimator, and a GREG estimator is common knowledge among statisticians in a statistical office, this is not the case between a gradient boosting algorithm, a random forest algorithm, and a CART algorithm. Notice that knowledge does not imply that a distributed isolated production organisation should be assumed; even when central units specialised on these techniques are considered, the knowledge should be extensive to all production staff.

Pilot experiences and proposals need to start their path to implementation in usual production conditions. A standard protocol to promote proofs of concept, minimal viable products, and experimental statistics to official statistics need to be also put in place hopefully following common international guidelines. For example, when should these new features be included (then made compulsory) in national and international legal regulations? (e.g. can it be made legally compulsory to deliver a short-term business statistics in, say, 15 days?)

Finally, as a prominent suggestion motivated by the high predictive power of these methods in terms of data at statistical units level, we think that the main focus in official statistics should be shifted from the quality in statistical outputs and aggregates as the primary concern of production to the availability of synthetic microdata to increase timeliness, granularity, cost efficiency, accuracy, and response burden reduction. Once high-quality microdata are in place through a combination of data collection and statistical learning methods, standard aggregation procedures can always be used to produce traditional outputs. Now the challenge is to guarantee the quality of these models and their input training data.

Acknowledgements We acknowledge the invaluable collaboration of colleagues from different statistical domains (Structural and Short-Term Business Statistics, Labour Market Statistics, Social Statistics) and the IT department of Statistics Spain (INE).

References

I. Arbués, P. Revilla, D. Salgado, An optimization approach to selective editing. J. Official Stat. **29**, 489–510 (2013)

G. Ascari, K. Blix, G. Brancato, T. Burg, A. McCourt, A. van Delden, D. Krapavickaitė, N. Ploug, S. Scholtus, P. Stoltze, T. de Waal, L.-C. Zhang, Quality of multisource statistics – the KOMUSO project. Surv. Statist. **81**, 36–51 (2020)

R. Avouac, T. Faria, F. Comte, A cloud-native data science platform for official statistics, in *Foundations and Advances of Machine Learning in Official Statistics*, ed. by F. Dumpert, Chap. 8 (Springer, Berlin, 2025)

S. Barragán, L. Barreñada, J. Calatrava, J.G.S. de Cueto, J.M. del Moral, E. Rosa-Pérez, D. Salgado, Early Estimates of the Industrial Turnover Index using Statistical Learning Algorithms (2022). Statistics Spain Working Paper 03/22. https://www.ine.es/GS_FILES/DocTrabajo/art_doctr032022.pdf

L. Barreñada, P. Dhiman, D. Timmerman, A.-L. Boulesteix, B.V. Calster, Understanding overfitting in random forest for probability estimation: a visualization and simulation study. Diagn. Progn. Res. **8**(1), 14 (2024)

S. Barragán, M. de Blas, C. Sáez, L. Sanguiao, G. Gorgas, An end-to-end statistical process to aggregate high-resolution temporal survey microdata: the monthly Labour Force Survey use case, in *Conference on New Technologies and Techniques in Statistics 2025, 10–12 March, Brussels* (2025)

S. Bates, T. Hastie, R. Tibshirani, Cross-validation: what does it estimate and how well does it do it? J. Am. Stat. Assoc. **119**(546), 1434–1445 (2024)

M. Beręsewicz, M. Wydmuch, H. Cherniaiev, R. Pater, Multilingual hierarchical classification of job advertisements for job vacancy statistics (2024). https://arxiv.org/abs/2411.03779

F. Beuter, J. Gussenbauer, E. Minther, V. Szabo, S. Wegner, Approaches to automated NACE coding of German business activity descriptions, in *Foundations and Advances of Machine Learning in Official Statistics*, ed. by F. Dumpert, Chap. 10 (Springer, Berlin, 2025)

G. Biau, L. Devroye, G. Lugosi, Consistency of random forests and other averaging classifiers. J. Mach. Learn. Res. **9**, 2015–2033 (2008)

BOE, Real Decreto 10/2025, de 14 de enero, por el que se aprueba la Clasificación Nacional de Actividades Económicas 2025 (CNAE-2025) (2025). https://www.boe.es/eli/es/rd/2025/01/14/10

S. Bohnensteffen, Selective data editing of continuous variables with random forests in official statistics. EMOS Master's Thesis, Complutense University of Madrid, 2020. https://eprints.ucm.es/63245

G. Brackstone, Discussion on "the Official Statistics Olympic Challenge: Wider, Deeper, Quicker, Better, Cheaper". Am. Stat. **61**(1), 9–12 (2007)

F. Breidt, J. Opsomer, Model-assisted survey estimation with modern prediction techniques. Stat. Sci. **32**(2), 190–205 (2017)

P. Daas, S. Ossen, M. Tennekes, L.-C. Zhang, C. Hendriks, K. Foldal Haugen, F. Cerroni, G. Di Bella, T. Laitila, A. Wallgren, et al., Report on methods preferred for the quality indicators of administrative data sources (2011). http://www.pietdaas.nl/beta/pubs/pubs/BLUE-ETS_WP4_Del2.pdf

DAMA International, *DAMA-DMBOK: Data Management Body of Knowledge*, 2nd edn. (Technics Publications, Sedona, 2017)

E. de Jonge, M. van der Loo, *Statistical Data Cleaning with Applications in R* (Wiley, Hoboken, 2018)

T. de Waal, Selective editing: a quest for efficiency and data quality. J. Official Stat. **29**, 473–488 (2013)

T. de Waal, J. Pannekoek, S. Scholtus, *Handbook of Statistical Data Editing and Imputation* (Wiley, Hoboken, 2011)

DGINS, Scheveningen Memorandum – Big Data and Official Statistics (2013). https://ec.europa.eu/eurostat/documents/13019146/13237859/Scheveningen-memorandum-27-09-13.pdf/2e730cdc-862f-4f27-bb43-2486c30298b6?t=1401195050000

DGINS, Bucharest Memorandum on Official Statistics in a datafied society (Trusted Smart Statistics) (2018). https://ec.europa.eu/eurostat/web/european-statistical-system/-/dgins2018-bucharest-memorandum-adopted

EC Council, EC Council Regulation No. 1165/98, of May 1998 EC Council Regulation No. 1158/05, of July 2005 (1998). https://eur-lex.europa.eu/legal-content/EN/TXT/PDF/?uri=CELEX:32019R2152

EDIMBUS, *Recommended practices for editing and imputation in cross-sectional business surveys*. ISTAT and CBS and SFSO and EUROSTAT (2007). https://ec.europa.eu/eurostat/documents/64157/4374310/30-Recommended%2BPractices-for-editing-and-imputation-in-cross-sectional-business-surveys-2008.pdf

ESS, 3rd Round of Peer Reviews (2024). https://ec.europa.eu/eurostat/web/quality/peer-reviews/current-round-2021-2023

Eurostat, Handbook of Rapid Estimates (2017). https://ec.europa.eu/eurostat/documents/3859598/8555708/KS-GQ-17-008-EN-N.pdf/7f40c70d-0a44-4459-b5b3-72894e13ca6d?t=1513758176000

Eurostat, Nomenclature statistique des activités économiques dans la Communauté européenne (2024). https://ec.europa.eu/eurostat/web/nace

T. Faria, T. Seimandi, Classifying companies in France using machine learning, in *UNECE Machine Learning for Official Statistics Workshop 2023. Geneva, 05-07 June, 2023* (2023). https://unece.org/sites/default/files/2023-05/ML2023_S1_France_Faria_Paper.pdf

L. Fiedler, B. Hofmann, K. Loogman, T. Scherl, Domain adaptation of a BERT model for analyzing job advertisements at the German Federal Employment Agency, in *Foundations and Advances of Machine Learning in Official Statistics*, ed. by F. Dumpert, Chap. 9 (Springer, Berlin, 2025)

Financial Times, How flawed data is leaving the UK in the dark (2025). February, 8. https://www.ft.com/content/dd5515cc-e628-4e17-a4fd-1a10cc9f81e4

N. Forteza, S. García-Uribe, A score function to prioritize editing in household survey data: a machine learning approach. J. Official Stat. **41**(1), 144–171 (2025)

J. Gleick, *Information: A History, a Theory, a Flood* (Pantheon Books, New York, 2011)

R. Groves, L. Lyberg, Total survey error: past, present, and future. Public Opin. Quart. **74**, 849–879 (2010)

O. Guerrero, H. Margetts, Are all policymakers data scientists now? data, data science and evidence in policymaking. LSE Public Pol. Rev. **3**(3), 1–10 (2024)

D. Hand, Statistical challenges of administrative and transaction data. J. R. Stat. Soc. A **181**, 555–605 (2018)

M. Hansen, Some history and reminiscences on survey sampling. Stat. Sci. **2**, 180–190 (1987)

H.O. Hartley, A. Ross, Unbiased ratio estimators. Nature **174**, 270–271 (1954)

D. Holt, The official statistics Olympic challenge: wider, deeper, quicker, better, cheaper. Am. Stat. **61**(1), 1–8 (2007)

A. Joulin, E. Grave, P. Bojanowski, T. Mikolov, Bag of tricks for efficient text classification, in *Proceedings of the 15th Conference of the European Chapter of the Association for Computational Linguistics: Volume 2, Short Papers* (Association for Computational Linguistics, Stroudsburg, 2017), pp. 427–431

G. Kalton, Models in the practice of survey sampling (revisited). J. Official Stat. **18**, 129–154 (2002)

U. Kamath, J. Liu, J. Whitaker, *Deep Learning for NLP and Speech Recognition* (Springer, Berlin, 2019)

Y. Liu, M. Ott, N. Goyal, J. Du, M. Joshi, D. Chen, O. Levy, M. Lewis, L. Zettlemoyer, V. Stoyanov, Roberta: A robustly optimized bert pretraining approach (2019). http://arxiv.org/abs/1907.11692

M.R. Mickey, Some finite population unbiased ratio and regression estimators. J. Am. Stat. Assoc. **54**(287), 594–612 (1959)

A. Nieto, S. Barragán, A. Rodríguez, S. Saldaña, D. Salgado, Measuring the quality of administrative sources: at macro level with novel indicators and micro level with distributions comparison. Statistics Spain Working Paper 10/24 (2024). https://www.ine.es/GS_FILES/DocTrabajo/art_doctr102024.pdf

J. Norwood, Discussion on "the Official Statistics Olympic Challenge: Wider, Deeper, Quicker, Better, Cheaper". Am. Stat. **61**(1), 13–15 (2007)

J. Pannekoek, S. Scholtus, M.V. der Loo, Automated and manual data editing: a view on process design and methodology. J. Official Stat. **29**, 511–537 (2013)

M. Puts, D. Salgado, P. Daas, Leveraging machine learning for official statistics, in *Foundations and Advances of Machine Learning in Official Statistics*, ed. by F. Dumpert, Chap. 2 (Springer, Berlin, 2025)

W. Radermacher, *Official Statistics 4.0* (Springer, Berlin, 2020)

J. Rao, Interplay between sample survey theory and practice: an appraisal. Surv. Methodol. **31**, 117–138 (2005)

M. Reister, Assuring quality in the new data ecosystem: mind the gap between data and statistics! Stat. J. IAOS **39**(2), 421–430 (2023)

L. Sanguiao-Sande, L.-C. Zhang, Design-unbiased statistical learning in survey sampling. Sankhya A **83**(2), 714–744 (2020)

C.-E. Särndal, S. Lundström, *Estimation in Surveys with Nonresponse* (Wiley, Hoboken, 2005)

C.-E. Särndal, B. Swensson, J. Wretman, *Model-Assisted Survey Sampling* (Springer, Berlin, 1992)

V. Satopää, J. Albrecht, D. Irwin, B. Raghavan, Finding a "kneedle" in a haystack: Detecting knee points in system behavior, in *2011 31st International Conference on Distributed Computing Systems Workshops* (2011), pp. 166–171

T. Smith, Sample surveys 1975-1990: an age of reconciliation? Int. Stat. Rev. **62**, 5–19 (1994)

Statistics Spain (INE), Encuesta de Población Activa. Diseño de la encuesta y evaluación de la calidad de los datos. Technical report, Statistics Spain (INE) (2009). http://www.ine.es/docutrab/epa05_disenc/epa05_disenc.pdf

Statistics Spain (INE), *Industrial Turnover Indices & Industrial New Orders Received Indices. Base 2015* (2018). https://www.ine.es/en/metodologia/t05/t0530053_2015_en.pdf

Statistics Spain (INE), Structural Business Statistics, methodology. Technical report, Statistics Spain (INE) (2023). https://www.ine.es/en/metodologia/t37/metodologia_eee2021_en.pdf

Statistics Spain (INE), Clasificación Nacional de Actividades Económicas (2025a). https://www.ine.es/dyngs/INEbase/es/operacion.htm?c=Estadistica_C&cid=1254736177032&menu=ultiDatos&idp=1254735976614

Statistics Spain (INE), CodIA: nuevo codificador automático del INE (2025b). https://www.ine.es/codia/inicio

B. Toth, Machine learning for survey estimation: bridging the rigor of traditional statistics with the power of machine learning, in *Conference on Foundations and Advances of Machine Learning in Official Statistics. Wiesbaden, April 03–05, 2024* (2024). https://www.destatis.de/EN/About-Us/Events/Machine-Learning/program.html

UNECE (ed.), *Statistical Data Editing: Methods and Techniques*, vol. 1 (United Nations, New York City, 1994)

UNECE (ed.), *Statistical Data Editing: Methods and Techniques*, vol. 2 (United Nations, New York City, 1997)

UNECE, Generic Statistical Data Editing Model v2.0 (2019). https://statswiki.unece.org/display/sde/GSDEM

UNECE, Machine Learning for Official Statistics. Technical report, UNECE (2021)

UNECE, CES Seminars on data ethics and timeliness, frequency and granularity of official statistics (2023). https://unece.org/info/Statistics/events/377894

UNECE, Generic Statistical Information Model (GSIM) v2.0 (2024). https://statswiki.unece.org/spaces/gsim/pages/597033371/Generic+Statistical+Information+Model

UNSD, International Standard Industrial Classification of All Economic Activities (2024). https://unstats.un.org/unsd/classifications/Econ/isic

U.S. Census Bureau, UNIVAC I (2024). https://www.census.gov/history/www/innovations/technology/univac_i.html

C. Wu, R. Sitter, A model-calibration approach to using complete auxiliary information from survey data. J. Am. Stat. Assoc. **96**, 185–193 (2001)

L.-C. Zhang, Topics of statistical theory for register-based statistics and data integration. Stat. Neerlandica **66**(1), 41–63 (2012)

L.-C. Zhang, L. Sanguiao-Sande, D. Lee, Design-based predictive inference. J. Official Stat. **41**(1), 404–432 (2025)

Chapter 14
Building a Retrieval-Augmented Generation Pipeline to Trace Administrative Data Use in Academic Papers

Sebastian Seltmann, Emily Kormanyos, and Hendrik Christian Doll

14.1 Introduction

In today's data-driven world, where data extraction and accumulation has been likened to the "next step in capitalism,"[1] the value of (administrative) data extends beyond mere numbers; it serves as a cornerstone for informed decision-making and societal advancement (e.g., Card et al. 2010). As highlighted in the editorial summary by Meng (2024) in the Harvard Data Science Review (HDSR) special issue on Democratizing Data, data's true worth emerges when it is accessible and utilized effectively for the public good. This perspective aligns with the ongoing discourse on evidence-based policymaking, which emphasizes the integration of empirical data into policy decisions to enhance their efficacy and public trust. The Foundations for Evidence-Based Policymaking Act of 2018 exemplifies this approach, mandating US federal agencies to utilize data in the development of effective public policies (Lane et al. 2024). Moreover, the establishment of research data centers (RDCs) has facilitated access to high-quality data, supporting researchers and policymakers in making informed decisions (Bender et al. 2024). However, the successful implementation of evidence-based policies also hinges on public trust

[1] Sadowski (2019), as summarized by Crawford (2021).

All views expressed in this chapter are personal views of the authors and do not necessarily reflect the views of Deutsche Bundesbank or the Eurosystem.

S. Seltmann · E. Kormanyos (✉) · H. C. Doll
Deutsche Bundesbank, Frankfurt am Main, Germany
e-mail: emily.kormanyos@bundesbank.de

F. Dumpert (ed.), *Foundations and Advances of Machine Learning in Official Statistics*, Society, Environment and Statistics,
https://doi.org/10.1007/978-3-032-10004-7_14

in scientific data and institutions. The prior literature shows that public confidence in scientists influences the acceptance and effectiveness of evidence-based policy decisions (e.g., Bundi and Pattyn 2023; Cologna et al. 2025). Therefore, fostering transparency and engagement between scientists and the public is crucial for the advancement of evidence-based policymaking.

Yet, as Meng (2024) notes, ensuring that data informs policy decisions effectively requires more than just accessibility. The ethical and efficient use of data demands robust frameworks that balance transparency with privacy concerns and align with societal values. For instance, the Federal Data Strategy's Data Ethics Framework provides guidelines to ensure that data-driven policymaking adheres to principles of fairness and accountability (Federal Data Strategy 2020). Similarly, the recent literature stresses the importance of AI ethics in shaping responsible data practices, advocating for frameworks that integrate (expert) human oversight into data-driven decision-making (e.g., Crawford 2021; Felländer et al. 2022; Burr and Leslie 2023; Joseph et al. 2024; Tariq 2025). These considerations are especially pertinent as advancements in artificial intelligence and machine learning (ML) increase the complexity and scope of data utilization in policymaking.

RDCs stand to gain significantly from implementing retrieval-augmented generation (RAG) pipelines, as demonstrated in our study. Bender et al. (2024) discuss practical lessons from social science RDCs, underscoring the challenges of data accessibility and researcher support. By adopting RAG pipelines, RDCs can automate the extraction of data citations from scholarly articles, thereby streamlining the discovery process and enhancing the visibility of datasets. This advancement not only reduces the manual effort required in data management but also fosters a more interconnected research ecosystem where data is more readily discoverable and reusable.

Finally, the access of researchers and policymakers to FAIR (findable, accessible, interoperable, and reusable) data, which enables trustworthy and evidence-based public policy, hinges on data search and discovery. Recent advancements utilize (public) big data sources and machine learning (ML) methods: For instance, Hausen and Azarbonyad (2024) present an ensemble machine learning approach to identify government-funded datasets within scientific literature, enhancing transparency and accountability in publicly funded research. Sostek et al. (2024) delve into the challenges faced by large-scale platforms like Google Dataset Search, offering recommendations to improve dataset discoverability on the web. Pallotta et al. (2024) focus on unstructured corpora, proposing techniques to uncover datasets and identify new research opportunities. Additionally, Zdawczyk et al. (2024) discuss the development of intuitive labeling systems for data search and discovery, aiming to make data usage more comprehensible and accessible to a broader audience. Collectively, these contributions highlight the critical role of advanced search methodologies like the one proposed in this chapter in democratizing data and empowering both researchers and policymakers.

14.2 Related Literature

The automated extraction of dataset mentions from research papers has been explored across various disciplines, often using rule-based methods, traditional natural language processing (NLP) models, or more recently, large language models (LLMs). However, the application of LLMs for dataset extraction remains relatively underexplored, with only a few recent studies such as Patiny and Godin (2023), Nishio et al. (2024), and Dagli et al. (2024) explicitly investigating this task.

Earlier approaches relied on lookup tables or structured metadata, which struggled with recall due to the diverse ways datasets can be cited in text. Figure 14.1 illustrates this challenge with real-world citation variations. While persistent identifiers such as digital object identifiers (DOIs) offer a potential solution, they are not consistently used in economics and finance research, necessitating NLP-based extraction methods.

LLMs provide a practical means for dataset citation extraction without requiring extensive programming expertise. Their accessibility makes them a promising tool for institutions such as research data centers (RDCs) and statistical agencies. Our study demonstrates how such institutions can leverage LLMs to streamline dataset identification in research literature.

Among comparable studies, Patiny and Godin (2023) and Dagli et al. (2024) focus on dataset extraction in chemistry and medical sciences, respectively. These domains differ from economics and finance, where datasets are often standardized and widely shared rather than created on a case-by-case basis. The closest work

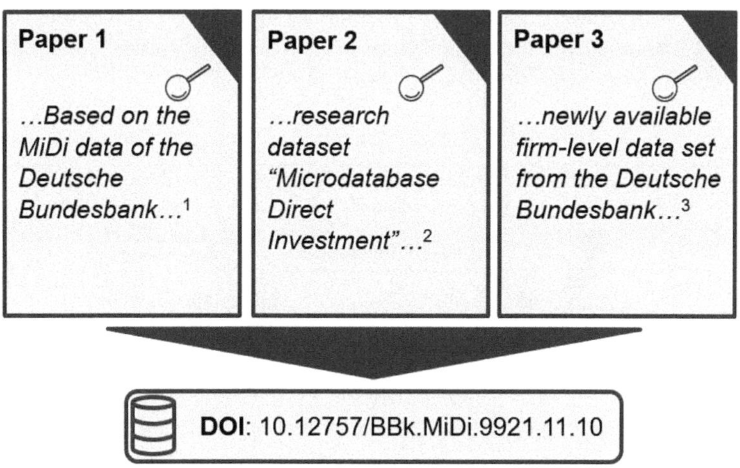

Fig. 14.1 Examples for diverse citations of the same dataset, highlighting the complexity of the extraction task. The dataset can be identified by the digital object identifier (DOI); however, the DOI is often not present or cited (Sources: Authors' own depiction based on actual data citations in (1) Lipponer 2006, (2) Blank et al. 2020, and (3) Buch et al. 2005)

to ours is Nishio et al. (2024), who explore dataset and methodology extraction for scientific recommender systems, and Polak and Morgan (2024), who introduce ChatExtract, an LLM-powered tool achieving high recall and precision in material sciences. However, both studies focus on different disciplinary contexts and objectives.

Prior to the rise of LLMs, research in automated information extraction focused on identifying citations (e.g., Saier and Färber 2020), key topics (Chen and Luo 2019), methodologies (Houngbo and Mercer 2012; Luan 2018), software usage (Boland and Krüger 2019), and researcher networks (Luan 2018; Färber et al. 2021). More recent work has shifted toward NLP-based approaches for dataset extraction (Kumar et al. 2021; Gemelli et al. 2023), including frameworks such as TDMS-IE (Hou et al. 2019) that extract dataset, methodology, and performance metrics from NLP papers.

A common limitation in prior studies is their focus on data science, ML, and bibliometrics literature, where methodological descriptions are often more structured than in economics and finance papers. By applying LLMs to these disciplines, our study contributes to understanding their effectiveness in extracting dataset mentions from research papers that may lack standardized metadata.

Related work such as Kumar et al. (2021) and Ikeda et al. (2020) explored dataset extraction for research transparency and dataset recommender systems, though these studies predate the widespread adoption of LLMs. Our approach extends beyond known dataset identification by addressing cases where dataset mentions are unknown a priori, making it applicable to researchers and RDCs seeking to analyze dataset usage trends at scale.

Our primary use cases include (i) recommending datasets to researchers based on related studies and (ii) assessing dataset impact through research output analysis. Figure 14.2 summarizes these applications. The following section details the data sources used in our extraction pipeline (Sect. 14.4).

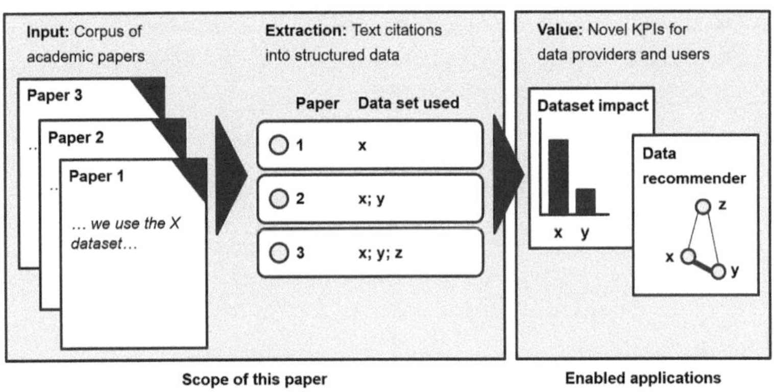

Fig. 14.2 Scope of this chapter and value proposition (Source: Authors' own depiction)

14.3 Data Provision Through Research Data Centers (RDCs)

RDCs are facilities where accredited researchers can securely access sensitive granular data. They are often located at the premises of data owners.[2] To tailor data provision services adequately and understand the downstream impacts of their work, RDCs might wish to be able to measure the usage of the data they provide. Therein, they can benefit by exploring data usage systematically in order to enable their mission of serving researchers in need of administrative data and to monitor the contribution of their provided data.

RDCs provide high-quality granular data from administrative source low cost or free of charge. Usually, they are publicly funded. As such, RDCs themselves or the public might have an interest in understanding the value created by the data provision. Currently, the impact of data provision on research output and data-driven decision-making remains hard to measure, as it requires tracing data citations in research papers. Figure 14.3 provides a model for this data value chain.

Consequently, there is a need shared by RDCs and researchers that often requires human readers to process descriptions of data usage and modifications in research papers. Since the reporting and discussion of these data sources do not follow a standardized textual form, this process is time-consuming even for relatively small numbers of research papers. Importantly, it is also prone to human error. Hence, an automated system which identifies and categorizes research data sources would provide much-needed information while increasing efficiency as well as accuracy.

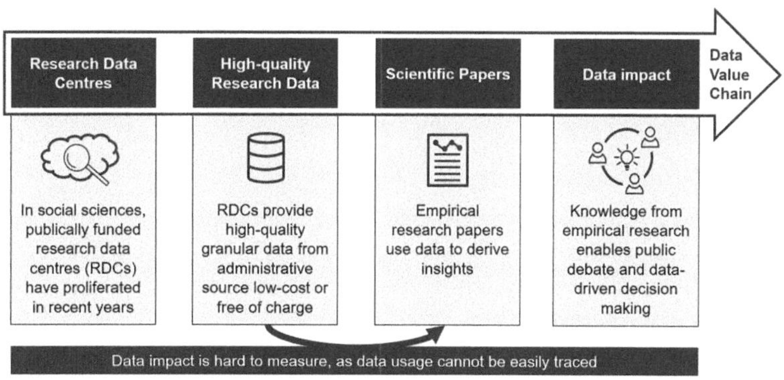

Fig. 14.3 The data value chain (Source: Authors' own depiction)

[2] In this context, data owners refer to data collectors or originators and comprise the institutions mandated with collecting the administrative raw data in question.

14.4 Data

14.4.1 Sample Collection

In order to provide a feasibility study for central banks, (national) statistics offices, and academia regarding obtaining data on structured citations from textual data mentions in papers, we use the data from the discussion paper series of Deutsche Bundesbank. This series comprises papers written by researchers including one or more Bundesbank employees.

Our input sample for the dataset extraction pipeline comprises all research papers released under the discussion paper series of Deutsche Bundesbank between February 1995 and July 2023. The series covers a variety of topics such as monetary policy, banking supervision, and household economic behavior. In total, this corpus includes 1,102 research papers at the time of our data collection in late 2023, published between 1995 and 2023 (see Table 14.1). This constitutes the corpus to be labeled. Subsequently, it is straightforward to extend our approach to label any corpus of academic papers.

From the corpus to be labeled, we randomly select 103 discussion papers as an evaluation sample. In order to obtain a baseline to evaluate the performance of our approach, we label these papers manually. By labeling in this context, we refer to reading a paper and manually noting the datasets used (if any). These manually labeled papers constitute our evaluation sample for the dataset extraction pipeline. This means that the performance evaluation metrics introduced and discussed in Sects. 14.5 and 14.6 are based on estimated averages across this evaluation sample.

14.4.2 Manual Labeling of Evaluation Sample

For manual labeling, we randomly assign 20 or 21 papers each to 5 human readers. Human readers scan the papers for specific dataset mentions. Figure 14.1 illustrates that dataset mentions differ widely, lacking a standardized format and standardized sections within papers. Therefore, the human readers need to read all papers thoroughly to find dataset citations.

Table 14.1 Data sample used including the corpus of research papers for automated extractions of data citations and the evaluation sample

	Source	Sampling	Number of papers	Sample period	Labeling
Corpus to be labeled	Bundesbank discussion paper series	Full universe	1,102	02/1995–07/2023	GPT 3.5
Evaluation sample	Bundesbank discussion paper series	Random sample	103	06/1995–04/2023	Manual

For all identified datasets in the evaluation sample, we record two types of labels: short labels and long (verbose) labels. The latter corresponds simply to the full text passage where each used dataset is discussed. In this step, we record only the first mention per dataset, omitting repeated (and differing) descriptions of the same dataset. Please see the box titled "Labeling instructions" for the step-by-step labeling guide used.

For short labels, we record all synonyms of the dataset names for each of the manually labeled papers. To illustrate what this means, consider the following example: The full name of the dataset commonly referred to as SHS is Securities Holdings Statistics. In addition, this dataset has different scopes with differing names which might or might not be explicitly mentioned in papers. Therefore, papers might refer to this dataset as either one of SHS, SHS-S, or Securities Holdings Statistics, among various other names.

Labeling Instructions

The following enumeration provides a labeling guide with step-by-step instructions used for human labeling of the evaluation sample:

1. Open the folder with the list of pdfs assigned to your name.
2. For each pdf:
    - Open the relevant pdf.
    - Read paper for mentions of dataset used for empirical analysis.
    - If mention found, record all synonyms used for the dataset(s) that you find in the text.
    - If mention found, record the entire passage that describes the dataset(s) that you find in the text (this can range from one sentence to an entire paragraph or more).
3. Save resulting json file.

If authors refer to the same dataset with different names within or across papers, recording only one strict definition as the ground truth will introduce a downward bias on the evaluation of the algorithm's performance. If we force our algorithm to find exact matches for the identified datasets, all datasets identified with another name than that specifically mentioned in the paper would be recorded as unidentified, i.e., negatives. This is more severe when definitions are long, since the chance of not finding each of, e.g., ten words in the correct order is generally higher than for, say, definitions of three words length.

In addition, the ground truth will likely not be represented in any but one paper, i.e., it is unlikely that several researchers use the same definition for the same dataset except for instances where dataset names are extremely short and standardized. This means that we would not be able to aggregate or compare results for the same dataset across papers if we do not undertake some sort of standardization of the dataset mentions.

We record these synonyms as equivalents for each paper. Considering again the above example, if a paper mentions any synonymous name for the SHS dataset, we note `SHS == SHS-S == Securities Holdings Statistics` for this paper. When we encounter new synonyms for datasets, we append them to this list of synonyms. In the performance evaluation pipeline, we subsequently loop through each of these synonyms and perform the dataset-level evaluation metrics on them.

When manually reading and labeling papers, we notice several ambiguities, demonstrating the complexity of the extraction task not only for machines but also for humans. Beyond the wide range of synonyms used to describe datasets, further particularities arise. While labeling, it is not always clear what constitutes a dataset name. This can be present in cases when there are descriptions in textual form instead of an actual dataset name (compare example in Fig. 14.1).

Issues on what constitutes an actual dataset further arise in the case of not well-defined sources. In these cases, we decide not to consider single time series or model calibration parameters as datasets. Furthermore, without domain knowledge, it is not always straightforward to decide what constitutes data usage for the respective paper as opposed to simple mentions of existing data that was used in other papers.

As a result of the manual labeling process, we end up with a corpus of papers to be labeled and an evaluation sample where we know (i) which datasets are used within, (ii) synonyms of the datasets used, if any, and (iii) the text passage where the dataset was mentioned. With this data ready, we proceed in the next section (Sect. 14.5) by outlining the methodology used, specifically the extraction pipeline we built.

14.5 Methodology

To extract data citations automatically from the corpus of papers obtained, we construct a pipeline for assistant-style LLMs. Our goal is to build a flexible process allowing us to (i) test different variations efficiently, (ii) track the results systematically, and (iii) easily transfer code to other interested parties. In the following, we outline the components of this pipeline, including the measures we use to judge performance.

14.5.1 Pipeline Overview

The pipeline takes as input a paper in the form of a PDF file and produces a list of datasets as output. Our pipeline can be segmented into five steps. First, we split papers into smaller segments. Second, we filter out irrelevant segments, before—third—combining each segment with the instruction for the LLM. In the fourth

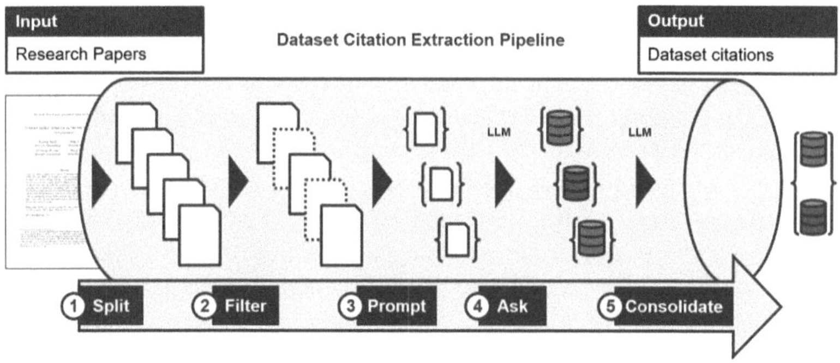

Fig. 14.4 The Dataset Citation Extraction Pipeline used for the purpose of this study. LLM: large language model (Source: Authors' own depiction)

step, we ask the LLM for the information contained, before finally consolidating the responses in the fifth step (compare Fig. 14.4 for a graphical depiction of the dataset citation extraction pipeline used). This is an application of a pipeline generally known as a retrieval-augmented generation (RAG) pipeline.[3]

In the first step, we split the document into smaller segments or snippets. The main goal here is to ensure that each snippet will be small enough to fit into the limited context window of the LLM. A context window is the zone in which text can be analyzed coherently by the LLM, so to speak the length of the "conversation" per paper we can engage with the model. In our case, we use GPT-3.5.[4] An intuitively simple approach could be, e.g., to treat each page of the document as one snippet.

While page splitting is easy to implement, this has the drawback of splitting an ongoing paragraph into two at the page boundary. In fact, instances where contextually connected paragraphs do not end at the end of a page are most frequent, causing the LLM to consider only part of the true context of the respective page-paragraph combination at a time. A mitigation of this issue is to define an overlap between the chunks. We achieve this with `langchain`, a solution tailored to application building with LLMs.

[3] An example of a RAG pipeline successfully used to generate structured data for central bank analysis from text can be found in Dimmelmeier et al. (2024) or BIS (2024).

[4] More precisely, we use the GPT-3.5 turbo model of the OpenAI API. It is technically possible to compare the performance to GPT-4 with a longer context window in the future. The model used at the time of writing allowed a context window of 4,000 tokens, also referred to as 4k (approximately five pages of text). A token can be thought of as a piece of a word, and one token is approximately four characters in English language. Newer versions of GPT-3.5 allow for a 16k token context window (approximately 20 pages). Recently, a 128k token context window was announced for GPT-4 (encompassing approximately 300 pages of text). In the interest of cost efficiency and technological availability at the time of writing, we relied on GPT-3.5 as a trade-off between performance and cost, but if implemented in production, our suggested pipeline could use larger snippets as well.

Alternatively, we evaluate splitting documents by paragraphs by using the occurrence of more than one new-line character as split separators. This yields a larger number of smaller snippets compared to a page split, which leads to higher application programming interface (API) costs but might improve performance.

The second, optional, step is to prefilter the snippets for informative content. Independently of the splitting approach used, the text conceivably contains large swathes of text without relevant information in the context of dataset extraction. Examples of such uninformative text snippets include (most) tables containing numbers, the references section, or small fragments without textual content.[5] In the interest of efficiency and cost considerations, we remove such snippets from consideration for further processing based on simple heuristics prior to the involvement of the LLM.[6]

Next, the text snippets from the paper have to be combined with instructions. This creates a prompt for a zero-shot task (compare, e.g., Sun et al. 2021). A prompt in this context is the plain English input we send to the model. The exact language model used determines the particulars of how to phrase the instructions. Depending on how the underlying model was designed and trained, it might allow a differentiation between "system" and "user" prompts or allow fill-in-the-middle prompts.

The quality of the generated responses also differs significantly depending on the phrasing of the instructions. As shown by, e.g., Reynolds and McDonell (2021), zero-shot prompts can outperform prompts with examples of task completion. Thus, our pipeline allows for systematic tracking of changes in output quality caused by alternative instructions. The prompts we use for this analysis can be found in the appendix. In short, we ask the model to identify a list of data sources or simply write "None" if it cannot find any.

In the fourth step, we send the prompt to the LLM, consisting of instructions and text content. At this point different sampling parameters can be set. Of particular interest to us is the "temperature" parameter, which rebalances the probability distribution for the next token in the generated sequence, with high temperature corresponding to a more uniform distribution.

High temperature is useful to generate more creative output and to prevent stiff, predictable writing style. In our use case, where we are interested in the simple retrieval of factual information from a given text, we set the temperature as low as possible. Our pipeline also takes care of any connectivity issues, for example, due to downtime or throttling. The result of this step is an answer to our inquiry for each individual text snippet in the form of a list of contained dataset names (or "None" if the LLM does not detect any mentions).

---

[5] Tables could be informative if they contain dataset information. This can be the case, for instance, when an appendix table includes information on further datasets used.

[6] *Simple heuristics* here refers to the logic we use to identify such uninformative text snippets. For instance, references do not contain dataset mentions and can be identified by the title "References" or "Bibliography." Removing such passages decreases cost and increases efficiency of the extraction pipeline since they do not have to be passed through the LLM API.

In the fifth and last step, the individual model replies per snippet must be consolidated into a single answer covering the entire input paper. Any single data source used will more often be mentioned multiple times across several sections of a research paper than not. The result of this pipeline is a single list of all—ideally unique—data sources used in the processed research paper. The identification of the union of all the individual answer sets generated at the text-snippet level is also left to the LLM. This part is achieved by a consolidation prompt (see appendix and the pipeline schematic in Fig. 14.4). Using the evaluation set of research papers, we subsequently evaluate in detail how close the automatically generated list is to what a human reader would expect (see Sect. 14.6 for the results, and recall Table 14.1 for the sample breakdown).

14.5.2 Performance Measures

Evaluating the consolidated results provides several insights. Firstly, LLMs such as GPT-3.5 have been reported to hallucinate. Secondly, our evaluation aims to compare two lists of free text (the consolidated model output list and the manually labeled evaluation sample), requiring fuzzy string comparisons.

LLMs are notorious for their propensity to hallucinate. More accurately than hallucination, this tendency might be rephrased as confabulation (compare, e.g., Smith et al. 2023). That is to say, they may make up incorrect information and present it as fact. This is a consequence of the way this kind of model is trained: "I do not know" is rarely if ever the "correct"/ best continuation after a query. Instead, the model assigns high probabilities to the sequences that present the most plausible answer.

While our application of fact retrieval from a given text is not particularly susceptible to confabulation, we nonetheless need to measure the quality of the responses that our pipeline produces. This is also crucial information to test the impact of changes to our processing pipeline.

Determining the performance of the model and pipeline by comparing its output to a list labeled by a human is not trivial. In essence, we would like to gauge the similarity of the two lists. The fact that the generated lists are free text, rather than a subset of predefined labels, complicates this task. While our approach allows the model to identify data sources beyond our current awareness, the output then necessitates the use of fuzzy string comparisons. In Fig. 14.5, we show the evaluation process pipeline.

We explore different versions of this fuzzy comparison, highlighting the trade-off between error types based on the strictness of the match. We estimate model recall (share of identified "true" data sources) and output similarity measured by various criteria.

The result of the comparison yields an incomplete confusion matrix. It is incomplete because the universe of datasets which are not in the list of true datasets and not identified by the model cannot be unambiguously defined. More specifically,

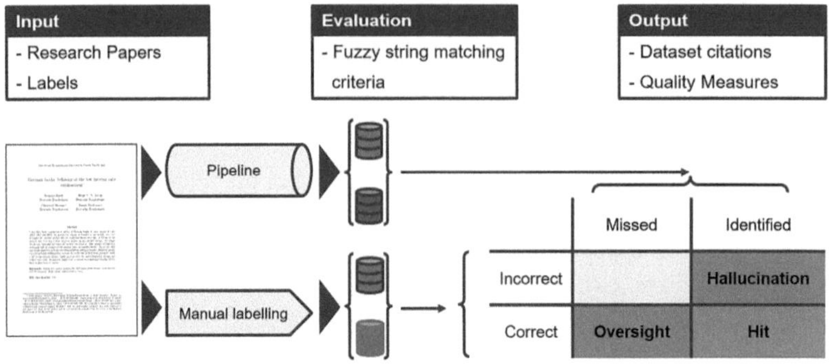

Fig. 14.5 Assessing the quality of the model's output (Source: Authors' own depiction)

we do not know the number of incorrect datasets. This set could either comprise all datasets in the complete text corpus, which we cannot know without labeling all papers (thereby defeating the purpose of this exercise). More importantly, however, even this list cannot truly include all possible datasets. Therefore, the true number of incorrect datasets, i.e., datasets not included in each paper, is infinite and can therefore not be used to estimate the performance of the dataset extraction pipeline.

For the task at hand, the model can make two kinds of errors. The first is oversight, where the model does not recognize a mentioned data source, resulting in a false negative (correct/missed in the confusion matrix). The second type of error consists of incorrect identifications, or false positives, either because the model misunderstood the output or confabulated answers (incorrect/identified in the confusion matrix). These two kinds of errors are not equally important in our case. For our use case, we prefer incorrect (confabulated) data source mentions to overlooking true ones, because domain experts, i.e., RDC employees, can easily correct implausible retrievals downstream. This could be achieved by, e.g., comparing the list of data sources managed and maintained by the RDC in question against the retrieved results or by comparing (aggregate) numbers of total datasets provided to those used in the resulting research output.

In evaluating the performance of our dataset extraction pipeline, we prioritize recall over precision, since RDCs maintain a known set of datasets. The latter fact means we have access to clear ground truth for our evaluation. Our primary objective is to assess the extent to which these datasets are acknowledged in research outputs, making it essential that our pipeline captures as many relevant mentions as possible. While precision measures the relevance of identified datasets, recall ensures that we do not overlook valid mentions, thereby providing a more comprehensive view of dataset impact. Given the role of RDCs in curating and providing these datasets, the risk of confabulation (incorrect identifications) is secondary to the risk of oversight (missed mentions), reinforcing our emphasis on maximizing recall. The performance metrics we use therefore focus on different modifications of recall estimates (compare Fig. 14.5 for the breakdown of instances of oversight/hallucination).

Finally, we study the costs associated with our dataset retrieval pipeline (Sect. 14.6.2). We present the monetary cost incurred when running on the OpenAI API platform at the time of writing and using GPT-3.5 turbo. On the platform, the current payment scheme charges users per token of input and output text. After evaluating model performance, we subsequently compare costs for each input, instructions, and the text itself (Sect. 14.6.2).

14.6 Results

14.6.1 Model Performance

Given the emphasis we place on recall, our evaluation of the pipeline performance provides a lower bound on the correct identification of datasets used in research papers. Since recall measures the proportion of actual dataset mentions that are successfully identified, our approach ensures that we capture most dataset mentions while allowing for some incorrect identifications. This conservative estimate reflects the minimum level of dataset recognition that our pipeline achieves, acknowledging that some true mentions may still be missed, and some identified mentions might be erroneous. However, as RDCs maintain a known set of datasets and our evaluation process focuses on maximizing recall, this approach provides a robust measure of how well the pipeline identifies dataset usage in research outputs while the true performance potential upon implementation might be even higher than suggested herein.

When evaluating recall of our approach (i.e., the share of correctly identified dataset citations as a fraction of all dataset citations), we apply several criteria to compare model predictions with our evaluation sample. We (i) let the LLM judge its own results as described above,[7] (ii) compare exact string matches, (iii) fuzzy string matches, (iv) fuzzy string matches including synonyms, and (v) the Jaccard similarity.

Figure 14.6 depicts recall results for all evaluation criteria used. As these criteria are not equally strict in comparing similarity, measures of recall vary widely. In the strictest scenario, when we only count exact matches, we reach a recall of only 12% (17% conditional on the fact that the model predicted any data to be contained at all). This result comes maybe unsurprisingly as it signifies the variety of dataset names, i.e., complexity of the extraction task.

When we ask the LLM to evaluate its own results, we obtain a 34% recall (62% when considering only papers where the model predicted any data). While elegant in the context of our pipeline to leverage the LLM for evaluation, this criterion too remains a black box subject to LLM reproduction variability, as the exact evaluation

[7] Please see the appendix for the corresponding recall prompt.

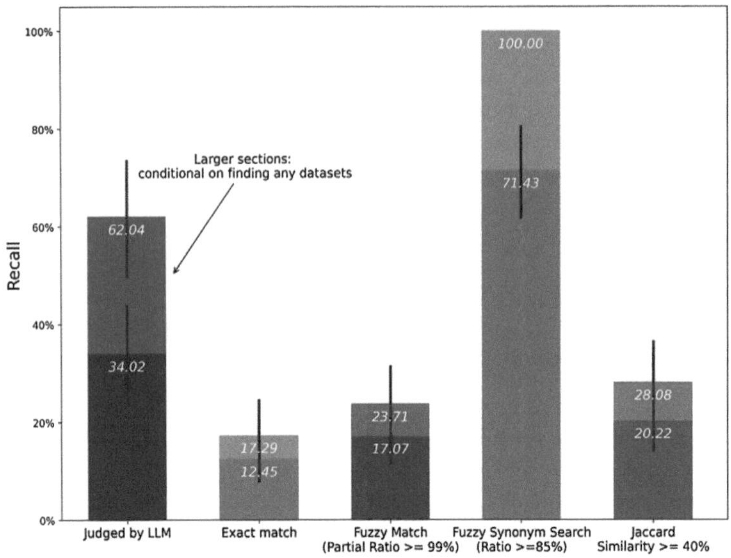

Fig. 14.6 Performance measures of model recall based on different criteria (Source: Authors' own calculations)

metric is subjective to the model and users are unaware a priori of the direction of a potential bias.

Arguably, the most lenient, or least strict, evaluation measure is the fuzzy synonym search, where any variations of synonyms of the dataset that are found are counted as a hit. Arguably, this may be a valid measure for our use case, as a synonym of a dataset name is often just as correct as any other. Using this criterion, our approach finds 71% of dataset citations in the papers provided (100% for papers where any dataset mention is extracted).

Note that fuzzy synonym search is not necessarily always feasible to evaluate results. If, for example, our pipeline were to be applied in another context, it requires different domain knowledge to identify synonyms of dataset names. However, providers of datasets often possess such domain knowledge. Further, note that such domain knowledge is only needed for the evaluation of synonyms, not algorithm performance itself.

As the recall numbers vary considerably depending on the exact criteria used, due to the loosely defined task of comparing lists of strings, we feel compelled to briefly comment based on our subjective feeling as employees of the Bundesbank's RDC: When looking at the similarity of lists ourselves, we were pleasantly surprised by the capability of the LLM to identify datasets correctly, even in the case of complicated names or infrequently used datasets.

Users of the automated dataset extraction pipeline might be interested in inspecting the performance of the LLM assistant on an individual-dataset level (Fig. 14.7). Here, we notice a very high ability of the algorithm to identify well-

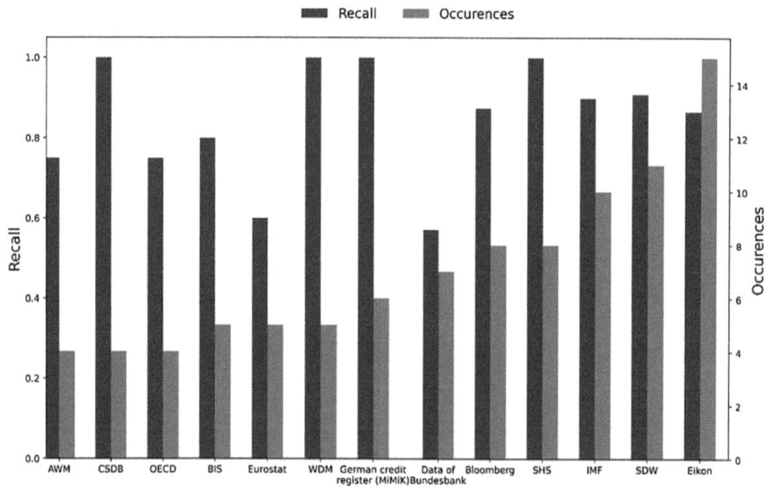

Fig. 14.7 Recall for selected datasets, judged by LLM (Source: Authors' own calculations)

structured and well-defined datasets, in our case due to the corpus of input papers particularly from administrative data sources (datasets originating from Deutsche Bundesbank) and commercial providers. A particularity of these datasets is that they are usually clearly defined, named, and separable entities.

Furthermore, the existence of relatively unique abbreviations seems to support the feasibility of our pipeline for automated dataset extraction (e.g., "CSDB" and "SHS"[8]). Figure 14.7 shows the LLM-based recall for a number of selected administrative datasets in our sample. Before deriving ambitious conclusions from this finding, the scope of the extraction would have to be enlarged, as the number of occurrences per dataset in our sample remains limited. However, we take this as an encouraging sign for the potential of our approach.

14.6.2 Costs

In order for data-providing institutions to assess their data usage with RAG pipelines such as the one proposed here, the question of costs to measure the impact of their data is of importance. This is all the more important in light of the fact that data providers may wish to justify public investment in data provision. If measuring the success comes as yet another cost driver, not much would be gained. Therefore, in addition to measuring performance, the impact of the decisions made in the

[8] "Central Securities Data Base" and "Securities Holdings Statistics," two large and well-documented European administrative datasets used for noncommercial research.

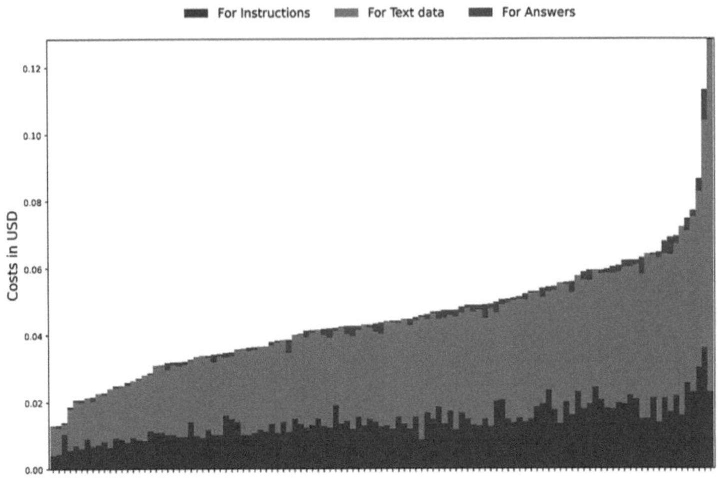

Fig. 14.8 API costs per paper—instruction and retrieval prompts. The bottom bars show instruction-prompt costs, whereas the middle and top bars show costs for the passed text snippets from which dataset mentions are to be extracted and response tokens (i.e., LLM replies), respectively (Source: Authors' own calculations)

pipeline on costs is also presented in Fig. 14.8. Luckily, overall monetary costs of the presented approach prove to be relatively low.

When considering the costs of our approach, we refer to the costs of API usage. In the context of this discussion, we exclude staff costs for developing and implementing our pipeline, costs for hardware, and data retrieval. The API is paid per token for prompts (i.e., textual input we feed to the model) and completions (i.e., the answers generated by the model). Therefore, the API costs for using the model for our use case can be divided into three categories: (i) costs for sending instructions to the LLM, (ii) costs for sharing the textual data, and (iii) costs for answers.

Overall, processing costs per paper seem to be very acceptable, at least for our presented use case at hand. Monetary costs for the API lie in the range of USD 0.02 to USD 0.08 for the overwhelming majority of papers. Costs are driven predominantly by instructions (approximately one third of API costs), and for the textual data, we share with the model (approximately two thirds of API costs). Answers make up only a small fraction of API costs (see Fig. 14.8 for a paper-by-paper overview).

As the textual data we share with the model makes up the largest cost driver in our approach, we are also interested in how the lengths of different splitting approaches change model performance in terms of recall (in this exercise using Jaccard similarities). We compare the model performance when we feed entire pages to the LLM and when we feed paragraphs ("page splitting" vs. "paragraph splitting").

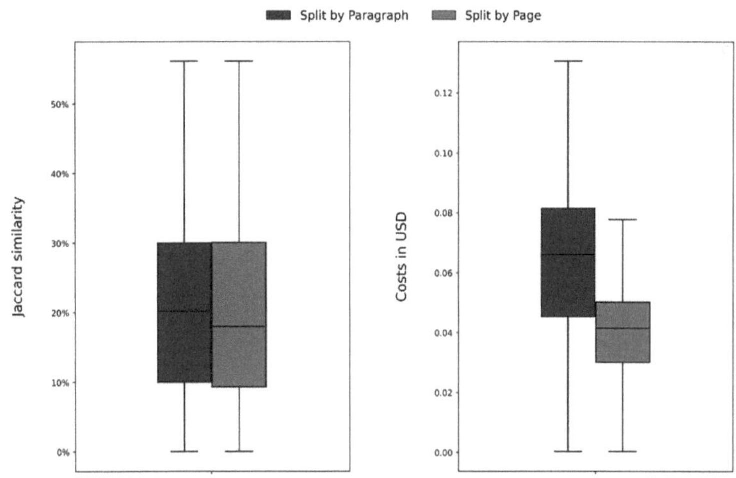

Fig. 14.9 Text splitting—performance vs. cost differences for page and paragraph splitting (Source: Authors' own calculations)

On the one hand, we find that providing larger contexts (i.e., page splitting) results in negligible changes in performance, if any. Figure 14.9 displays Jaccard similarities as an exemplary measure of model performance. On the other hand, we note that larger contexts significantly reduce overall costs. This is driven by the fact that for smaller contexts, the overhead of sending instructions every time increases proportionally.

The upper limit of the context size is bound by the token length the API accepts. Maximum token length in turn is limited by the underlying model. Furthermore, easy performance gains can be achieved by rule-based prefiltering of snippets (see Fig. 14.4). Prefiltering becomes less efficient in the case of very large contexts—consider the case of a context length of the entire paper. In this case, nothing should be filtered out. This adjustment to larger contexts does not produce noteworthy differences in performance, irrespective of the evaluation criterion considered (see Fig. 14.10).

When evaluating costs of our approach, we note that overall monetary costs for API usage of the presented approach tend to be small in the context described. To use this pipeline for a range of research work on a similar scale to ours, the cost of API usage can be expected to be in the low four-figure Euro range. While pricing models can change and newer models tend to be more costly, we deem monetary costs for API usage to be small in comparison to labor costs for setting up such a pipeline and to costs in the alternative scenario when attempting to run a similar LLM model on-site.

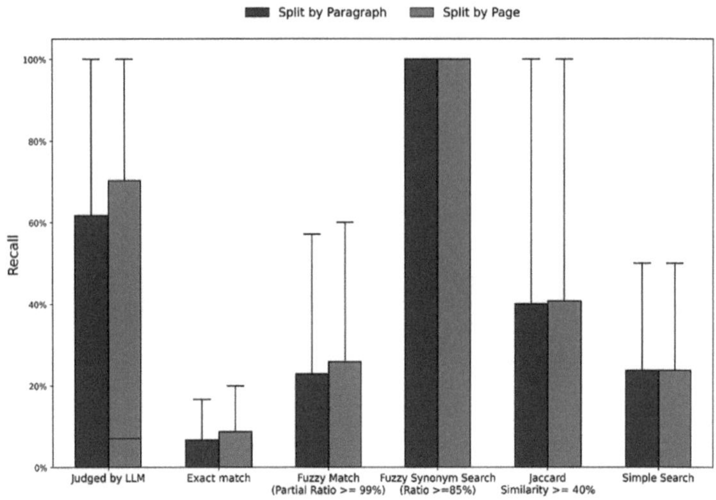

Fig. 14.10 Text splitting—performance differences for page and paragraph splitting, considering multiple evaluation criteria. Recall is depicted for those cases where any datasets are predicted to be included (Source: Authors' own calculations)

14.6.3 Reproducibility of Results

Our dataset extraction pipeline is designed to improve reproducibility compared to purely generative AI approaches by incorporating structured processing steps, deterministic filtering, and predefined dataset references. Instead of relying solely on an LLM's open-ended text generation, our pipeline first processes and filters text passages to isolate relevant content before prompting the model. This structured approach reduces variability in input data and ensures that dataset mentions are extracted systematically. Additionally, dataset name consolidation and recall evaluation steps introduce further layers of control, minimizing the impact of inconsistencies in individual LLM responses. By incorporating multiple recall measures—ranging from exact matches to fuzzy synonym-based recognition with similarity thresholds of 90% for fuzzy matching, 85% for synonym recognition, and 40% for broader similarity—we create a robust framework for identifying dataset mentions while maintaining consistency across different research papers (compare Figs. 14.6 and 14.7 of Sect. 14.6.1). Preprocessing steps, such as PDF-to-text conversion and snippet-based text filtering, further help standardize input data before it is processed by the LLM, reducing the likelihood of missing dataset mentions due to extraction errors. Moreover, by comparing results against a predefined set of datasets maintained by our division, we ensure that recall estimates are evaluated against a known reference, reinforcing consistency in our findings (compare the pipeline schematic in Fig. 14.4).

Despite the structured nature of our pipeline, some inherent variability remains due to the stochastic nature of LLMs and other external dependencies such as text extraction quality and threshold-based matching techniques. While our pipeline significantly reduces the impact of randomness compared to a direct LLM-based approach, it does not fully eliminate variability across repeated runs. One key limitation is that LLM outputs, even when using controlled temperature settings, may differ slightly due to the probabilistic nature of token generation. Additionally, small changes to the fuzzy matching similarity thresholds may alter recall outcomes, as minor variations in dataset naming conventions could affect whether a mention is identified. The consolidation step, where an additional LLM call merges and removes redundant mentions, introduces another layer of potential inconsistency, as the model determines which mentions to retain based on patterns in the provided lists. Furthermore, errors in PDF-to-text conversion, such as missing characters or formatting issues, may cause dataset mentions to be lost before reaching the LLM, affecting the overall recall performance.

The reproducibility of our pipeline is enhanced by its structured multistep approach, which combines LLM-based dataset extraction with deterministic filtering, consolidation, and recall evaluation. By grounding our analysis in a predefined set of datasets maintained by our division, we mitigate the open-ended variability commonly associated with generative AI. Additionally, multiple recall measures ensure robustness against different dataset naming conventions, while the use of structured prompts and postprocessing steps further reduces inconsistencies. However, despite these safeguards, some variability remains due to the random elements of LLM responses that often cannot be controlled by the user (e.g., by specifying a seed), sensitivity to fuzzy matching thresholds, and potential inconsistencies in PDF-to-text conversion. While our approach significantly improves reproducibility compared to direct LLM-based methods, results may exhibit minor variations across runs.

14.7 Value Added for Official Statistics

If the baseline of dataset extraction tasks is human extraction, which can be costly and prone to human errors, the results presented in this chapter seem to suggest that it is efficient to streamline the process of identifying data usage in papers with the use of LLM-based RAG pipelines. This holds true even if model performance itself is deemed insufficient for certain use cases, as long as prelabeling by the model increases efficiency of subsequent human labeling. Hence, using our pipeline for known datasets in an RDC or as a data provider seems to yield a fairly reliable gauge of data impact in research, with or without human supervision.

The data citations we extract form a network which shows how datasets are used in actual research papers. In Fig. 14.11, we depict the resulting network graph. In the network, the size of the nodes represents the number of occurrences of any one dataset in our sample, and the thickness of the edges represents the frequency of

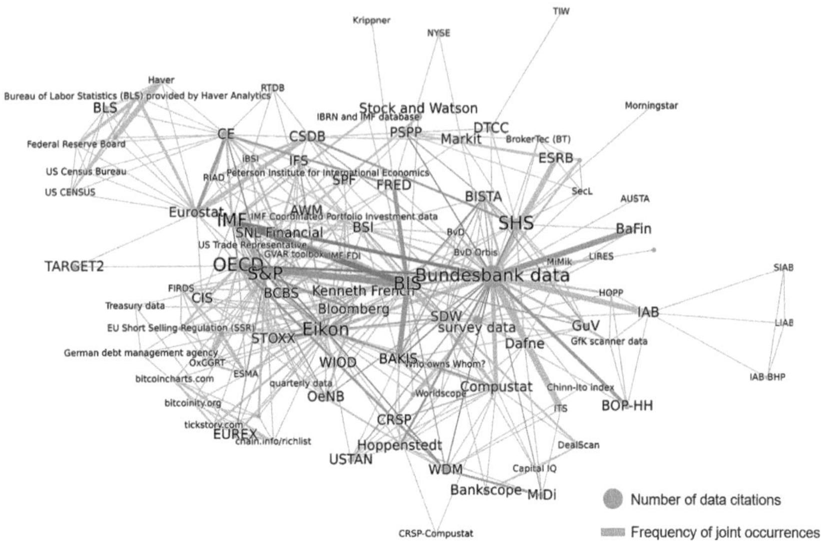

Fig. 14.11 Data usage network based on automatically identified data usage in research papers (Source: Authors' own calculations)

joint usage of datasets in research papers. To construct the network, we consider all available extracted labels, in total 218 for the quality-assured sample of 129 papers. The network is drawn as a force-directed graph with as few crossing edges as possible.

A network graph such as the one presented here can be used by data-providing institutions to tailor their data provision services. This holds especially true for providers of official statistics as this can be their only means of user-centric feedback, whereas commercial providers can ultimately gauge their data provision services to increase sales.

We see that a few nodes dominate the central part of the network with many connections to the surrounding environment. Often, the names of data-providing institutions exhibit high centrality, such as "Eurostat," "OECD," "BIS,"[9] "Bundesbank," or "IAB."[10] Since these are not necessarily dataset names in the traditional sense, this finding presumably stems from the fact that data citations often occur in the form of "We use data from Eurostat to [. . .]." Since our sample comprises the entire Deutsche Bundesbank discussion paper series, we see a large cluster around "Bundesbank."

The network depicted in Fig. 14.11 offers interesting starting points for human visual analysis and for network analysis. From a first glance, we note the joint data

---

[9] Bank for International Settlements.

[10] "Institut für Arbeitsmarkt- und Berufsforschung," the German official institute for employment research, a large provider of administrative research data.

usage clusters of US American datasets to the upper left, proprietary datasets in the lower left, and administrative datasets toward the right of the network with a "Bundesbank" and an "IAB" cluster. These joint data usage clusters have strong connections among their nodes and to a degree less intensive connections to other areas of the network.

With administrative data, a reasonable assumption could be that usage restrictions play a role. Since confidential administrative data tends to be provided only on the premises of the data-providing institutions, usage restrictions can render joint usage of two confidential datasets from two different providers mutually exclusive. This could be a reason that proprietary datasets tend to connect clusters in the network depicted above. This brief analysis points to obvious improvements in joint data usability to be obtained if usage of administrative data from multiple sources could be enabled.

Knowing the usage of datasets in research papers yields several important use cases for data providers beyond the anecdotal analysis above. Exemplary applications include (i) focusing resources on value-creating data, (ii) promoting popular datasets, (iii) recommending tailored datasets to researchers while additionally considering researcher information, (iv) providing frequently jointly used datasets together, and (v) identifying underused data by removing potential usage hurdles (see Fig. 14.12).

Extracting datasets with the approach presented in this chapter enables data providers to know which datasets are the most-value creating in terms of usage in research with high impact. Data providers can then focus resources on these top performing datasets specifically, e.g., optimize time-to-market for these datasets. This in turn ensures data-driven data provision and therefore improved dissemination of insights from the data.

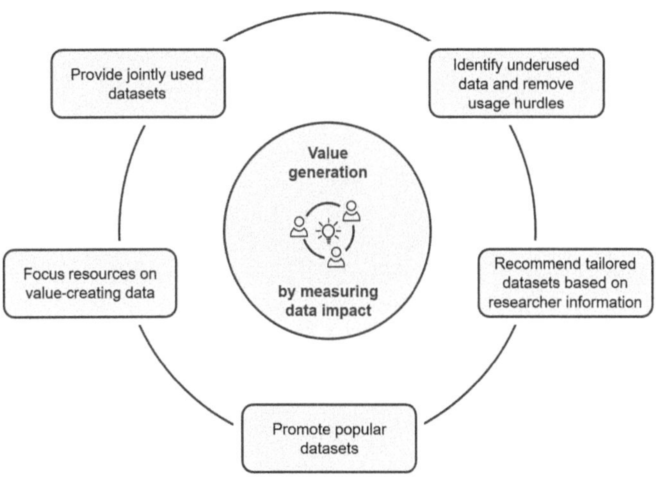

Fig. 14.12 Value proposition based on achieved results (Source: Authors' own depiction)

Related to this aspect, popular datasets can be promoted based on their usage. If data providers find that particular datasets are commonly used in a field (a piece of information that is currently not structurally available), new researchers arriving at an RDC (such as PhD candidates) can be provided with a guide to how other researchers utilize the data in question. When searching and citing research papers, it is common practice to use scholarly search engines. However, these engines often fail to identify data usage in research papers due to the unstructured nature of data citations. The use case at hand provides an avenue toward similarly searchable data usage.

Taking the promotion of popular datasets one step further, it is possible to recommend tailored datasets to researchers by additionally considering researcher characteristics. Since only registered researchers usually obtain data access at RDCs, information on prior research, co-authors, and areas of interest is often available to the RDCs. This allows tailored recommendations in the style of online retailers. Recommendations could take the form "Researchers like you often used dataset X" or "Have you considered also using dataset Y, since it is often used in conjunction with dataset X?".

If RDCs find that several datasets are often used together, they can streamline provision processes to grant access to datasets which frequently tend to be used jointly in one single step. This can remove usage hurdles and reduce overhead effort for researchers having to apply or register twice and for the data providers, while the providers themselves can reduce administrative effort.

Analogously to the potential of promoting popular datasets outlined above, usage hurdles can also be identified to promote underused data. Combined with implicit knowledge of the data provider staff, measuring the data impact can help identify underused data. This can be of interest in cases where a supposedly valuable dataset is rarely used in research. RDC staff in this case can understand, investigate, and remove usage hurdles, e.g., by contacting researchers who successfully published their work using the datasets in question.

Taken together, the value propositions outlined above demonstrate the potential for data providers to tailor their data provision on actual data-driven feedback. Central banks and national statistical offices can streamline data provision for research (i) in terms of costs both by optimizing processes based on data usage and (ii) by improving data impact through optimized usability. With these value propositions in mind, a next step would be to incorporate a larger corpus of research papers and implement the pipeline at other RDCs to leverage the potential of LLMs for value-driven research output measurement using automated extraction pipelines such as the one proposed in this work.

14.8 Conclusions

Tracing dataset usage in research is valuable for both data providers and researchers, enabling impact assessment and systematic exploration of data citation patterns. However, dataset mentions in academic papers remain largely unstructured, limiting their use for such analyses. Recent advancements in NLP, particularly large language models (LLMs), offer a way forward.

We present a retrieval-augmented generation (RAG) pipeline using GPT-3.5 to extract dataset citations from research papers, evaluating its performance on a manually labeled sample. Our findings indicate that while recall is promising—especially for well-known datasets with predefined synonyms—variability remains due to the challenges of string-matching evaluation. Despite this, the approach provides a practical improvement over manual tracing, offering a scalable prescreening tool to support data impact assessments.

The cost of implementation is relatively low, with our analysis showing that a page-level input structure balances cost and accuracy effectively. Moreover, as LLMs continue to evolve, model performance and context window size are expected to improve, enhancing the feasibility of this approach.

Our method is transferable to other institutions interested in dataset tracking and can be adapted to different LLMs, including on-premise models for handling confidential texts. Future developments could extend the pipeline to integrate dataset citation tracking into broader research evaluation frameworks, improving discoverability and strategic data management for research data centers.

Acknowledgments We wish to express our gratitude to Shir Frank, Kilian Graef, and Caspar Schauhoff for their excellent student research support, as well as to Carolin Schaaff for revisions.

Appendix

Prompts

Final prompt used

```
A dataset is a collection of structured or unstructured data
that is organized and grouped together for a specific
purpose. It typically consists of multiple data points or
observations related to a particular topic or subject. A
dataset can include various types of information such as
numerical values, text, images, audio, video, or any other
form of data. It is often used in the context of data
analysis, machine learning, and statistical research, where
```

(continued)

the data is utilized to extract insights, train models, or draw conclusions. Datasets can be generated through various means, including surveys, experiments, observations, or by gathering existing data from different sources.

The text after the empty lines is a scientific paper excerpt.

According to the definition on the first line, search for any mentions of datasets or data-sources used in the paper's research.

If you find any, please compile a list of all mentioned datasets in this excerpt.

If there aren't any, please reply with 'None'.

Consolidation prompt

You will be provided one or more lists of datasets.
Each new list starts with '=>'.

You need to combine all the lists into a single one by removing redundant entries. Delimit the final list with simple '-' bullet points.

Recall prompt

''You will be provided with text delimited by triple quotes that is supposed to be a list of datasets. Check if the following true datasets are directly contained in the answer:

For each of these true datasets perform the following steps:

1 - Restate the true dataset
2 - Write 'yes' if the true dataset is mentioned in the answer, otherwise write 'no'

Finally, provide a count of how many 'yes' findings there are. Provide this count as {"count":<insert count here>}"

References

S. Bender, J. Blaschke, C. Hirsch, A practical use case: lesson learned from social science research data centers. Harvard Data Sci. Rev. (2024). https://doi.org/10.1162/99608f92.8a2f4507

BIS, Project Gaia – Enabling climate risk analysis using generative AI. Bank for International Settlements (BIS) Innovation Hub (2024). https://www.bis.org/publ/othp84.pdf

S. Blank, A. Lipponer, C.-J. Schild, D. Scholz, Microdatabase Direct Investment (MiDi)–A full survey of German inward and outward investment. Ger. Econ. Rev. **21**(3), 273–311 (2020)

K. Boland, F. Krüger, Distant supervision for silver label generation of software mentions in social scientific publications, in *BIRNDL@SIGIR* (2019), pp. 15–27

C.M. Buch, J. Kleinert, A. Lipponer, F. Toubal, R. Baldwin, Determinants and effects of foreign direct investment: evidence from German firm-level data. Econ. Policy **20**(41), 52–110 (2005)

P. Bundi, V. Pattyn, Trust, but verify? Understanding citizen attitudes toward evidence-informed policy making. Public Adm. **101**(4), 1227–1246 (2023)

C. Burr, D. Leslie, Ethical assurance: a practical approach to the responsible design, development, and deployment of data-driven technologies. AI Ethics **3**(1), 73–98 (2023)

D. Card, R. Chetty, M.S. Feldstein, E. Saez, Expanding access to administrative data for research in the United States. American economic association, ten years and beyond: Economists answer NSF's call for long-term research agendas (2010)

H. Chen, X. Luo, An automatic literature knowledge graph and reasoning network modeling framework based on ontology and natural language processing. Adv. Eng. Inf. **42**, 100959 (2019)

V. Cologna, N.G. Mede, S. Berger, J. Besley, C. Brick, M. Joubert, E.W. Maibach, S. Mihelj, N. Oreskes, M.S. Schäfer, et al. Trust in scientists and their role in society across 68 countries. Nat. Hum. Behav. **9**(4), 1–18 (2025). https://pubmed.ncbi.nlm.nih.gov/39833424/

K. Crawford, *The Atlas of AI: Power, Politics, and the Planetary Costs of Artificial Intelligence* (Yale University Press, New Haven, 2021)

M.M. Dagli, Y. Ghenbot, H.S. Ahmad, D. Chauhan, R. Turlip, P. Wang, W.C. Welch, A.K. Ozturk, J.W. Yoon, Development and validation of a novel AI framework using NLP with LLM integration for relevant clinical data extraction through automated chart review. Sci. Rep. **14**(1), 26783 (2024)

A. Dimmelmeier, H. Doll, M. Schierholz, E. Kormanyos, M. Fehr, B. Ma, J. Beck, A. Fraser, F. Kreuter, Informing climate risk analysis using textual information–A research agenda, in *Proceedings of the 1st Workshop on Natural Language Processing Meets Climate Change (ClimateNLP 2024)* (2024), pp. 12–26

M. Färber, A. Albers, F. Schüber, Identifying used methods and datasets in scientific publications, in *SDU@AAAI* (2021)

Federal Data Strategy, Data ethics framework (2020). https://resources.data.gov/assets/documents/fds-data-ethics-framework.pdf

A. Felländer, J. Rebane, S. Larsson, M. Wiggberg, F. Heintz, Achieving a data-driven risk assessment methodology for ethical AI. Digit. Soc. **1**(2), 13 (2022)

A. Gemelli, E. Vivoli, S. Marinai, CTE: a dataset for contextualized table extraction (2023). https://arxiv.org/abs/2302.01451

R. Hausen, H. Azarbonyad, Discovering data sets through machine learning: an ensemble approach to uncovering the prevalence of government-funded data sets. Harvard Data Sci. Rev. (2024). https://doi.org/10.1162/99608f92.18df5545

Y. Hou, C. Jochim, M. Gleize, F. Bonin, D. Ganguly, Identification of tasks, datasets, evaluation metrics, and numeric scores for scientific leaderboards construction (2019). https://arxiv.org/abs/1906.09317

H. Houngbo, R.E. Mercer, Method mention extraction from scientific research papers, in *Proceedings of COLING 2012* (2012), pp. 1211–1222

D. Ikeda, K. Nagamizo, Y. Taniguchi, Automatic identification of dataset names in scholarly articles of various disciplines. Int. J. Inst. Res. Manage. **4**(1), 17–30 (2020)

S. Joseph, T.M. Kolade, O. Obioha Val, O.O. Adebiyi, O.S. Ogungbemi, O.O. Olaniyi, AI-powered information governance: Balancing automation and human oversight for optimal organization productivity. Asian J. Res. Comput. Sci. **17**(10), 110–131 (2024)

S. Kumar, T. Ghosal, A. Ekbal, Dataquest: an approach to automatically extract dataset mentions from scientific papers, in *Towards Open and Trustworthy Digital Societies: 23rd International Conference on Asia-Pacific Digital Libraries, ICADL 2021, Virtual Event, December 1–3, 2021, Proceedings 23* (Springer, Berlin, 2021), pp. 43–53

J. Lane, A. Spector, M. Stebbins, An invisible hand for creating public value from data. Harvard Data Sci. Rev. (2024). https://doi.org/10.1162/99608f92.03719804

A. Lipponer, Microdatabase direct investment–MiDi. A brief guide. Technical report, Bundesbank working paper, Frankfurt (2006). https://www.bundesbank.de/resource/blob/604692/50a67498fac4bde377c1c762834dc247/mL/2011-midi-documentation-data.pdf

Y. Luan, Information extraction from scientific literature for method recommendation (2018). https://arxiv.org/abs/1901.00401

X.-L. Meng, Data democratization: an ecosystemic contemplation and coordination. Harvard Data Sci. Rev. (2024). https://doi.org/10.1162/99608f92.f2a44cc0

S. Nishio, H. Nonaka, N. Tsuchiya, A. Migita, T. Banno, T. Hayashi, H. Sakaji, T. Sakumoto, K. Watabe, Extraction of research objectives, machine learning model names, and dataset names from academic papers and analysis of their interrelationships using LLM and network analysis, in *2024 IEEE International Conference on Industrial Engineering and Engineering Management (IEEM)* (IEEE, Piscataway, 2024), pp. 1377–1381

N. Pallotta, J. Lane, J. Locklear, X. Ren, V. Robila, A. Alaeddini, Discovering data sets in unstructured corpora: discovering use and identifying new opportunities. Harvard Data Sci. Rev. (2024). https://doi.org/10.1162/99608f92.77bfa1c9

L. Patiny, G. Godin, Automatic extraction of FAIR data from publications using LLM. chemRxiv Org. Chem. (2023). Available at https://chemrxiv.org/engage/chemrxiv/article-details/656c34ab29a13c4d478b2a12

M.P. Polak, D. Morgan, Extracting accurate materials data from research papers with conversational language models and prompt engineering. Nat. Commun. **15**(1), 1569 (2024)

L. Reynolds, K. McDonell, Prompt programming for large language models: beyond the few-shot paradigm. in *Extended Abstracts of the 2021 CHI Conference on Human Factors in Computing Systems* (2021), pp. 1–7

J. Sadowski, When data is capital: datafication, accumulation, and extraction. Big Data Soc. **6**(1), 2053951718820549 (2019)

T. Saier, M.Färber, unarXive: a large scholarly data set with publications' full-text, annotated in-text citations, and links to metadata. Scientometrics **125**(3), 3085–3108 (2020)

A.L. Smith, F. Greaves, T. Panch, Hallucination or confabulation? Neuroanatomy as metaphor in large language models. PLOS Digital Health **2**(11), e0000388 (2023)

K. Sostek, D.M. Russell, N. Goyal, T. Alrashed, S. Dugall, N. Noy, Discovering datasets on the web scale: challenges and recommendations for google dataset search. Harvard Data Sci. Rev. (2024). https://doi.org/10.1162/99608f92.4c3e11ca

X. Sun, J. Gu, H. Sun, Research progress of zero-shot learning. Appl. Intell. **51**, 3600–3614 (2021)

M.U. Tariq, Navigating the ethical frontier-human oversight in AI-driven decision-making system, in *Enhancing Automated Decision-Making Through AI* (IGI Global Scientific Publishing, Hershey, 2025), pp. 425–448

C. Zdawczyk, J. Lane, E. Rivers, M. Aydin, Searching for how data have been used: intuitive labels for data search and discovery. Harvard Data Sci. Rev. (2024). https://doi.org/10.1162/99608f92.f1cbbfbb

MIX
Papier aus verantwortungsvollen Quellen
Paper from responsible sources
FSC® C105338

If you have any concerns about our products,
you can contact us on
ProductSafety@springernature.com

In case Publisher is established outside the EU,
the EU authorized representative is:
Springer Nature Customer Service Center GmbH
Europaplatz 3, 69115 Heidelberg, Germany

Printed by Libri Plureos GmbH
in Hamburg, Germany